Springer Series in Synergetics

Editor: Hermann Haken

Synergetics, an interdisciplinary field of research, is concerned with the cooperation of individual parts of a system that produces macroscopic spatial, temporal or functional structures. It deals with deterministic as well as stochastic processes.

Hermann Haken (Ed.)

Neural and Synergetic Computers

Proceedings of the International Symposium
at Schloß Elmau, Bavaria,
June 13–17, 1988

With 139 Figures

Springer-Verlag Berlin Heidelberg New York
London Paris Tokyo

Professor Dr. Dr. h. c. mult. Hermann Haken

Institut für Theoretische Physik und Synergetik, Universität Stuttgart, Pfaffenwaldring 57/IV,
D-7000 Stuttgart 80, Fed. Rep. of Germany and
Center for Complex Systems, Florida Atlantic University,
Boca Raton, FL 33431, USA

ISBN 978-3-642-74121-0   ISBN 978-3-642-74119-7 (eBook)
DOI 10.1007/978-3-642-74119-7

© Springer-Verlag Berlin Heidelberg 1988
Softcover reprint of the hardcover 1st edition   1988

2154/3150-543210 – Printed on acid-free paper

Preface

These proceedings contain written versions of the talks given by the invited speakers at an international workshop on "Neural and Synergetic Computers" at Schloss Elmau, Bavaria, in June 1988.

It was a great pleasure for me to be able to welcome there most of the early pioneers of this field, in particular Shun-ichi Amari, Eduardo Caianiello, Stephen Grossberg, and Werner Reichardt. The field of neurocomputers is presently mushrooming. It will lead, I am sure, to important new concepts in computers and will provide us with new ideas about how brains work, although in both cases a good deal of work still lies ahead of us. The reason why I included "synergetic computers" in the title of the workshop is explained in my contribution to these proceedings.

I wish to thank Mrs. Ursula Funke cordially for the perfect organization of this workshop and the Volkswagenwerk Foundation for its efficient support of the synergetics project and, in particular, for the support of this workshop. Last but not least, I wish to thank Springer-Verlag for the excellent cooperation that has become a tradition.

Stuttgart, July 1988   *H. Haken*

Contents

Part I

Synergetics, Self-Organization, Pattern Recognition

Synergetics in Pattern Recognition and Associative Action

H. Haken

Institut für Theoretische Physik und Synergetik,
Universität Stuttgart, Pfaffenwaldring 57/IV,
D-7000 Stuttgart 80, Fed. Rep. of Germany and
Center for Complex Systems, Florida Atlantic University,
Boca Raton, FL 33431, USA

1. Why "synergetic" computers?

Biology abounds in examples for the self-organization of structures and behavioral patterns. There is increasing evidence that the mammalian brain is, to a rather large extent, self-organizing. Thus when we wish to implement the abilities of biological systems into the construction of new types of computers, it appears quite natural to try to use principles derived from the study of biological systems, in particular the brain. This is certainly one of the motivations which has led to the mushrooming field of neurocomputers [1]. Whether a present day neural computer can be considered as a sensitive model of the neural network of a brain is, to put it mildly, an open question. Indeed, opinions range from great optimism to a profound scepticism (see for instance Kohonen [2]).

Yet, there is still another avenue to self-organization as has been revealed by the interdisciplinary field of synergetics [3], [4]. Namely, self-organization may take place in the much simpler systems of physics and chemistry. Starting from such systems we were able to find general principles underlying self-organization, and a number of models dealing with morphogenesis or behavioral patterns can be shown to be based on such principles.

This leads us to the question whether synergetics can provide us with a novel access to the construction of highly parallel computers. Since this question can be answered in the positive, I chose the title of these proceedings as neural _and_ synergetic computers. Hopefully, there will be a convergence of ideas coming from the bottom-up approach that starts from the properties of individual model-neurons and their connections and the top-down approach offered by synergetics. Indeed, there are strong indications that this will be the case. My contribution is organized as follows:

Section 2 presents a brief reminder of some typical examples and basic concepts of synergetics. Section 3 recapitulates our standard model for pattern recognition that is an algorithm which can be implemented on a serial computer (cf. the contribution by Fuchs and Haken to these proceedings) or on a "neural net" in a straightforward manner (cf. section 5). Section 4 contains some generalization of that model to deal with ambiguous figures and hysteresis effects in perception. Section 6 shows how the approach of synergetics leads to construction principles of parallel computers and their learning procedure in a rather natural way. Section 7 discusses some possible "hardware" realizations. Finally, section 8 presents some ideas on the development of a synergetics of cognition and behavior. This section is intended to further the dialogue with physiologists and neurobiologists in particular.

Springer Series in Synergetics Vol. 42: **Neural and Synergetic Computers**
Editor: H. Haken   ©Springer-Verlag Berlin Heidelberg 1988

2. Synergetics: A Reminder

The interdisciplinary field of synergetics [3], [4] deals with the spontaneous formation of patterns or structures in systems far from thermal equilibrium via self-organization. Here I list only a few typical examples:

1. The convection instability which was discovered by Benard around the turn of the century. When a fluid in a vessel is heated from below, a temperature difference between the lower and upper surface is built up. When this temperature difference exceeds a critical value, a macroscopic motion in the form of rolls (or hexagons) may start (Fig.1). When we have a circular vessel and look from above at the top of it, we recognize a system of such rolls. In a subsequent experiment we may find the same roll system but oriented in a different direction (Fig.2). Thus the fluid may show multistability.

2. The laser
When a laser is pumped only weakly, it acts like a lamp and emits incoherent wave tracks. When the pump power into the laser is increased, it may form the coherent laser wave, or in other words, a temporal pattern or structure is formed.

3. Chemical reactions When specific chemicals, e.g. in the Belousov-Zhabotinski reaction, are poured together, specific spatio-temporal patterns can be formed, such as concentric rings or moving spirals.

Let us analyse what is happening in this case more closely by decomposing the total volume into small volume elements (Fig.3). Then in each of these volume elements reactions are going on which can be described by multiplication, addition, or subtraction of concentrations of molecules. In addition to this, diffusion between different volume elements takes place which can be described as transmission of information. Thus the chemical reaction in the whole volume can be considered as a parallel computation which, eventually, leads to the formation of spatial patterns. The question arises whether such kind of processes can be inverted, i.e. whether instead of the formation of patterns we may recognize patterns by such mechanisms.

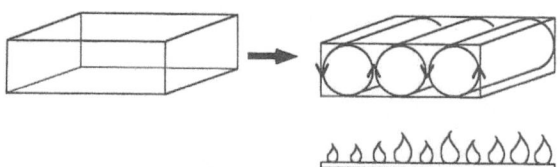

Fig. 1:   A fluid layer heated from below may spontaneously form rolls.

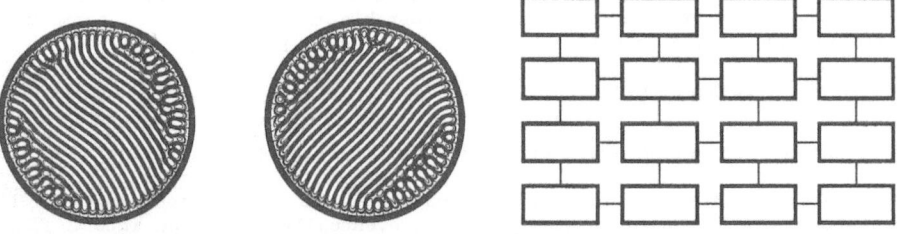

Fig. 2:   A fluid layer in a circular vessel may form roll systems in various orientations.

Fig. 3:   Decomposition of a spatially extended chemical reaction into individual volume elements.

Let us briefly recall what is done in synergetics to deal with pattern formation. To this end we introduce a state vector $q(x,t) = (q_1,\ldots, q_N)$ where each component depends on space and time and where the components may mean the velocity components of a fluid, its temperature field, or in the case of chemical reactions, concentrations of chemicals. The state vector obeys an evolution equation which has the general form

$$dq/dt = \dot{q} = L(\alpha) \, q + \underset{\sim}{N}(q) + \underset{\sim}{F}(t), \qquad (1)$$

which contains on the right hand side a linear operator L, a nonlinear part $\underset{\sim}{N}(q)$ and randomly fluctuating forces $\underset{\sim}{F}(t)$ which may stem from external or internal sources. In general L, and in some cases $\underset{\sim}{N}$, depend on control parameters α, which may be the temperature difference in the fluid, the pump power into the laser, or concentrations of chemicals added to the reaction.

Let us consider the spectral decomposition of the linear operator L into its eigenvectors $\underset{\sim}{v}_j$ which are defined by

$$L \bullet \underset{\sim}{v}_j = \lambda_j \bullet \underset{\sim}{v}_j.$$

Then equation (1) acquires the form

$$\dot{q} = \sum_u \lambda_u \underset{\sim}{v}_u \, (\underset{\sim}{v}^u \, q) + \sum_s \lambda_s \underset{\sim}{v}_s \, (\underset{\sim}{v}^s \, q) + \underset{\sim}{N}(q) + \underset{\sim}{F}(t) \, . \qquad (2)$$

The adjoint vectors $\underset{\sim}{v}^j$ are defined by

$$\underset{\sim}{v}^j \bullet \underset{\sim}{v}_j{}' = \delta_{jj'} \, . \qquad (3)$$

The first sum in (2) refers to the so-called unstable modes v where $\lambda_u \geq 0$. The second sum refers to the stable modes where $\lambda_s < 0$ and the remaining terms are the same as in equation (1). The slaving principle of synergetics allows us to eliminate the degrees of freedom which refer to the stable modes. In the leading approximation, equation (1) can be transformed into the following equations:

$$\dot{q} = \sum_u \lambda_u \underset{\sim}{v}_u \, (\underset{\sim}{v}^u \, q) - \sum_{u \neq u'} (\underset{\sim}{v}^u \, q)^2 \, \underset{\sim}{v}_u(\underset{\sim}{v}^u \, q) - |q|^2 \, q + \underset{\sim}{F}(t) \, . \qquad (4)$$

Here I have chosen a specific form for the nonlinearity which applies to those systems where a competition between patterns occurs. The first term on the right hand side stems again from the linear operator but including only the unstable modes; the second term is the nonlinearity which as we shall see below, describes discrimination between patterns and the third term describes the so-called saturation. The state vector q can be decomposed into the eigenvectors of L, where the leading terms are given by the unstable modes, the small correction terms contain the stable modes:

$$q = \sum_u d_u(t) \, \underset{\sim}{v}_u + \text{corrections} \, . \qquad (5)$$

The patterns are determined essentially by the first part of equation (5). The amplitudes $d_u(t)$ are, in the parlance of synergetics, the <u>order parameters</u>. When an initial state $q(0)$ occurs e.g. caused by a fluctuation, it will be driven into a time-dependent state $q(t)$ that eventually is driven into a final state that is identical with one of the patterns described by the vectors v_u. In earlier work in the field of synergetics [3], [4] the transition from $\lambda_u < 0$ to $\lambda_u > 0$ was studied and it was found that this transition has all the typical properties of a phase transition. But because in the present case it happens away from thermal equilibrium it was called a <u>nonequilibrium phase transition</u>. When we project the equation (4) on the eigenvectors $\underset{\sim}{v}_u$, very simple equations for the order parameters d_u are obtained, namely

$$\dot{d}_u = (\lambda_u - D) \, d_u + d_u^3 + F_u(t) \, ,\qquad\qquad\qquad (6)$$

where

$$D = \sum_{u'} d_{u'}^2 \, , \quad F_u = \underset{\sim}{v}^u \bullet \underset{\sim}{F}(t) \, .\qquad\qquad\qquad (7)$$

As can be shown, this kind of equation guarantees that there are no spurious states, i.e. all stable states coincide with one of the patterns $\underset{\sim}{v}_u$.

3. Our standard model for pattern recognition [5]

Some years ago I established an analogy between pattern formation and pattern recognition [6]. To study such an analogy, consider Fig.4. On the left hand side we study <u>pattern formation</u> which, as we have seen, is governed by order parameters d_u. If one order parameter is selected, the behavior of the subsystems is fixed by means of the vectors $v_u(x)$ and even the fine structure is fixed by means of the slaving principle. Once a few subsystems are in the state which belongs to the order parameter d_u, the slaving principle guarantees that all other subsystems are pulled into that specific ordered state. In <u>pattern recognition</u> a specific pattern is described by the order parameters d_u to which a specific set of features belongs. Once some of the features are given which belong to the order parameter d_u, the order parameter will complement the other features so that the whole system acts as <u>associative memory</u>.

Let us now show how we can exploit this analogy explicitly. Let us consider as an example a face to be recognized. We put a grid on the prototype faces and take as the features the tones of grey in each pixle. Equivalently we could use other features, e.g. Fourier transforms. Then we adopt our above set of equations for pattern recognition (4), but we attribute a new meaning to this equation and its individual terms on the right hand side of this equation. λ_u plays, as we shall see below, the role of an attention parameter. The term $\underset{\sim}{v}_u \bullet \underset{\sim}{v}^u$ is part of the learning matrix which is wellknown from theories on association [7] and which in particular occurs in the Hopfield model [8], although there the states can have only integer components, 1 or -1. The next term, the double sum over u' and u serves for pattern discrimination, the third term serves for saturation and the last term, $\underset{\sim}{F}$, takes care of fluctuating forces which may or may not be included depending on the procedure. In contrast to our above procedure

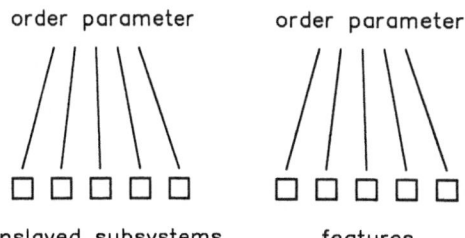

order parameter   order parameter

enslaved subsystems   features

Fig. 4:   Analogy between pattern formation (l.h.s.) and pattern recognition (r.h.s.).

5

of decomposing the linear operator L, the first sum on the right hand side does not necessarily stem from such a spectral decomposition. Rather λ_u and v_u are now quantities fixed by ourselves, the v_u representing the stored prototype patterns. Equation (4) derives from a potential V, i.e.

$$\dot{q} = - \partial V / \partial q^+ + F(t) .\tag{8}$$

In complete analogy to the dynamics valid for pattern formation, the following happens when a test pattern $q(0)$ is offered to the system. The dynamics will pull $q(0)$ via $q(t)$ into a specific final state v_u, i.e. it identifies the test patterns with one of the prototype patterns. The evolution of $q(t)$ can be identified with the overdamped motion of a particle with the (multidimensional) position vector q in a potential $V(q)$. Fig. 5 provides us with a two-dimensional example of such a potential. We have applied this formalism to the recognition of faces in the following way: First we stored a number of faces in the computer so that they formed the vectors v_u (Fig. 6). Then a part of the face was given to the computer as initial state. Via the dynamics (4) the computer pulled the initial state into the completed pattern so that the full face was restored ("associative memory") (Fig. 7). The accompanying

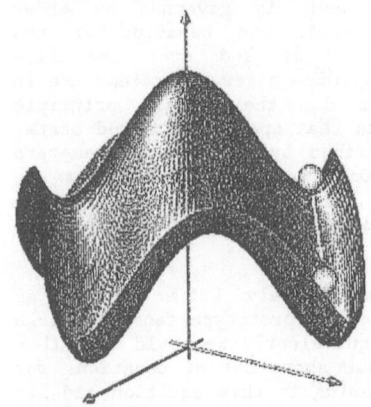

Fig. 5: The potential V in two dimensions spanned by the prototype vectors q_1 and q_2.

Fig. 6: Examples of faces stored in the computer. The faces were digitized on 60 x 60 pixels with 4 bits for each pixel (After [9]).

Fig. 7: Reconstruction of a stored face from an initially given part of that face. Note that the reconstruction is done in the presence of all other stored patterns (After [9]).

Fig. 8: Example of a scene. (After [10], [11]; see also [5]).

paper by Fuchs and Haken [10] generalizes this procedure to include names that are encoded as letters on the same picture. Thus, if a name (letter) is presented to the computer, it finds the corresponding face by restoring the pattern: "letter + face". If, on the other hand, part of a face is presented to the computer, it not only restores the face but provides us with the person's name.

By a suitable preprocessing of the images and a suitable new interpretation of the test vector q and the prototype vectors $\underline{v}$ the recognition process was made invariant against simultaneous translation, rotation and scaling within the plane. For an explicit representation of this procedure I refer the reader to the accompanying paper [10]. We then offered composite scenes, such as Fig.8, to the computer using translational invariance only, but no different scales or angles. It then identified the person in front. Then we let the computer put the corresponding attention parameter, λ, equal to 0 so that this front pattern could no more be identified with the corresponding prototype pattern. Then the computer indeed identified the second partly hidden face. In this way up to five prototype patterns could be identified subsequently (for details cf. Fuchs and Haken, [10], [11]). While this procedure worked well if the faces had the same size as that of the prototype faces, the computer failed when the procedure had been made invariant against scales. In order to check whether human perception is scale-invariant, consider Fig. 9. At a first glance we recognize the face of Einstein. But when we study this picture at a smaller scale, we suddenly recognize three bathing girls. Thus, human perception is not scale invariant. Rather we make hypotheses, mostly subconsciously, on the meaning of possible interpretations of images.

Clearly, such hypothesis formation must be implemented into future computers for pattern recognition (including those for speech recognition!). Once the scales are set adequately, we may construct faces by choosing parts from reservoirs and we may attempt to recognize faces in this manner (cf. Fig. 10).

Fig. 9: Compare text.
This picture was drawn
by Sandro Del-Prete
and reproduced from [12].

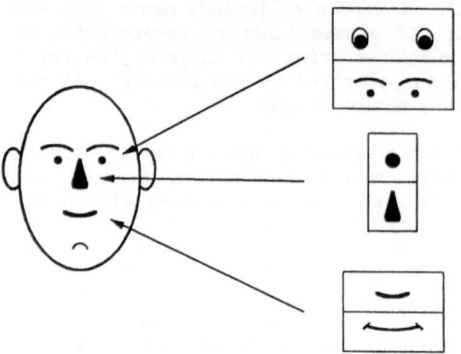

Fig. 10: Composition of faces from individual parts such as eyes etc.

4. Oscillations and hysteresis in perception

In order to let the computer identify ambiguous patterns, such as that of Fig. 11, we introduced a term which describes the saturation of the attention parameter once the corresponding order parameter increases, e.g. we implemented equations of the following form:

$$\dot{\lambda}_j = - \gamma \lambda_j + 1 - d_j . \tag{9}$$

Then the solution of a coupled set of order parameter equations (6) and attention parameters showed an oscillatory behavior (Fig. 12) as it is actually

8

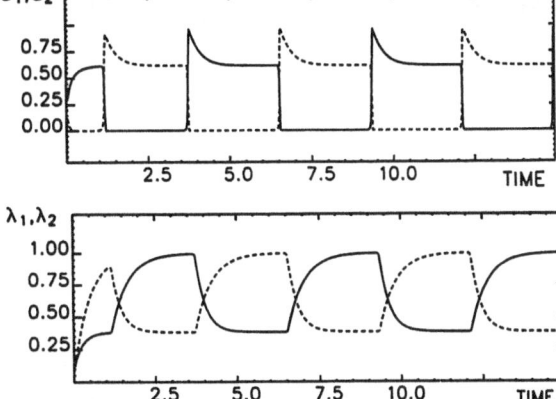

Fig. 11:   Vase or faces ?

Fig. 12

Fig. 12:   Oscillations of perception. Upper part: The dashed curve refers to the recognition of a vase,
the solid line to the recognition of a face. The oscillation can clearly be seen. Lower part:
Oscillation of attention parameters (After H. Haken and T. Ditzinger, unpublished).

Fig. 13:   Hysteresis·in perception.

found in experiments on the recognition of ambiguous figures by humans. It seems
possible to choose the constants in (4) and (9) so that the experimentally
observed switching times between perceived patterns can be matched. This may open
a way to a quantitative modeling of this kind of perception phenomena. Let us
discuss a second experiment which this time shows hysteresis of perception. When
we look at the individual parts of Fig. 13 starting from the left upper corner we
shall recognize a face. This perception persists in the first and second part in
the second row when it suddenly switches to the perception of a girl. When we
follow the individual parts in the reverse sequence, a jump from girl to face
happens only somewhere in the first row. Thus the jump of perception from face to
girl or girl to face depends on the history. This phenomenon can be incorporated
into our general model by the following consideration. Looking at the different
parts of Fig. 13 leads to a learning process in which the ridge separating the
valleys "face" and "girl" is shifted so that a pattern "face (or "girl") is
accepted even if the feature vectors already have a stronger overlap with the
other prototype vector (Fig. 14). This rotation by an angle α can be achieved by
the following term:

$$V = \frac{C}{r^4}\, d_1^2\, d_2^2 (1 - \frac{4\alpha}{r^2}\, (d_2^2 - d_1^2))^2, \qquad r^2 = d_1^2 + d_2^2 ,$$

which replaces the second sum on the r.h.s. of (4) in the case of two prototype
patterns. The angle α depends on the learning process and has only a rather
short term memory. A change of attention parameters λ may also be included.

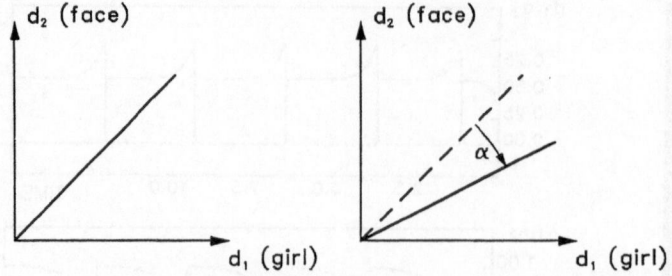

Fig. 14: Two-dimensional plot of the ridge of the potential curve in the unbiased case (l.h.s.) and the biased case after learning (r.h.s.).

5. Some possible realizations by neural nets

It is rather simple to realize our abstract model by means of neural networks. When we write out the individual equations for the features, i.e. for the components q_j, we find coupled differential equations which on the right hand sides are linear or cubic in the individual feature terms, q_k. Then it is obvious that we can identify q_k with the activity of a formal neuron of kind k, and these neurons are connected by connectivities $\lambda_{jk}...$ containing λ's with two and four indices (Fig. 15). The order parameter concept allows us to replace this completely parallel network by a three-layer network where the first layer contains the features at time $t = 0$ [5]. This network then projects via the components of the vectors v_j^u to a network of order parameters , d_u (Fig. 16). Finally in a third layer the order parameters project on the output which may contain again the features q_j (auto-associative memory) or other vectors in the case of a hetero-associative memory. The order parameter equations (6) allow an interpretation as network as shown in Fig. 17. Here no more direct connections are present between the neurones representing the order parameters, but the neurones interact only via a common reservoir $D = \Sigma\, d_u^2$. This leads us to suggest a

$\underset{\sim}{q}(t)$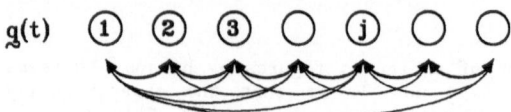

Fig. 15: Visualisation of equation (4) as a neural network (after [5]).

$\underset{\sim}{q}(0)$

$d_k(t)$

$\underset{\sim}{v}^{(k')}$

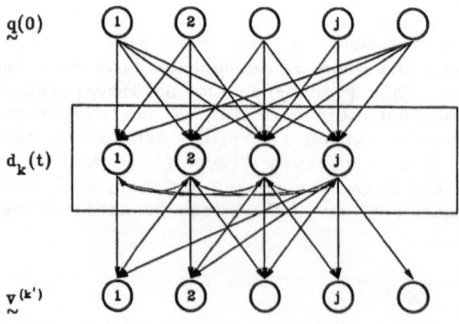

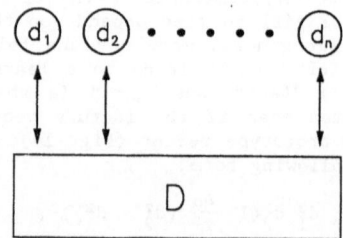

Fig. 16: A three layer network utilizing the order parameters d_u (after [5]).

Fig. 17: Network connections for the order parameters (after [5]).

new kind of learning in such nets which may be called non-Hebbian learning. Namely, procedures must be found to minimize the number of synapses, which has obviously occurred when we go from the original network of Fig. 15 to the final network of Fig. 17. In contrast to conventional neurocomputers of the spin-glass type, the neurones presented here are not threshold elements but rather perform simple multiplications and additions.

6. Learning of patterns, processes and associative action [13]

The methods of synergetics provide us with learning rules in a very natural way. In order to treat pattern formation including the effects of fluctuations, the general procedure is this: We start from evolution equations of the form (1). They are then transformed into order parameter equations to which the Fokker-Planck equation is established. In a variety of cases the condition of "detailed balance" is fulfilled which allows us to write down the stationary solution f of the Fokker-Planck equation explicitly. From it, we may read off the most probably macroscopic patterns the system may acquire. In pattern recognition we precisely follow the reverse path. We first make guesses on f, and we eventually deduce the explicit form of the equation

$$\dot{\underline{q}} = L\underline{q} + \underline{N}(\underline{q}) + \underline{F}(t),$$

from which we may read off the wiring of the neural net.

So let us briefly discuss the construction principle when we do not know the offered patterns a priori but wish to let the system do the job. When we offer patterns $P^{(1)}$, $P^{(2)}$ ... again and again we may, at least in principle, measure a probability distribution $f(P)$. The maxima of this distribution define the offered patterns. In practice we need specific feature detectors which measure features which are quantified by numbers q_1, q_2 ... . These numbers then form the components of the pattern vector $\underline{q}$. The question arises how we may reconstruct the probability distribution for these vectors $\underline{q}$ when we know the output of the feature detectors. As usual in physics this can be done by measuring moments of the form

$$<q_i\, q_j>, \;\cdot\cdot\;,\;\; <q_i\, q_1\, q_m\, q_n> , \tag{10}$$

which must be taken into account at least up to the fourth order. Then the maximum entropy principle of Jaynes [14] allows us to make unbiased estimates on the probability distribution function f. The constraints (10) are taken care of by adequate Lagrange parameters. The result can be written in the form

$$\bar{f}(\underline{q}) = \exp[- \lambda - \sum_{ij} \lambda_{ij}\, q_i\, q_j - \sum_{ijkl} \lambda_{ijkl}\, q_i\, q_k\, q_1\, q_m\,] , \tag{11}$$

where λ, λ_{jk}, λ_{jklm} are the Lagrange multipliers. We may interpret $\bar{f}$ as the stationary solution of a Fokker-Planck equation which belongs to a set of constant noise sources, $F_j(t)$. This Fokker-Planck equation is then equivalent to Langevin equations of the form

$$\dot{q}_j = 2 \sum_i \lambda_{ij}\, q_j + 4 \sum_{ikl} \lambda_{ijkl}\, q_k q_1\, q_m + F_j(t) . \tag{12}$$

As is evident, the synaptic connections between the neurones are identical with the Lagrange multipliers. Actually a learning procedure can be implemented now in which the information gain

$$K = \int \hat{f}(\underline{q}) \ln(\hat{f}(\underline{q})/\bar{f}(\underline{q})) \; d^N q$$

is minimized. This is achieved by using a distribution function $\hat{f}$ which has the same form as $\bar{f}$, but where the Lagrange multipliers are still to be determined. When we adopt an optimization procedure using a gradient strategy, we obtain the equations

$$\dot{\lambda}_{ij} = <q_i q_j>_{\hat{f}} - <q_i q_j>_{\bar{f}} \tag{13}$$

and

$$\dot{\lambda}_{ijkl} = <q_i q_j q_k q_l>_{\hat{f}} - <q_i q_j q_k q_l>_{\bar{f}} \; . \tag{14}$$

These equations may be considered as a generalization of those of Ackley, Hinton and Sejnovsky [15] to the present case.

The use of a Fokker-Planck equation allows us even to write down the construction principles for an autonomous network which performs <u>specific processes</u>. This procedure then includes perception-motor-control. We assume in the following that the processes to be learned by the system are Markovian. Then its joint probability

$$P_n \; (\underline{q}_n, \; t_n; \; \underline{q}_{n-1}, \; t_{n-1}; \; \cdots \; \underline{q}_o, \; t_o) \tag{15}$$

can be written in the form

$$P_n = \prod_{m=o}^{n-1} P(\underline{q}_{m+1}, \; t_{m+1} \; | \; \underline{q}_m, \; t_m) \; P_o(\underline{q}_o, \; t_o) , \tag{16}$$

where the individual terms P represent the conditional probability. To apply the maximum entropy principle we need the conditional moments $<q_i(t+\tau)>_{\underline{q}(t)}$ and $<q_j(t+\tau) \; q_k(t+\tau)>_{\underline{q}(t)}$. Applying the maximum entropy principle we may make an unbiased guess on the conditional probability which we study in the limit $\tau \to 0$. This conditional probability allows us to write down the corresponding Ito-Fokker-Planck equation which is equivalent to the Ito-Langevin equation of the form

$$\dot{q}_j = N_j(\underline{q}) + F_j(\underline{q}, \; t) \; . \tag{17}$$

We need the Ito calculus because in general the noise sources F now become $\underline{q}$-dependent. In addition, it turns out that the network is no more symmetric, at least in general. For a detailed representation of the procedure described in this section I refer the reader to [13].

7. Some possible realizations

Finally I wish to discuss possible realizations of the standard model of section 3. At a different instance I reported about a possible realization by a chemical parallel computer. This system is, however, rather complicated. More realistic devices seem to be lasers. In a laser, different modes may occur. When we describe the modes by their photon-numbers $n_j = E_j^2$ where the E_j are the corresponding field amplitudes, the equations of motion read [16]

$$\dot{n}_j = (G_j - \kappa_j - D_j) \; n_j \; , \tag{18}$$

where

$$D_j = \sum_{j'} D_{jj'} \, n_{j'} \, , \qquad\qquad (19)$$

G_j is the gain of mode j, κ_j its loss and D_j describes the saturation. Quite evidently these equations are more general forms of the order parameter equations (6), where d_u corresponds to E_j. Thus by an adequate choice of atom field interaction we may arrive at bringing the general equation into the form of the order parameter equation (6) or into forms equivalent to them. This suggests a three-layer treatment, again an input layer of data, which for instance by holographic mappings can be mapped onto the laser, then the laser with its modes serving as a decision making device and finally an output layer in which the surviving mode triggers new sets of features. In this respect the results of Lugiato and Narducci on multistable spatial patterns (cf. these proceedings) seem to be promising for a realization of the decision making layer.

8. Towards a synergetics of cognition and behavior

One of the central concepts of synergetics is that of the order parameter. As we have seen before it may stand for a total pattern such as a specific face or it appears in theories on the formation of behavioral patterns as was suggested at various previous occasions [17] and [18] which is dealt with in the article by Kelso in these proceedings. For various reasons it appears to be in order to discuss the order parameter in more detail. Namely, what is the very nature of the order parameter from the ontological point of view? According to synergetics the order parameter describes the macroscopic order and it gives orders to the subsystems. In the context of biological systems this requires some discussion. For instance, Kelso's experimental findings can be interpreted by using phase (or phase angle) as order parameter. But how can a phase give orders to the subsystems in a biological system? Fist of all it must be said that in the mathematical approach of synergetics the concept of order parameters emerges in a quite natural manner, when a complex system is studied close to its point of instability, i.e. where an old structure or an old behavioral pattern is replaced by a new one. So at first sight the order parameter seems to be a mathematical construct which is as real as the number π or any other mathematical concept. In any specific case of a physical or biological system the order parameters have specific properties that can be measured. For instance, the phase occurring in Kelso's experiments can be measured as a function of time, as well as its specific changes between different states and its fluctuations. That there is much more to the concept of order parameter from the ontological point of view can best be visualized by the consideration of physical systems. In the well-known physical example of the laser, the behavior of the laser atoms is governed by the field strength which acts as order parameter. Here, the order parameter has a material quality. In addition, we observe circular causality because the individual atoms by their emission of light generate the light field which in turn reacts on the atoms and determines their behavior. When in a fluid layer heated from below rolls are formed, the pattern of the rolls is an immaterial quality. Similarly, when swimmers in a pool start to move in circles to minimize their mutual hindrance, the circular swimming is an immaterial concept. Nevertheless, a swimmer entering the swimming pool will be automatically forced to join the general swimming motion. In addition, in the case of the fluid, the order parameter has specific measurable properties and allows us to predict the motion of the fluid.

In a way the order parameter can be thought of as an abstraction of a consensus among the individuals of a system who found a collective motion. However, as the fluid or the laser reveals, this interpretation as a consensus may have a too strong anthropomorphic flavor. In a way it is a philosophical question whether we say the individual parts of a system behave as if they were governed by an order parameter, or whether the order parameter is a real quantity. To say the least, it may be used as a well founded working hypothesis. But in my interpretation order parameters in biological systems are at least as real as our thoughts. In whatever way we wish to interpret the order parameter concept, it has far reaching consequences on the interpretation of phenomena of biology because we realize that an ensemble of individual cells may produce a joint macroscopic action described and prescribed by order parameters.

Another question which, in my opinion, is still not entirely settled concerns the localisation of cells or neurons which produce their order parameters which then in turn may influence or steer other cells like those of muscles. For instance in Kelso's experiments on involuntary changes of hand movements as well as in the movements of decerebrate cats, it is likely that the neurons are situated in the spinal cord. Their action seems to be more similar to that of a self-organized central pattern generator than that of a fixed motor program. However, afferences may play an important role in the establishment of patterns governed by order parameters. While the experiments by Erich von Holst indicate that afferences play a minor or no role in the motion of fins of fishes, afferences may be important in animals not subjected to a homogeneous surrounding such that of water, i.e. for animals living on the earth or birds. A further question is whether rhythmic movements and single event movements such as grasping for an object may be modelled in a similar fashion. I think that there is good reason to believe that our brain contains a model of our surrounding and after a task is set the brain forms a "potential" in which the state of the system moves, under additional constraints, along a specific trajectory to the final attractor state. Thus in contrast to rhythmic movements we rather have to deal with transients towards a specific fixed point attractor. Again, order parameters determining the motion come into play.

It is my profound conviction that the concepts of order parameters and slaving play a crucial role in cognition also and that the same principles come into play at different hierachical levels. Our approach on pattern recognition outlined above demonstrates that we may cast specific tasks of pattern recognition in precise mathematical terms and that we may model various phenomena such as association, hysteresis, and oscillations. In addition, the results obtained by Fuchs and Haken [10], [11] show that linear preprocessing may play an important role with respect to invariance properties such as translation, rotation, and scaling. First tests also show that deformations can be taken care of to some extent, for instance deformation of faces. There exists a solid frame composed of nose, eyes and partly eyebrows, whereas other parts can undergo some deformations which can be taken care of by smooth mathematical mappings. It appears also that a few typical views of a face are sufficient to characterize it, and that a machine or our brain may reconstruct any other position by means of specific rotations not only in the plane of the observer but also in other planes. While linear processing or preprocessing helps us to identify a specific face by association, its discrimination against all other faces is done by nonlinear decision making as is at least demonstrated by our model outlined in section 3. This nonlinear desicion making is taking place at a rather global and universal level, namely here a majority decision is made over the features connected with one object (face) or another one.

While the concept of order parameters and slaving will help us to penetrate more deeply into the secrets of cognition, at least one warning should be added. By all means the brain of even a lower animal is a complex system. Complex systems are practically inexhaustible or, to use another interpretation, one may ask many different questions of a complex system, which may answer them in a variety of ways. In view of the rich variety to ask questions and to get answers it appears to me that we may set up not only experiments but also _specific_ models which apply to _specific_ behavioral patterns or to _specific_ processes of cognition, but I strongly doubt that we shall ever be able to set up a complete model of our brains. (For an interesting discussion of complexitiy see [19]).

I wish to thank J.A.S. Kelso, H.-P. Koepchen und O. Meijer for highly stimulating discussions on the subject of section 8. This work was supported by the Volkswagenwerk Foundation, Hannover.

References

[1] M. Candill, S. Butler eds., IEEE First Int. Cong. on Neural Networks, Vols. 1-4, SOS Printing San Diego (1987) with many further references

R. Eckmiller, Ch.v.d. Malsburg, eds., Neural Computers, Springer, Berlin, New York (1987),
W.v. Seelen, G. Shaw, U.M. Leinkos, eds., Organization of Neural Networks, VCH, Weinheim, (1988)

[2] T. Kohonen, Neural Networks $\underline{1}$, 3 (1988)

[3] H. Haken, Synergetics. An Introduction. 3rd ed., Springer, Berlin, New York (1983)

[4] H. Haken, Advanced Synergetics, 2nd corr. print., Springer, Berlin, New York (1987)

[5] H. Haken, in Computational Systems - Natural and Artificial, H. Haken, ed., Springer, Berlin, New York (1987)

[6] H. Haken, in Pattern Formation by Dynamical Systems and Pattern Recognition, H. Haken, ed., Springer, Berlin, New York (1979)

[7] K. Steinbuch, Kybernetik $\underline{1}$, 36 (1961)

[8] J.J. Hopfield, Proc. Nath. Acad. Sci. USA, $\underline{79}$, 2554 (1982)

[9] H. Haken, A. Fuchs, in Dynamic Patterns in Complex Systems, J.A.S. Kelso, A.J. Mandell, M.F. Shlesinger, eds., World Scientific, Singapore (1988)

[10] A. Fuchs, H. Haken, these proceedings

[11] A. Fuchs, H. Haken, Biol. Cybernetics, in press

[12] Sandro Del-Prete, Illusorismen, Illusorismes, Illusorisis s, Beuteli Verlag, Bern, 3rd. ed. (1984)

[13] H. Haken, Information and Self-Organization, Springer, erlin, New York (1988)

[14] E.T. Jaynes, Phys. Rev. $\underline{106}$, 4, 620 (1957); $\underline{108}$, 171 (1.57)

[15] D.H. Ackley, G.E. Hinton, T.H. Sejnovsky, Cogn. Sci. 9, 147 (1985)

[16] H. Haken, Laser Theory, corr. print., Springer, Berlin, New York (1984)

[17] H. Haken, in Synergetics of the Brain, Eds., E. Basar, H. Flohr, H. Haken, A.J. Mandell, Springer, Berlin, New York (1983)

[18] J.A.S. Kelso, G. Schoener, in Lasers and Synergetics, Ed., R. Graham, A. Wunderlin, Springer, New York (1987)

[19] R. Rosen, in Dynamic Patterns in Complex Systems, J.A.S. Kelso, A.J. Mandell, M.F. Shlesinger, eds., World Scientific, Singapore (1988)

Computer Simulations of Pattern Recognition as a Dynamical Process of a Synergetic System

A. Fuchs and H. Haken

Institut für Theoretische Physik und Synergetik, Universität Stuttgart,
Pfaffenwaldring 57, D-7000 Stuttgart 80, Fed. Rep. of Germany

1. Introduction

In this paper we want to present computer simulations of an algorithm of pattern recognition and associative memory, which was introduced by Haken [1] at the International Symposium at Schloß Elmau, Bavaria in 1987. This was the first specific formulation of a synergetic system that exhibited the analogy between pattern formation and pattern recognition that Haken suggested in 1979 [2]. The crucial point is that the ability of a system to form patterns is strongly related to its ability to recognize patterns or to act as an associative memory. While these basic ideas are discussed in the article of Haken (cf. these proceedings), we shall focus our attention on explicit simulations, and on a discussion of possible generalizations of this approach [3,4]. We shall proceed as follows: In Section 2 we introduce the mathematical assumptions that are nessecary, and show how an associative memory can be realized by a synergetic system [5,6]. In Section 3 we shall reduce the degrees of freedom of the system by introducing order parameters describing the interaction between the macroscopic states of the system. Then in the Sections 4 and 5 generalizations of our approach are discussed. We show that the approach can be formulated in a translationally invariant form which will lead us to the ability to decompose complex patterns or scenes [7]. In Section 5, a further generalization with respect to rotation and scaling is introduced, by use of the complex logarithmic map which also exists in the visual systems of mammals [8,9] as the connection between the retina and the visual cortex.

2. A Dynamical System Acting as an Associative Memory

In the following we shall understand by "pattern" a two-dimensional array of colors or grey-values of pixels, that are represented as components of a vector. Equivalently a pattern is represented by the vector built up of expansion coefficients which are obtained by the projection of the pattern onto a set of basis vectors. An example for such vectors is provided by the faces shown in fig. 1 which form a set of M basic vectors or prototype patterns $\mathbf{v}_{(k)}$ having N components, and which are normalized, $|\mathbf{v}_{(k)}| = 1$. Of course, in general, such vectors do not form an orthogonal set so that we introduce a set of adjoint vectors $\mathbf{v}^{(k)}$ in a way that the relations

$$\mathbf{v}_{(k)} \, \mathbf{v}^{(k')} = \delta^k_{k'} \tag{1}$$

hold. In general, the number of components N of the vectors $\mathbf{v}_{(k)}$ will be (much) larger than the number of prototype patterns N and there are infinitely many possibilities to fulfill relation (1). To obtain a unique definition of the vectors $\mathbf{v}^{(k)}$ we

Springer Series in Synergetics Vol. 42: **Neural and Synergetic Computers**
Editor: H. Haken   ©Springer-Verlag Berlin Heidelberg 1988

Fig. 1: Set of basis vectors

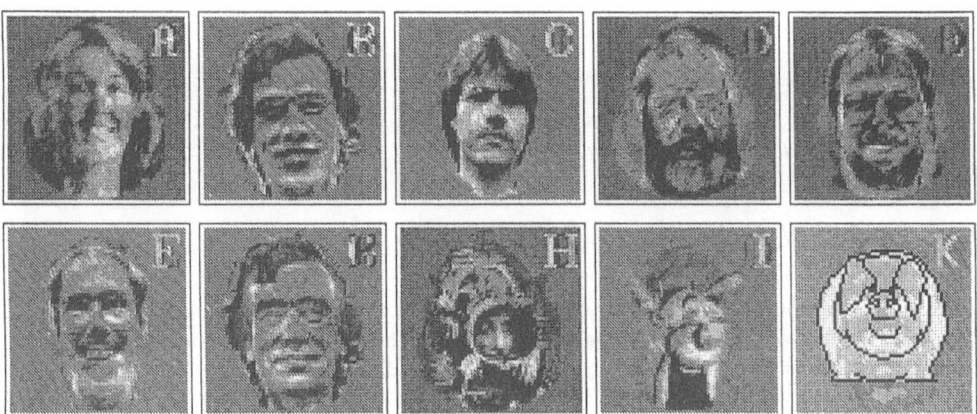

Fig. 2: The adjoint vectors corresponding to the set of fig. 1

restrict them to the subspace defined by the prototype patterns i.e. we write the $\mathbf{v}^{(k)}$ as linear combinations of the $\mathbf{v}_{(k)}$

$$\mathbf{v}^{(k)} = \sum_{k'=1}^{M} a^{kk'} \mathbf{v}_{(k')} \quad . \tag{2}$$

When we multiply eq. 2 by $\mathbf{v}_{(j)}$, we obtain a linear system of equations that determines the coefficients $a^{kk'}$, and therefore the vectors $\mathbf{v}^{(k)}$ uniquely. The "adjoint" faces which correspond to the patterns of fig. 1 are shown in fig. 2. To get a complete set of N basis vectors we add (N-M) vectors $\mathbf{w}_{(k)}$ and their adjoint ones $\mathbf{w}^{(k)}$ to the prototype patterns in a way that the metric tensor acquires the following form:

17

$$g_{kk'} = \mathbf{v}_{(k)} \, \mathbf{v}_{(k')} \qquad g^{kk'} = \mathbf{v}^{(k)} \, \mathbf{v}^{(k')} \qquad\qquad \text{if} \quad k, \, k' \leq M$$

$$g_{kk'} = \mathbf{w}_{(k)} \, \mathbf{v}_{(k')} = g^{kk'} = \mathbf{w}^{(k)} \, \mathbf{v}^{(k')} = 0 \qquad \text{if} \quad k > M, \, k' \leq M \qquad (3)$$

$$g_{kk'} = \mathbf{w}_{(k)} \, \mathbf{w}_{(k')} = g^{kk'} = \mathbf{w}^{(k)} \, \mathbf{w}^{(k')} = \delta^k_{k'} \quad \text{if} \quad k, \, k' > M \; .$$

This can be achieved, for instance, by use of Schmidt's orthogonalization procedure.

Now we are able to define a dynamical system by a potential function that depends on the prototype patterns and a test pattern given by the vectors $\mathbf{q}$ and $\mathbf{q}^+$, where $\mathbf{q}^+$ is related to $\mathbf{q}$ by the metric tensor $\underline{\underline{\mathbf{G}}}$.

With the notation used above the potential function given by Haken [1] reads

$$V = V_1 + V_2 + V_3$$

$$V_1 = -\frac{1}{2} \sum_{k=1}^{M} \lambda_k \, (\mathbf{q}^+ \mathbf{v}_{(k)}) \, (\mathbf{v}^{(k)} \mathbf{q}) = \mathbf{q}^+ \, \underline{\underline{\mathbf{J}}} \, \mathbf{q}$$

$$\underline{\underline{\mathbf{J}}} = \sum_{k=1}^{M} \lambda_k \, \mathbf{v}_{(k)} \, \mathbf{v}^{(k)} \qquad\qquad (4)$$

$$V_2 = \frac{1}{4} B \sum_{k \neq k'}^{M} (\mathbf{q}^+ \mathbf{v}_{(k)}) \, (\mathbf{v}^{(k)} \mathbf{q}) \, (\mathbf{q}^+ \mathbf{v}_{(k')}) \, (\mathbf{v}^{(k')} \mathbf{q})$$

$$V_3 = \frac{1}{4} C \, (\mathbf{q}^+ \mathbf{q})^2 \; .$$

The interpretation of these terms and a visualization of the potential function is given in the contribution of Haken in these proceedings. The evolution equations in time for the test patterns are determined by the negative gradient of the potential.

$$\dot{\mathbf{q}} = -\frac{\partial V}{\partial \mathbf{q}^+} \qquad\qquad \dot{\mathbf{q}}^+ = -\frac{\partial V}{\partial \mathbf{q}}$$

$$\qquad\qquad (5)$$

$$\dot{\mathbf{q}} = \sum_{k=1}^{M} \lambda_k \, (\mathbf{v}^{(k)} \mathbf{q}) \, \mathbf{v}_{(k)} - B \sum_{k \neq k'}^{M} (\mathbf{v}^{(k)} \mathbf{q})^2 \, (\mathbf{v}^{(k')} \mathbf{q}) \, \mathbf{v}_{(k)} - C \, |\mathbf{q}|^2 \, \mathbf{q} \; .$$

For our numerical simulation we choose all $\lambda_k = \lambda$ for the moment, and we shall return to the interpretation of these parameters later in section 5. If λ is smaller than zero, the trajectories of the equations (5) have only one stable fixedpoint, $\mathbf{q} = 0$. If λ becomes positive, each one of the prototype patterns generates two stable fixed points, corresponding to the pattern and its negative, because the potential is symmetric with respect to $\mathbf{q}$ and $-\mathbf{q}$. A plot of the trajectories for a test pattern which is a linear combination of two prototypes is given in fig. 3 where for the case λ greater than zero the fixedpoint at $\mathbf{q} = 0$ has become unstable. Figure 4 shows a numerical simulation of the equations (5) where as initial condition the label of one of the faces i.e. the capital letter in the upper right corner of the pattern was used. During the evolution in time the face corresponding to the label is restored – the system acts as an associative memory.

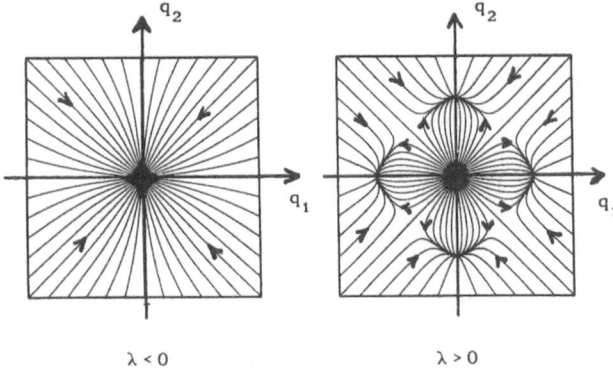

$\lambda < 0$ $\qquad\qquad\qquad$ $\lambda > 0$

Fig. 3: Trajectories of eq. (5) below and above the threshold value of λ

Fig. 4: Associative recall of a face from its label

3. Reduction of the Degrees of Freedom

To this end we expand the test vector into the basis set $\mathbf{v}_{(k)}$ and $\mathbf{w}_{(k)}$

$$\mathbf{q} = \sum_{k=1}^{M} q^k \, \mathbf{v}_{(k)} + \mathbf{w} \qquad\qquad \mathbf{q}^+ = \sum_{k=1}^{M} q_k \, \mathbf{v}^{(k)} + \mathbf{w}^+ \qquad (6)$$

and insert this ansatz into the basic equations (5). From there the equations of motion for the expansion coefficients q^j and the residual vector $\mathbf{w}$ can be derived.

$$\dot{q}^j = \lambda_j q^j - B \sum_{k \neq j}^{M} q^k q_k q^j - C \{ \sum_{k=1}^{M} q^k q_k + |\mathbf{w}|^2 \} \, q^j$$

$$\dot{\mathbf{w}} = - C \{ \sum_{k=1}^{M} q^k q_k + |\mathbf{w}|^2 \} \, \mathbf{w} \; . \qquad\qquad (7)$$

In fig. 5 the evolution in time of the different q^j and $|\mathbf{w}|$ is given for the example mentioned in the previous section. The q^j that corresponds to the pattern which is recognized is plotted in a solid line, the others are dotted.

Because $|\mathbf{w}|$ appears only in the saturation term of equation (7), and decreases exponentially in time, it can be neglected from the very beginning. So we end up with equations for the expansion coefficients q^j only:

$$\dot{q}^j = \lambda_j q^j - B \sum_{k \neq j}^{M} q^k q_k q^j - C \sum_{k=1}^{M} q^k q_k q^j \; . \qquad\qquad (8)$$

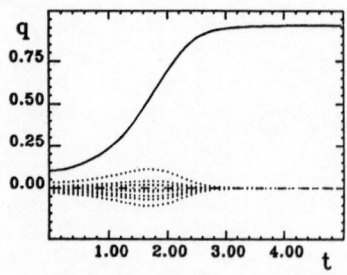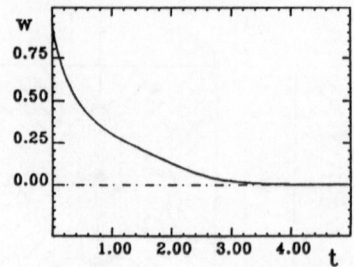

Fig. 5: Time evolution of the expansion coefficients q^j, and the magnitude of the residual vector **w**

These are equations which describe the interaction of the internal macroscopic states of the system i.e. the prototype patterns. In the synergetic theory of pattern formation [10,11] similar equations, which represent the competition of the macroscopic modes, are treated. They describe the dynamics of the order parameters and are wellknown as generalized Ginzburg-Landau equations. This system of M equations is, in general, low dimensional compared with the dimension N, given by the number of pixels of each pattern - so we have reduced the degrees of freedom of the dynamical system dramatically.

4. Translational Invariance and Decomposition of Scenes

In this section we want to generalize the former approach of associative memory to a formulation which is invariant with respect to spatial shifts of the patterns. This will provide us with the ability to extract single patterns from complex scenes. For what follows it is advantageous to introduce continuous variables in space. Then the vectors representing the patterns have to be substituted by functions

$$\mathbf{q}(t) \rightarrow |q(x,y,t)\rangle \qquad \mathbf{v}_{(k)} \rightarrow |v_k(x,y)\rangle \ , \tag{9}$$

the scalar product is to be replaced by an integral over the whole space

$$(\mathbf{v}^{(k)} \ \mathbf{q}) \rightarrow \int dx \int dy \ v^k(x,y) \ q(x,y,t) = \langle v^k(x,y)|q(x,y,t)\rangle \ , \tag{10}$$

and, finally, the gradient of the potential function which defines the equations of motion becomes the functional derivative of $V(|q\rangle, \langle q|)$

$$|\dot{q}\rangle = - \frac{\delta V}{\delta \langle q|} \quad . \tag{11}$$

Now we can apply a Fourier transformation to the prototypes and to the test pattern

$$\tilde{v}(k_x, k_y) = \int_{-\infty}^{\infty} dx \int_{-\infty}^{\infty} dy \ v(x,y) \ e^{i(k_x x + k_y y)}$$

$$\tilde{q}(k_x, k_y) = \int_{-\infty}^{\infty} dx \int_{-\infty}^{\infty} dy \ q(x,y,t=0) \ e^{i(k_x x + k_y y)} \quad . \tag{12}$$

These complex functions can be split into a magnitude, and a phase factor

$$\tilde{v}(k_x,k_y) = |\tilde{v}(k_x,k_y)|\, e^{i\varphi} \qquad \text{with} \qquad \varphi = \arctan\frac{\text{Im}(\tilde{v})}{\text{Re}(\tilde{v})} \quad . \tag{13}$$

If we now use only the magnitudes $|\tilde{v}(k_x,k_y)|$, and $|\tilde{q}(k_x,k_y)|$ for the dynamics, we have a formulation of the algorithm which is invariant with respect to spatial shifts [12] of the patterns. In fig. 6 a test pattern is shown which is part of one of the prototypes, but now shifted from its normalized location. The dynamics is started with the coefficients $\tilde{q}^j$ given by

$$\tilde{q}^j = \int_{-\infty}^{\infty}dk_x \int_{-\infty}^{\infty}dk_y\, |\tilde{v}^{\,j}(k_x,k_y)|\, |\tilde{q}(k_x,k_y)| \quad . \tag{14}$$

It is shown in the plot (fig. 6) of the temporal evolution of the different $\tilde{q}^j$, that their behavior is very similar to the behavior we obtained in the previous section by the integration of the expansion coefficients q^j (fig. 5). When the dynamics has reached its stationary state, the pattern is identified, and its spatial shift with respect to its prototype can be calculated. This can be done by searching for the values x_0 and y_0, where the correlation function $K(x_0,y_0)$ reaches its absolute maximum.

$$K(x_0,y_0) = \int_{-\infty}^{\infty}dx \int_{-\infty}^{\infty}dy\, v^j(x-x_0,y-y_0)\, q(x,y,t=\infty)$$

$$= \int_{-\infty}^{\infty}dk_x \int_{-\infty}^{\infty}dk_y\, \tilde{v}^{\,j*}(k_x,k_y)\, \tilde{q}(k_x,k_y,t=\infty)\, e^{i(k_x x_0 + k_y y_0)} \tag{15}$$

$$= \mathfrak{F}^{-1}\Big\{\mathfrak{F}^*\{v^j(x,y)\}\, \mathfrak{F}\{q(x,y,t=\infty)\}\Big\} ,$$

where $\mathfrak{F}$ stands for the Fourier transform, and the asterisk means complex conjugation. In fig. 7 a contour plot of $K(x_0,y_0)$ is shown, and the maximum corresponds to the spatial shift of the test pattern with respect to its prototype.

With the translational invariant formulation, and the interpretation of the values λ_j in equations (8) as attention parameters, it is possible to decompose scenes that consist for instance of a prototype which partially covers another one as shown in fig. 8. This procedure acts as follows:

 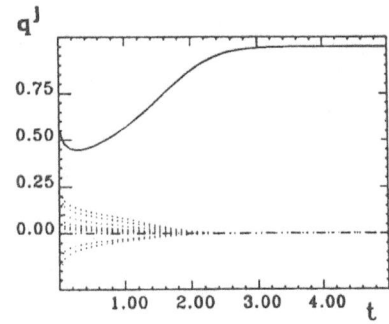

Fig. 6: Initial condition and time evolution of the coefficients for a shifted pattern

21

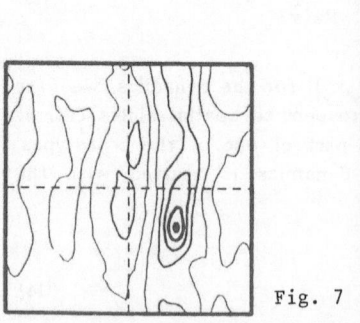

Fig. 7   Fig. 8

Fig. 7: Contour plot of the correlation function K showing the spatial shift of the initial pattern with respect to its prototype

Fig. 8: A scene of two faces, where one of them is partially covered by the other

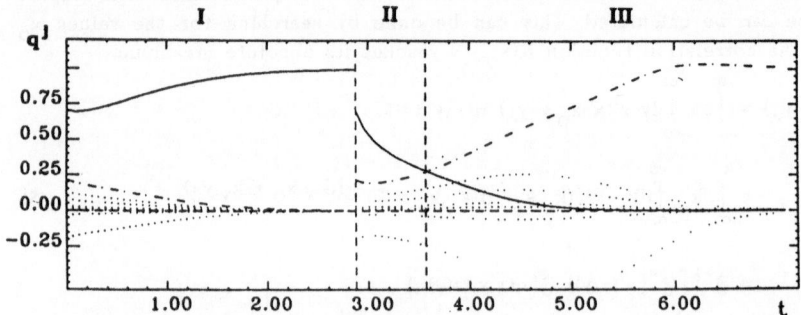

Fig. 9: Decomposition of the scene (fig. 8). In region I the first pattern is identified. The corresponding λ_j is changed to zero and the dynamics is restarted (reg. II). The second face is identified (region III).

1. A scene (fig. 8) is offered to the system and the pattern with the biggest $\tilde{q}^j$ is identifed by the computer (fig. 9).
2. When the dynamics has reached its stationary state (fig. 9), the λ_j that corresponds to the identified pattern is changed to zero, and the dynamics is started again with the original initial pattern.
3. The other pattern gets identified.

With this stepwise procedure it it possible to decompose complex scenes as shown in the figures 10.

Fig. 10: A complex scene, and three steps of the decomposition procedure. Left column: Time evolution of the expansion coefficients. Middle column: Contour plot of the correlation function showing the shift. Right column: Reconstructed pattern

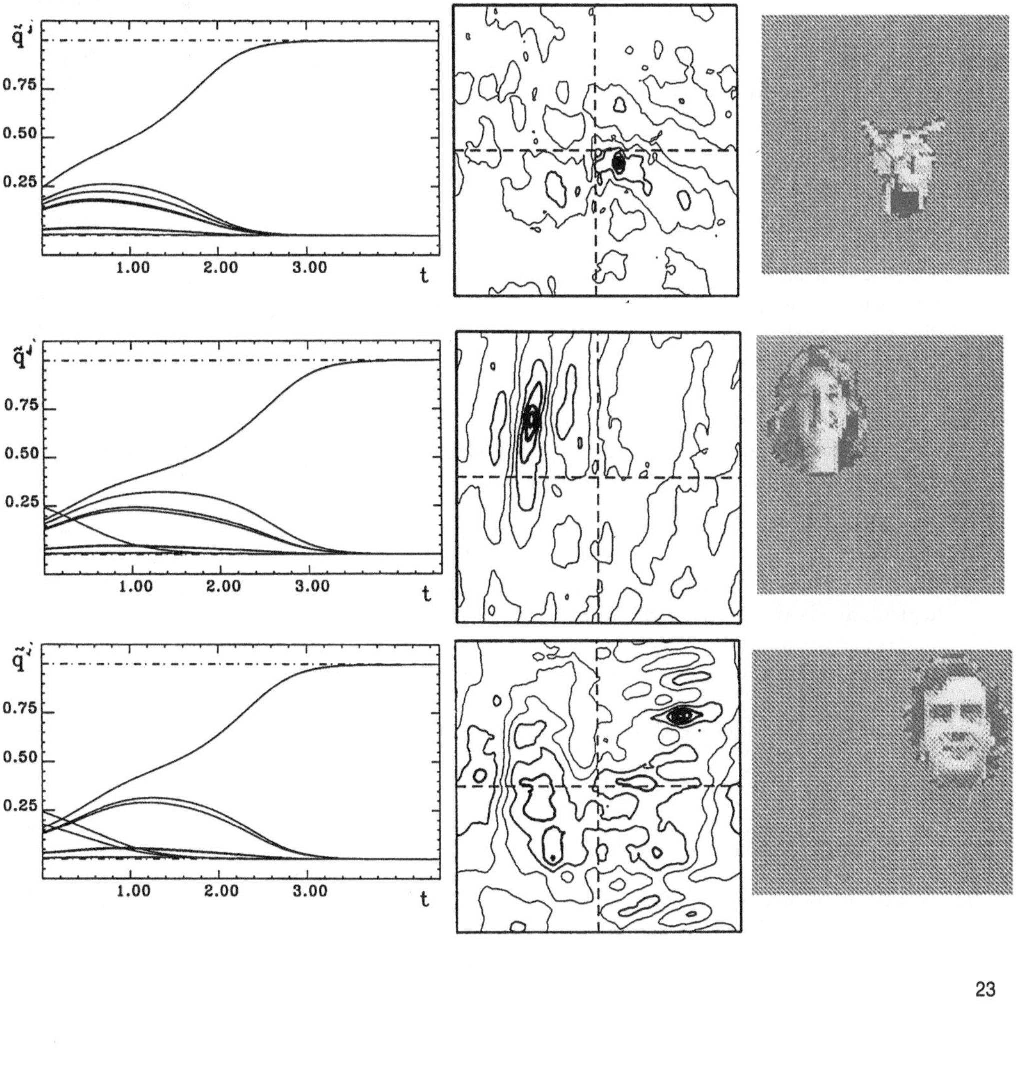

5. Simultaneous Invariance with Respect to Translation, Rotation, and Scaling

Now we show how our procedure can be formulated in a fashion which is simultaneously invariant under translation, rotation, and scaling. Above, the patterns $|\tilde{v}(k_x,k_y)|$ were shown to be invariant with respect to translation. Rotation and scaling of an image in real-space leads to rotation and scaling in k-space. It is well known that there exists a mapping from the retina to the visual cortex of mammals that transforms scaling and rotation of an image on the retina into a spatial shift at the visual cortex. Such a transformation can be realized by a conformal map with a complex logarithm. In order to incorporate translational invariance, we perform this transformation in k-space rather then in real-space, however. To this end we write the arguments (k_x,k_y) as complex numbers in the following form:

$$(k_x,k_y) = \mathbf{k} = k_x + i k_y =: r\, e^{i\varphi}$$

$$|\tilde{v}(k_x,k_y)| \rightarrow |\tilde{v}(\mathbf{k})| \rightarrow \hat{v}(r,\varphi) \quad .$$

(16)

We now apply a complex logarithmic mapping to the vector $\mathbf{k}$

$$\mathbf{k}' = \ln(\mathbf{k}) = \ln(r) + i\,\varphi \quad .$$

(17)

If $\mathbf{k}$ is represented in the form

$$\mathbf{k} = \frac{r}{r_0}\, e^{i(\varphi + \varphi_0)}$$

(18)

then $\mathbf{k}'$ reads

$$\mathbf{k}' = \ln(\frac{r}{r_0}) + i(\varphi + \varphi_0) = \ln(r) - \ln(r_0) + i(\varphi + \varphi_0),$$

(19)

which means a shift of the logarithm of the scaling factor r_0 along the real-axis and a shift of φ_0 along the imaginary axis. It is evident that this map diverges at the origin $r=0$, so that, in practical calculations, a circle around this point has to be neglected [13,14]. If this transformation is applied to an image in real-space, the eliminated region lies in the center of the image which is normally the most important one. This is not the case in k-space where the central region contains the coefficients of the constant, and slowly varying brightness which can be neglected without changing essential features of the image.

The procedure which is invariant under spatial transformations goes the following way:

1. A Fourier-transformation is applied to the original pattern and the phase factors are neglected: $v(x,y) \rightarrow |\mathfrak{F}\{v(x,y)\}| = |\tilde{v}(k_x,k_y)|$.

2. The magnitude is mapped by means of a complex logarithmic function:

$$|\tilde{v}(k_x,k_y)| \rightarrow \hat{v}(r,\varphi) \quad .$$

3. Another Fourier-transformation is now performed on $\hat{v}(r,\varphi)$:

$$\hat{v}(r,\varphi) \rightarrow |\mathfrak{F}\{\hat{v}(r,\varphi)\}| = |\tilde{\hat{v}}(r,\varphi)| .$$

The resulting values fulfill the required invariance properties. Numerical examples of this procedure are shown in the colored figures.

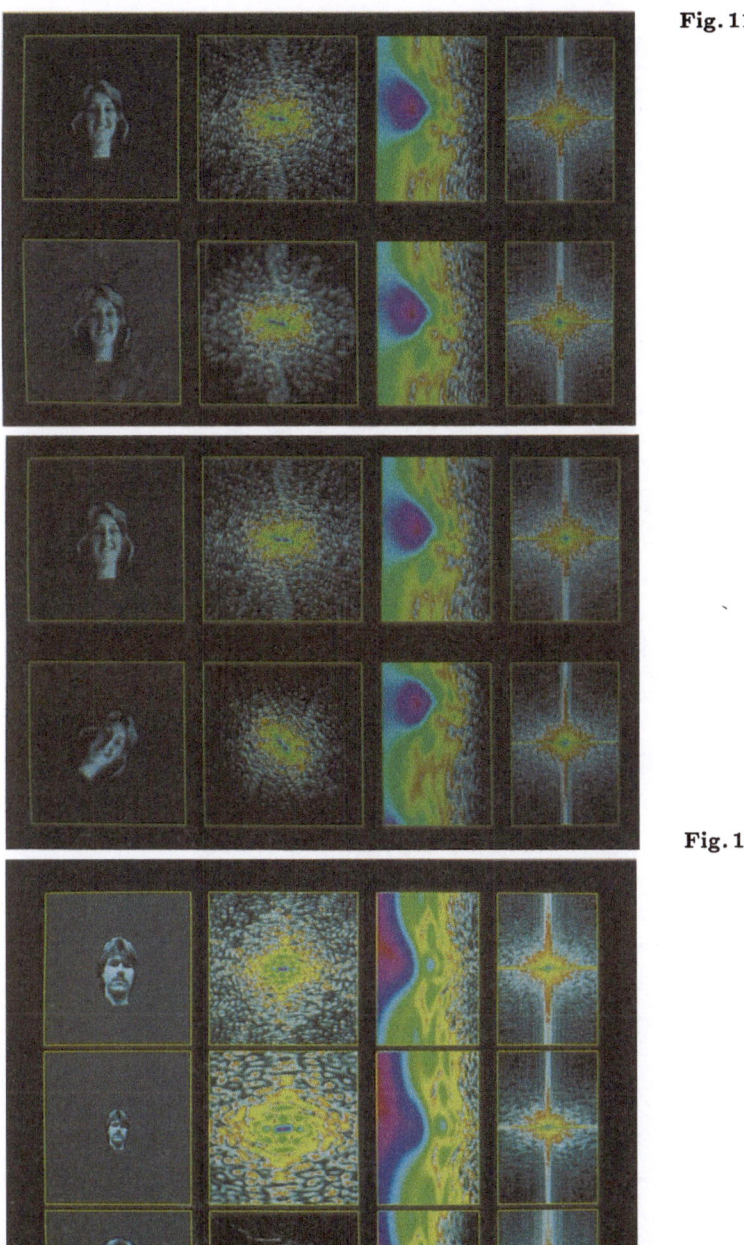

Fig. 11

Fig. 12

Fig. 13

Fig. 11. The steps which lead to the invariant pattern. (*Upper row*) The original pattern, its Fourier transformation, after applying the complex logarithmic map, after the second Fourier transformation. (*Lower row*) The same as above, but now calculated from the right to the left

Fig. 12. A pattern rotated in real space and ist different transforms

Fig. 13. A pattern scaled in real space and its different transforms

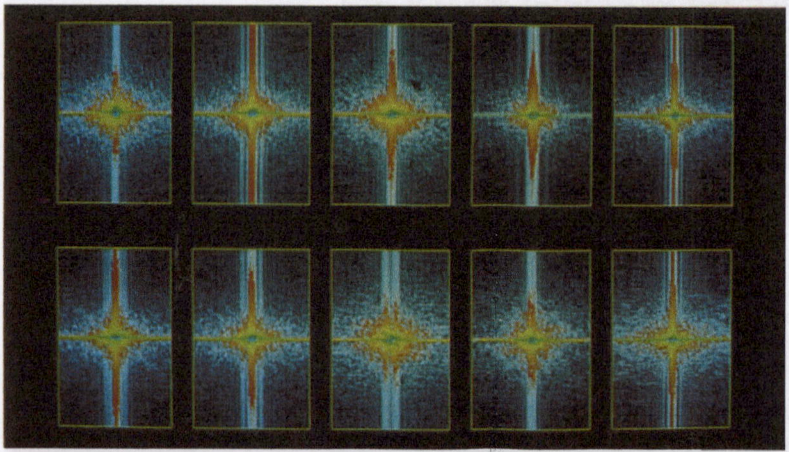

Fig. 14

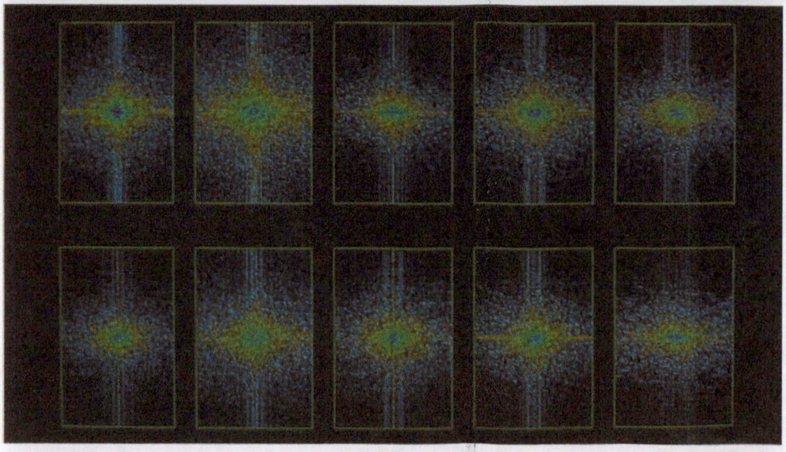

Fig. 15

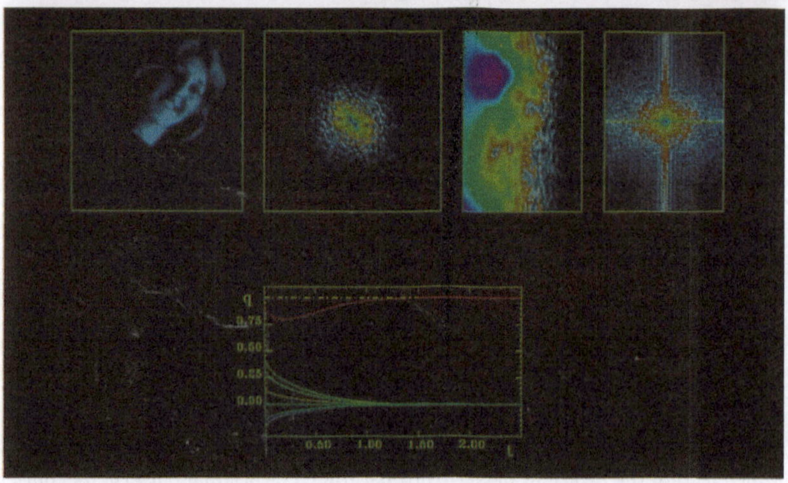

Fig. 16

Fig. 14. The basis patterns of Fig. 1 in the invariant representation

Fig. 15. Adjoint patterns corresponding to Fig. 14

Fig. 16. Associative recall of a pattern which is simultaneously scaled, shifted, and rotated in real space, by use of the invariant representations

Conclusion

We have shown that synergetic systems can act as associative memory. Our specific system is able to restore patterns which are offered only partially. The procedure was formulated in a manner invariant with respect to spatial transformations. It turned out that if a test pattern was built from prototypes which were only shifted but partially hidden, we succeeded in decomposing complex scenes using a procedure that identified the different prototypes one after the other.

Acknowledgement

We thank the Volkswagen Foundation, Hannover, for financial support within the project on Synergetics.

References:

[1] Haken, H. : Synergetic Computers for Pattern Recognition and Associative Memory.
In: Computional Systems, Natural and Artificial,
H. Haken, Ed., Springer, Berlin, (1987)

[2] Haken, H.: Pattern Formation and Pattern Recognition – An Attempt at a Synthesis.
In: Pattern Formation by Dynamical Systems and Pattern Recognition,
H. Haken, Ed., Springer, Berlin, (1979)

[3] Haken, H.: Nonequilibrium Phase Transitions in Pattern Recognition and
Associative Memory. Z. Phys. B – Condensed Matter 70, (1988), 121-123

[4] Fuchs, A. & Haken, H: Nonequilibrium phase transitions in pattern recognition
and associative memory. Numerical results.
Z. Phys. B – Condensed Matter 73, (1988), in press

[5] Fuchs, A. & Haken, H.: Pattern Recognition and Associative Memory as Dynamical
Process in Nonlinear Systems. Submitted to: IEEE International Conference on
Neural Nets, (1988)

[6] Haken, H. & Fuchs, A.: Pattern Formation and Pattern Recognition as Dual
Processes. In: Dynamic Patterns in Complex Systems, J.A.S. Kelso, A.J. Mandell,
M.F. Schlesinger, Ed., World Scientific, Singapore, (1988)

[7] Fuchs, A. & Haken, H.: Pattern Recognition and Associative Memory as Dynamical
Process in Synergetic Systems. Translational Invariance, Selective Attention,
and Decomposition of Scenes. To be published in Biol. Cybern., (1988)

[8] Fischer, B.: Overlap of receptive field centers and representation of the visual
field in the cat's optic tract. Vision Res. 13, (1988), 2113-2120

[9] Schwarz, E.L.: Spatial mapping in the primate sensory projection: analytic structure
and relevance to perception. Biol. Cybern. 25, (1977), 181-194

[10] Haken, H.: Synergetics, an Introduction, 3rd. ed. , Springer, Berlin, (1983)

[11] Haken, H.: Advanced Synergetics, 2nd. corr. printing, Springer, Berlin, (1987)

[12] Altes, R.A.: The Fourier-Mellin transform and mammalian hearing.
J. Acoust. Soc. Am. **63**, (1978), 174-183

[13] Reitboeck, H. J. & Altmann, J.: A Model for Size- and Rotation-Invariant
Pattern Processing in the Visual System. Biol. Cybern., **51**, (1984), 113-121

[14] Braccini, C. et. al.: A Model of the Early Stage of Human Visual System:
Functional and Topological Transformations Performed in the Peripherical
Visual Field. Biol. Cybern., **44**, (1982), 47-58

Variational Principles in Pattern Theory

W. Güttinger and G. Dangelmayr

Institute for Information Sciences, University of Tübingen,
D-7400 Tübingen, Fed. Rep. of Germany

1. Introduction

The understanding of pattern formation and its dual, pattern recognition, is one
of the most exciting areas of present research. It is the question of how complex
systems can generate coherent global structures and how systems are designed which,
by means of sensory and perceptional mechanisms, can construct internal representations of patterns in the outside world. The field represents a remarkable con-
fluence of several different strands of thought.

Pattern formation takes place in excitable media through nonlinear interactions of
competing forces and under the influence of boundary constraints. On the other hand,
when measurements group together and begin to form clusters in some feature space,
one tends to observe that a pattern is developing. Both phenomena represent qualitative properties of a system. By qualitative properties we mean invariants under
differentiable changes of coordinates, as opposed to quantitative properties which
are invariant only under linear changes of coordinates. The clustering of system
components or pixel intensities into coherent, visible global structures, on the
other hand, is the result of energy concentration in parts of space. It is therefore natural to identify and select dominant features in both pattern formation
and recognition from this unifying point of view. The resulting low-dimensional
description of complex phenomena through optimality principles is the subject of
this paper. In section 2 we discuss feature selection in pattern formation by means
of projections. Section 3 describes probabilistic nonlinear inversion in pattern
recognition. Feature selection by projection and pattern classification are discussed in sections 4 and 5.

2. Feature Selection in Pattern Formation

The formation and evolution of spatio-temporal patterns is the result of sudden
qualitative changes, known as bifurcations. Bifurcation occurs in a nonlinear system
if an externally controllable parameter passes through critical values so that a
balance existing between competing system-immanent forces breaks down. As a conse-

quence, an initially structureless system becomes unstable and, in a sequence of bifurcations, restabilizes successively in ever more complex space-and time-dependent configurations. Describing pattern formation by bifurcation requires an intuitive selection and identification of significant global state variables or order parameters in which most of the energy of the system is concentrated. These quantities measure the degree and type of ordering or structure which is built up or destroyed in the vicinity of a bifurcation point. Spontaneous structural changes are then characterized by the appearance of singularities in these observables. Analyzing just these amounts to a reduction of the system's degrees of freedom to a few basic ones [9].

Let

$$\frac{\partial u}{\partial t} = F(u,\lambda) \equiv L(\lambda)u + N(u,\lambda) \tag{2.1}$$

be a system of evolution equations for u with linear part L and nonlinear part N, $u(x,t)$ being an element of a Hilbert space and λ an externally controllable parameter. A bifurcation occurs if the real part of an eigenvalue of L passes through zero from negative values. A zero eigenvalue leads to stationary solutions whereas a pair of imaginary eigenvalues gives rise to time-periodic solutions. Suppose that for a critical value $\lambda = \lambda_c$, L possesses m eigenmodes ϕ_i whose eigenvalues have zero real parts. Then, when λ goes through λ_c, the nonlinear term N generates an interaction between the modes ϕ_i in the solution vector

$$u = \sum_{i=1}^{m} c_i \phi_i + w(c,\lambda) \ ,$$

where w is orthogonal to the ϕ_i and $c = (c_1, \ldots c_m)^T$. As a consequence, the amplitudes c_i of these modes would increase indefinitely due to resonance and u ceases to exist unless N is orthogonal to ϕ_i (Lyapunov-Schmidt reduction of (2.1)), i.e.,

$$G_i(c,\lambda): = (N(u,\lambda),\phi_i) = 0 \tag{2.2}$$

for all i, where $(\cdot,\cdot)$ denotes the scalar product. The stationary solutions of (2.1) are determined by the system of algebraic equations (2.2) whose solution curves are the bifurcation diagrams in (c,λ)-space.

Applications of the ensuing theory of pattern formation by mode interaction cover a wide spectrum ranging from nonlinear optics to solidification systems [9]. The basic patterns and transitions between them are selected by projecting the nonlinear system onto eigenmodes. Assuming that (2.1) is of gradient type this selection is summarized in the variational principle

$$R(u): = \frac{1}{2}(Lu,u) - f(u) = \text{Extremum}, \tag{2.3}$$

with N being the Frechet-derivative of f, whose solution $\partial R/\partial c_i = 0$ reduces to that of (2.2).

Feature selection by eigenmode decomposition is also familiar from image processing [6,25]. Given m picture vectors $u^k(x)$, with u the gray level at $x \in R^2$, consider the two-point spatial correlation function

$$K(x,y) = (1/m) \sum_1^m u^k(x)u^k(y).$$

Representing $u^k = \Sigma c_{ki}\phi_i$ in terms of the eigenpictures ϕ_i of the covariance matrix K, $K\phi_i = \mu_i\phi_i$, the total "energy" $R = (Ku,u) = \Sigma\mu_i$ is dominated by the first few eigenvalues (the sample variances). Thus, (2.3) may be regarded as a generalization of the Karhunen-Loève decomposition to the nonlinear situation. In setting up a dynamical system for pattern recognition, the projection operator K (which corresponds to $J = \Sigma_i \xi_i^\alpha \xi_j^\alpha$ in the binary Hopfield model [13]) provides the appropriate starting point [12].

In a dimensional reduction of the full dynamical system (2.1) the organized motion features are selected as follows [17,22]. Let now $\phi_i(x)$ be the orthonormal eigenfunctions of the covariance matrix

$$K(x,y) = <u(x,t)u^*(y,t)>$$

$$= \lim_{T\to\infty} \frac{1}{T} \int_0^T u(x,t)u^*(y,t)dt , \qquad (2.4)$$

where the usual ergodicity hypothesis is assumed and expand $u = \sum_i^\infty A_i(t)\phi_i(x)$ in terms of the coherent structures ϕ_i. Then the modes become decorrelated, $<A_iA_k> = \mu_i\delta_{ik}$ and $K = \Sigma\mu_i\phi_i(x)\phi_i(y)$. The μ_i,ϕ_i follow from the variational principle $\mu = max(Ku,u)$ subject to the constraints $(u,\phi_j) = 0$ for $j<i$. Here, the energy of the field is again dominated by the first few modes, $1\le i\le m$. Truncating the expansion of u and projecting onto ϕ_i yields

$$\dot{A}_i(t) = (F(\sum_1^m A_j\phi_j,\lambda),\phi_i) . \qquad (2.5)$$

This is the desired low-dimensional dynamical system which describes the interaction of the organized structures.

Classical pattern-forming systems are described by evolution equations (2.1) with polynomial interactions N(u) [9]. In complex systems such as spin-glass models of neural networks with many attractors present [1,19] (sec. 4), the dynamics (2.1) are governed by an interaction N which is a transcendental function of u and may also depend on the space variable x, e.g., $N(x,u) = tanh(g(x)u)$. Let $L = (1/\lambda)L_0$

and assume that L_0 and N do not depend on λ. In the case of a continuous stationary field ($\dot{u} = 0$), equ. (2.1) can be converted into the nonlinear integral equation

$$u(x) = \lambda \int J(x,y)N(y,u(y))dy \qquad (2.6)$$

for $u(x)$, J being the Green's function corresponding to L_0. Define $f(x,u)$ by $N = \partial f/\partial u$. Then the energy functional

$$R(\Psi) = \int dxdy J(x,y)\Psi(x)\Psi(y) + 2\int dx f(x,\int dy J(x,y)\Psi(y)) \qquad (2.7)$$

becomes extremal if $\Psi(x) = -N(x,u(x))$, where $u(x)$ is a solution of (2.6). By applying the projection technique it follows that u possesses an eigenfunction decomposition $u = \Sigma c_i \phi_i$ with the c_i being solutions of the "bifurcation equations" $c_i = -(\lambda/\mu_i)(N(y,u),\phi_i)$. It can then be shown that the number of bifurcating solutions and energy minima increases indefinitely as λ increases if N is bounded. In mean field theories of associative memory models [1,19] the c_i may be identified with the overlaps of input and stored patterns. Thus, the pattern recognition process is closely related to the formation of patterns through bifurcations. A similar relationship is discussed in [4] for remote sensing.

3. Probabilistic Nonlinear Inversion

The problem of pattern recognition is that of classifying or labelling a group of objects on the basis of certain requirements. We assume that each pattern or object is described by a set of real numbers $y_1,..,y_n$, arranged in a pattern vector $y = (y_1,..,y_n)^T \in R^n$, $n \leq \infty$. If $n = \infty$ this includes also continuously varying patterns in which case R^n is replaced by a Hilbert space and the y_i represent, e.g., appropriately chosen Fourier coefficients. The patterns are considered as random vectors with probability distribution $p(y)$. An important class of pattern recognition problems is that of ill-posed inverse problems. Here a pattern y has to be reconstructed from a set of incomplete and possibly noisy observed data $x_1,...,x_m$, $m \leq n$. We represent this process by

$$x = f(y) + v, \qquad (3.1)$$

where $x = (x_1,...,x_m)^T \in R^m$, f is a mapping from R^n to R^m with components $f_i(y)$ and $v = (v_1,...,v_m)^T$ is a realization of the noise. It is assumed that y and the components v_i are independent and that the statistics of all v_i are determined by the same normal distribution

$$p(v_i) = (\frac{2\pi}{\sigma})^{1/2} \exp\{-(v_i - \bar{v}_i)^2/2\sigma\}. \qquad (3.2)$$

As an example consider image restauration. Here the y_i may represent Fourier coefficients of a continuous intensity distribution on a screen and x collects a finite set of pixel intensities so that f is a linear mapping. Noise is induced by uncertainties of the measurement process. Alternatively, y and x may represent pixel intensities over the same grid with x being a degraded image of y. A common assumption for the degradation model is the combination of a blurring matrix H and of a nonlinear transformation F, f(y) = F(H(y)) [7]. The task of image restauration then is an optimal estimation of the original image y from measured degraded image data x. Another class of problems that leads to (3.1) is that of feature selection. If measured data have to be processed further it is often unavoidable to compress the dimensionality of the problem in order to reduce its complexity to a manageable size. In this case f represents certain discriminatory features of the original patterns which have to be chosen appropriately. The main problem of feature selection is with the optimal choice of the features $f_i(y)$ [6]. We address this problem below and for the moment assume that f is a given mapping.

The most natural approach to estimating y after measuring a data vector x consists of maximizing the conditional probability p(y/x) with respect to y for a given x. This is known as maximum a posteriori Bayesian estimation [7]. Let V(y) be a potential or energy function and suppose that y possesses a Gibbs distribution

$$p(y) = \frac{1}{Z} e^{-\beta V(y)}. \tag{3.3}$$

Then, by virtue of $p(x,y,v) = \delta(x-f(y)-v)\, p(y)p(v)$, p(y/x) is also a Gibbs distribution,

$$p(y/x) = \frac{1}{Z(x)} \exp\{-W_f(y;x)\}, \tag{3.4}$$

with a new normalization constant Z(x), and

$$W_f(y;x) = \beta V(y) + (1/2\sigma)|x - f(y) - \overline{v}|^2. \tag{3.5}$$

Actually, the noise need not be Gaussian. If $p(v) \sim \exp(-V_1(v))$, the second term in (3.5) is replaced by $V_1(x - f(y))$. Maximizing p(y/x) is obviously equivalent to the problem of determining

$$\min_y W_f(y;x) \text{ for given x.} \tag{3.6}$$

The minimization in (3.6) may, e.g., be performed by simulated annealing [3,8,15]. In the noiseless case ($\overline{v} = 0$, $\sigma \to 0$) (3.6) reduces to determining

$$\min_y V(y) \tag{3.7a}$$

subject to the constraint

$$x = f(y). \tag{3.7b}$$

The conventional approach to solving ill-posed problems is based on deterministic regularization techniques [18,24]. Let x = f(y) represent measured data of a pattern y and assume that f has no inverse. The regularization method to reconstruct y from measured data x makes use of the introduction of a cost function which is denoted also by V(y). For example, in early vision [21] y may represent a three-dimensional velocity field, x a two-dimensional projection of it and V(y) a measure of the smoothness of the field. Then y is determined by minimizing the modified cost function

$$\alpha V(y) + \frac{1}{2}|x - f(y)|^2 \tag{3.8}$$

for some regularization parameter α. Identifying the cost function V with the potential of a Gibbs distribution and setting $\alpha=\beta\sigma$, we see that the deterministic approach to solving ill-posed problems can be reinterpreted as a maximum a posteriori estimate with zero noise mean. The parameter α can be interpreted as a measure for the noise to signal fluctuations. If V(y) is a quadratic function, e.g., $V(y) = |Qy|^2$ with a linear operator Q and if f is also linear, then (3.8) becomes a quadratic polynomial for y. Minimization problems of this type can be solved in terms of electrical, chemical or mechanical networks [14,20].

As pointed out before, the main problem of feature selection is with the choice of the mapping f so that it represents the most important features of the original patterns. We propose the following choice. For a given mapping f define $y_f(x)$ by

$$W_f(y;y_f(x)) = \min_y W_f(y;x)$$

and assume $y_f(x)$ to be uniquely defined. Consider now a fixed pattern y and let p(x/y) be the conditional probability for measuring x, given y. Then

$$\int dx |y_f(x) - y|^2 p(x/y)$$

is the mean square error corresponding to the pattern y. A natural criterion for the optimum choice of f is that the average mean square error is minimal, i.e., that the functional

$$M[f] : = \int dy \int dx |y_f(x) - y|^2 p(y,x), \tag{3.9}$$

with p(y,x) = (1/Z)exp($-W_f$) becomes a minimum in the space of admissible mappings f. In the noiseless case M[f] reduces to

$$M[f] = \int dy |y_f(f(y)) - y|^2 p(y), \tag{3.10}$$

where $y_f(x)$ is determined by (3.7a,b). The minimization problem corresponding to (3.9) or (3.10) provides a rather general approach to feature selection on the basis of Bayesian estimates. However, minimizing M[f] is often complicated, even within the class of linear mappings. Furthermore, $y_f(x)$ should be uniquely defined what imposes restrictions on f which in general will depend on $V(y)$.

4. Inversion and Feature Selection Through Projections

In this section we consider the inverse problem (3.1) for the case where $f(y)$ is an orthogonal projection in the pattern space R^n, confining ourselves to the noiseless case. Letting $\{\Phi_1,\ldots,\Phi_m,\Psi_{m+1},\ldots,\Psi_n\}$ be an orthonormal basis of R^n we represent a given pattern $y \in R^n$ in the form

$$y = \sum_{i=1}^{m} x_i \Phi_i + \sum_{j=m+1}^{n} r_j \Psi_j , \qquad (4.1)$$

where $x_i = (y,\Phi_i)$, $r_j = (y,\Psi_j)$, with $(.,.)$ denoting the Euclidean scalar product in R^n. The projection of y onto the first term on the r.h.s. of (4.1) induces the linear mapping

$$x = f(y) \equiv ((y,\Phi_1),\ldots,(y,\Phi_m))^T , \qquad (4.2)$$

i.e., $f(y) = \Phi y$ where Φ is the mxn-matrix whose rows consist of the basis vectors Φ_i. The minimization problem (3.7) reduces now to the determination of the minimum of

$$V(r,x): = V(\sum_i x_i \Phi_i + \sum_j r_j \Psi_j) \qquad (4.3)$$

with respect to $r = (r_{m+1},\ldots,r_n)^T$ for a given measured feature vector x. An essential requirement for the basis vectors Φ_i is that a unique minimum $r_\Phi(x)$ of V exists for each x. Assuming that this condition is satisfied, we can decompose $V(r,x)$ according to

$$V(r,x) = V_0(x) + V_1(r,x) \qquad (4.4)$$

where $V_1(r_\Phi(x),x) = 0$ and $\nabla_r V_1(r_\Phi(x),x) = 0$. The conditional probability $p(r/x)$ can be written in the form

$$p(r/x) = \frac{1}{Z_1(x)} e^{-\beta V_1(r,x)} , \qquad (4.5)$$

i.e., as a Gibbs distribution with an x-dependent normalization $Z_1(x)$. Since V_1 has a global minimum at $r = r_\Phi(x)$ we may approximate V_1 by a quadratic form,

$$V_1(r,x) \cong \frac{1}{2}(K_1(x)[r-r_\Phi(x)],r-r_\Phi(x)) . \qquad (4.6)$$

Here, $K_1(x)$ is the Hessian matrix of V_1, evaluated at the global minimum $r_\Phi(x)$. In order that the approximation (4.6) makes sense, $K_1(x)$ must be non-degenerate.

If $p(y)$ is a Gaussian with zero mean, the potential V has the form $V(y) = \frac{1}{2}(Ky,y)$ where $\beta K = C^{-1}$, C being the covariance matrix. We denote by A the orthogonal nxn-matrix obtained by augmenting the rows of Φ by the Ψ_j. Setting

$$AKA^T = \begin{bmatrix} K_{11} & K_{12} \\ K_{12} & K_{12} \end{bmatrix}^T$$

where K_{11}, K_{12} and K_{22} are mxm, (n-m)xm and (n-m)x(n-m)-matrices, respectively, $V(r,x)$ takes the form

$$V(r,x) = \frac{1}{2}(K_{11}x,x) + (K_{12}x,r) + \frac{1}{2}(K_{22}r,r).$$

The minimum of V with respect to r is $r_\Phi(x) = -K_{22}^{-1}K_{12}x$ and V_0 reduces to $V_0(x) = (K_{11}x,x)/2$. The potential $V_1(r,x)$ is given by (4.6) with $K_1(x) = K_{22}$. If the Φ_i are chosen as eigenvectors of K we obtain $K_{12} = 0$ and hence $r_\Phi(x) = 0$.

Consider now the "neural network" potential

$$V(y) = \frac{1}{2} \sum_{i,j} K_{ij}y_iy_j - \sum_i h_i(y_i) , \tag{4.7}$$

with a positive definite symmetric matrix K_{ij}. The h_i are chosen such that their derivatives $h_i'(u) \equiv dh_i(u)/du$ are sigmoid functions, e.g., $h_i'(u) = \tanh(s_i u)$. Potentials of the type (4.7) are typical for mean field theories of spin glasses [19, 23] and for analogue networks [14,20]. The stationary points of V satisfy the equations

$$y_i = \sum_j J_{ij}h_j'(y_j) \quad (1 \le i \le n) , \tag{4.8}$$

where $J_{ij} = K_{ij}^{-1}$. We consider the continuum limit of (4.7) in which i is replaced by a continuous variable $\xi \in [0,1]$ and y_i becomes a function $y(\xi)$. Then (4.7) takes the form

$$V(y) = \frac{1}{2}(Ky,y) - \int_0^1 d\xi\, h(\xi,y(\xi)) , \tag{4.9}$$

where we have used the notation $Ky \equiv \int K(\xi,\eta)y(\eta)d\eta$ and $(\cdot,\cdot)$ denotes here the L_2-inner product. The stationarity equation (4.8) becomes now a nonlinear integral equation of the type (2.6),

$$y(\xi) = \int_0^1 d\eta\, J(\xi,\eta)h'(\eta,y(\eta)) . \tag{4.10}$$

Let $\Phi_i(\xi)$, $i = 1,2,\ldots$ be the eigenfunctions of K corresponding to the eigenvalues λ_i, $K\Phi_i = \lambda_i\Phi_i$, and assume that the λ_i form an increasing sequence, $\lambda_1 \leq \lambda_2 \leq \ldots$. We choose as pattern features the first m eigenfunctions $\Phi_i(\xi)$, $1 \leq i \leq m$. This means that $y(\xi)$ is split according to

$$y(\xi) = \sum_{i=1}^{m} x_i\Phi_i(\xi) + r(\xi) \equiv y_m(\xi;x) + r(\xi)$$

and that the projections $x_i = (y,\Phi_i)$ are assumed to be known. The remainder term $r(\xi)$ is orthogonal to the Φ_i, $(r,\Phi_i) = 0$ for $1 \leq i \leq m$. The potential $V(y) \equiv V(x,r)$ takes the form

$$V(x,r) = \sum_{i=1}^{m} \lambda_i\{\frac{1}{2}x_i^2 + (\Phi_i,r)x_i\} + \frac{1}{2}(Kr,r) - H(x,r), \quad (4.11a)$$

where

$$H(x,r) = \int_0^1 d\xi\, h(\xi,y_m(\xi;x)+r(\xi)) . \quad (4.11b)$$

We have to minimize $V(x,r)$ with respect to r for given features x. A necessary condition for a minimum is that $r(\xi)$ satisfies the nonlinear integral equation

$$r(\xi) = \int_0^1 d\xi J_m(\xi,\eta)\, h'(\eta,y_m(\eta;x)+r(\eta)) , \quad (4.12)$$

subject to the constraints $(r,\Phi_i) = 0$ $(1 \leq i \leq m)$, where J_m denotes the reduced kernel

$$J_m(\xi,\eta) = J(\xi,\eta) - \sum_{i=1}^{m} \lambda_i^{-1}\Phi_i(\xi)\Phi_i(\eta) .$$

Let $|\partial h'(\eta,u)/\partial u| \leq A$ and define k by the condition $\lambda_k \leq A < \lambda_{k+1}$. One can show that (4.12) possesses a unique solution $r(\xi;x)$ that corresponds to a global minimum of $V(x;r)$ for any x, provided that $m \geq k$. On the other hand, if $m < k$ (4.12) may and in general does possess several solutions in virtue of bifurcations. Consequently, if we require a unique Bayesian estimate for r we need to select at least k features. In neural networks the maximum slope of h' becomes very large for small temperatures, i.e., k is large, too. This reflects the presence of many minima of $V(y)$ [14].

The feature selection problem corresponding to (4.1) consists of finding the optimal choice for the basis vectors Φ_i. The functional $M[f]$ introduced in section 3 becomes now a function $M[\Phi]$, $\Phi = \{\Phi_i : 1 \leq i \leq m\}$, given by

$$M[\Phi] = \int dx\, e^{-\beta V_0(x)} \int dr |r_\Phi(x) - r|^2\, e^{-\beta V_1(r,x)} . \quad (4.13)$$

We note that (4.13) does not depend on the particular choice of the complementary basis vectors $\{\Psi_j\}$. Minimizing (4.13) is still a complicated problem for general potentials $V(y)$. For normal distributions one can show that a global minimum of $M[\Phi]$ is attained if the Φ_i are chosen as eigenvectors corresponding to the largest eigenvalues of the covariance matrix. Thus in this case optimal feature selection reduces to the standard Karhunen-Loeve expansion [6,17,25].

For large β in (4.5) one can prove the following. Let us assume that $V(y)$ has a unique and non-degenerate global minimum at y_0 and let H be the Hessian matrix of V, evaluated at y_0. Then, if β is sufficiently large, the minimum of $M[\Phi]$ is attained if the Φ_i are chosen as the eigenvectors of H corresponding to the m smallest eigenvalues ($p(y)$ becomes essentially a normal distribution). This optimal feature selection is clearly sensitive to bifurcations. If V depends on a parameter λ, then the global minimum may degenerate and new minima are formed when λ passes through a critical value λ_c. For such cases HAKEN [11] suggested to choose as features the eigenvectors corresponding to the degenerate directions of the Hessian matrix. For situations in which $p(y)$ is the steady state distribution for a Langevin-equation, e.g., $\dot{y} = -\nabla_y V + F(t)/\beta$ where $F(t)$ represents fluctuating forces, this choice is closely related to the pattern formation process in the averaged equation $\dot{y} = -\nabla_y V$ [10]. The Bayesian estimate can then be identified with the Lyapunov-Schmidt reduction procedure discussed in section 2.

5. Pattern Classification

In pattern recognition feature extraction is followed by classification. Consider a number of processes, each generating a sequence of patterns. Each generator then has a statistical description with it. A class w_ν ($1 \leq \nu \leq M$) is defined as the generator that produces an observed set of patterns. In this section we assume that an observed pattern is described by vectors of data $x = (x_1,\ldots,x_m)^T \varepsilon\ R^m$ that describe certain discriminatory features of the original patterns, as explained in the preceding sections. The joint probability distribution over all measurements and all classes is denoted by $p(x,w_\nu)$. More important are the marginal distributions $p(x)$, $p(w_\nu)$ and the conditional probabilities $p(x/w_\nu)$ and $p(w_\nu/x)$, the latter denoting the probability that x is generated by w_ν and the probability that w_ν has generated a measured x. The pattern classification problem can be formulated as a decision problem: given an input x, decide to which pattern class x belongs.

A standard approach to classifying x is the use of decision or discriminant functions $d_\nu(x)$ ($1 \leq \nu \leq M$) [6,25]. A vector x is associated with the class w_ν if $d_\nu(x) > d_\mu(x)$ for all $\mu \neq \nu$. This corresponds to a decomposition of R^m into regions Ω_ν such that x is classified to belong to class w_ν if and only if $x \varepsilon\ \Omega_\nu$. The boundary $\partial\Omega_{\nu\mu}$ between regions Ω_ν and Ω_μ is given by

$$\partial\Omega_{\nu\mu} : = \{x: d_\nu(x) = d_\mu(x) > d_\rho(x) \text{ for all } \rho \neq \mu,\nu\} . \quad (5.1)$$

In general, the decision functions are restricted to a certain class of functions. We assume that each function in this class can be expressed in the form $d(x,q)$ where q is a vector of parameters. For example, any linear function can be written as $d(x,q) = (x,c) - \theta$ with $c \in R^m$, $\theta \in R$, i.e., $q = (c,\theta) \in R^{m+1}$. The decision function $d_\nu(x)$ is then obtained by fixing the parameter vector, $d_\nu(x) = d(x,q_\nu)$. In the case of linear decision functions we have $d_\nu(x) = (x,c_\nu) - \theta_\nu$ and the regions Ω_ν are separated by pieces of hyperplanes. If $\theta_\nu = 0$ and $|c_\nu| = 1$ for all ν, then c_ν belongs to Ω_ν and may be regarded as a prototype vector for the class w_ν.

The main task of pattern classification is to determine the decision functions which in our case reduces to the determination of the enlarged parameter vector $Q = (q_1, ...,q_M)$. Our basic assumption is the existence of loss functions [6,25] $L_\nu(x,Q)$, $1 \leq \nu \leq M$, which may depend on the parameters to be determined. $L_\nu(x,Q)$ is the loss that will be incurred if the system classifies pattern x as belonging to class w_ν. Then

$$R_{\nu\mu}(Q) = \int_{\Omega_\nu} L_\nu(x,Q) \, p(x/w_\mu)dx \,,$$

where $\Omega_\nu = \Omega_\nu(Q)$ is determined by the decision functions $d(x,Q_\nu)$ as described before, is the risk due to a misclassification of patterns from class w_μ to class w_ν. Averaging $R_{\mu\nu}(Q)$ over all classes yields

$$R_\nu(Q): = \sum_\mu p(w_\mu) \, R_{\nu\mu}(Q)$$

$$= \int_{\Omega_\nu} L_\nu(x,Q) \, p(x) \, dx \,.$$

The total risk $R(Q)$ is obtained by summing the individual risks $R_\nu(Q)$, i.e.,

$$R(Q) = \sum_{\nu=1}^{M} \int_{\Omega_\nu} L_\nu(x,Q) \, p(x) \, dx \,. \tag{5.2}$$

The parameter vector Q is then obtained by minimizing $R(Q)$. The global minimum of $R(Q)$ may undergo bifurcations if $p(x)$ changes its shape in virtue of variations of further parameters that represent external influences. We note that there are stochastic versions of steepest descent algorithms, e.g., the Robbins-Monro algorithm [6], which converge to a local minimum of $R(Q)$. Here at each step k a vector x(k) is received with probability $p(x)$. The input vector x(k+1) is then used to determine an improvement Q(k+1) of the previously determined Q(k). Thus no knowledge of $p(x)$ is necessary. The disadvantage of the algorithm is that there is no guarantee of convergence to a global minimum. It is, therefore, desirable to design simulated annealing algorithms [8] combined with evolutionary optimization (and networks realizing these) so that the approach to a global minimum is ensured.

Once the regions Ω_ν have been determined in terms of decision functions, one has to design a pattern classifier that automatically classifies x to the class w_ν when x is in Ω_ν. This means that all vectors in Ω_ν are identified with each other. Alternatively we max fix an appropriately chosen vector x_ν in Ω_ν, e.g., the class-mean vector $x_\nu = \int_{\Omega\nu} xp(x)dx$, and regard x_ν as a prototype vector for the class w_ν in the sense that any $x \in \Omega_\nu$ is identified with x_ν. A pattern classifier would then consist of a system that receives an input vector x and produces x_ν as output if $x \in \Omega_\nu$. This leads naturally to the theory of dynamical systems. Suppose that $\dot{x} = X(x)$ is a dynamical system in R^m with the property that x_ν is an attractor and that its basin of attraction is Ω_ν. If $x \in \Omega_\nu$ is fed into the system as an initial condition, then $x(t) \to x_\nu$ for $t \to \infty$, thus yielding the desired input-output behaviour. Constructing dynamical systems with prescribed attractive behaviour is the basic idea underlying recent developments in associative memory recalls which has its origin in the work of CAIANIELLO [2], LITTLE [16] and HOPFIELD [13,14] (cf., also [5]). However, the discrete Hopfield model [13] cannot be used for our purpose because the data are continuous, whereas the continuous version [14] cannot be designed so that arbitrary patterns are stored. A more promising approach is provided by Haken's model [12], which allows one to store an arbitrary set of continuous patterns. It appears, however, difficult to adjust the basins of attraction in a prescribed manner.

We have demonstrated, in terms of representative examples, that pattern formation and pattern recognition are amenable to a low-dimensional unifying description using projection and variational principles. This provides a basis for understanding the qualitative analogies linking both fields. The robustness of pattern-forming and -recognizing systems and the observed sudden changes between collective patterns, however, lead one to ask for broader organization principles that generalize those discussed above to a statistical mechanics of invariant pattern recognition.

References

1. D. Amit, H. Gutfreund & H. Sompolinsky, Phys. Rev. A 32 (1985), 1007
2. E. R. Caianiello, *Neural Networks*, Springer 1968; and in *Progress in Quantum Field Theory*, H. Ezawa & S. Kamefuchi (eds.), North Holland 1986
3. V. Cerny, J. Optimization Theory and Appl. 45 (1985), 41
4. G. Dangelmayr & W. Güttinger, Geophys. J. R. Astr. Soc. 71 (1982), 79
5. R. Eckmiller & Ch. v.d. Malsburg (eds.), *Neural Computers*, Springer 1988
6. K. S. Fu, *Sequential Methods in Pattern Recognition and Machine Learning*, Academic Press 1968
7. S. Geman & D. Geman, IEEE Trans. PAMI 6 (1984), 721
8. S. Geman & C.-R. Hwang, SIAM J. Contr. and Opt. 24 (1986), 1031
9. W. Güttinger & G. Dangelmayr (eds.), *The Physics of Structure Formation*, Springer 1987
10. H. Haken, Z. Phys. B 61 (1985), 329; B 61 (1985), 335

11. H. Haken, in *Physics of Cognitive Processes*, E. R. Caianiello (ed.), World Scientific 1987

12. H. Haken, in *Computational Systems - Natural and Artificial*, H. Haken (ed.), p.2, Springer 1987

13. J. J. Hopfield, Proc. Natl. Acad. Sci. USA $\underline{79}$ (1982), 2554

14. J. J. Hopfield, Proc. Natl. Acad. Sci. USA $\underline{81}$ (1984), 3088

15. S. Kirkpatrick, J. Stat. Phys. $\underline{34}$ (1984), 975

16. W. A. Little, Math. Biosci. $\underline{19}$ (1974), 101

17. J. L. Lumley, *Stochastic Tools in Turbulence*, Academic Press 1970

18. M. Z. Nashed (ed.), *Generalized Inverses and Applications*, Academic Press 1976

19. C. Peterson & J. R. Anderson, Complex Systems $\underline{1}$ (1987), 995; $\underline{2}$ (1988), 59

20. T. Poggio & C. Koch, Proc. R. Soc. Lond. $\underline{B\ 226}$ (1985), 303

21. T. Poggio & V. Torre, Artif. Intelligence Lab. Memo., no. $\underline{773}$, M.I.T. Cambridge, Massachusetts

22. L. Sirovich, in Proc. Int. Conf. Fluid Mech., Beijing 1987

23. D. J. Thouless, P. W. Anderson & R. G. Palmer, Phil. Mag. $\underline{35}$ (1977), 593

24. A. N. Tikhonov & V. Y. Arsenin, *Solutions of Ill-Posed Problems*, Winston & Sons 1977

25. S. Watanabe, *Pattern Recognition: Human and Mechanical*, Wiley 1985

Self-Organizing Neural Network Architectures for Real-Time Adaptive Pattern Recognition

G.A. Carpenter [1,2] *and S. Grossberg* [2]

[1]Department of Mathematics, Northeastern University, Bosten, MA 02115, USA
[2]Center for Adaptive Systems, 111 Cummington Street, Boston University, Boston, MA 02215, USA

1. Active Regulation of Self-Organized Learning and Recognition by Attention and Expectation

Many of the most important properties of biological intelligence arise through a process of self-organization whereby a biological system actively interacts with a complex environment in real-time. The environment is often noisy and nonstationary, and intelligent capabilities are learned autonomously and without benefit of an external teacher. For example, children learn to visually recognize and manipulate many complex objects without being provided with explicit rules for how to do so.

This chapter summarizes some recent results concerning the process whereby humans learn recognition codes in real-time through a process of self-organization. This analysis requires consideration of a number of basic issues with which we are all familiar through our own experiences.

Why do we pay attention? Why do we learn expectations about the world? In particular, how do we cope so well with unexpected events, and how do we manage to do so as well as we do when we are on our own, and do not have a teacher as a guide? How do we learn what combinations of facts are useful for dealing with a given situation and what combinations of facts are irrelevant? How do we recognize familiar facts so quickly even though we may know many other things? How do we join together knowledge about the external world with information about our internal needs to quickly make decisions that have a good chance of satisfying these needs? Finally, what do all of these properties have in common?

2. The Stability-Plasticity Dilemma and Adaptive Resonance Theory

One answer to these questions has been found through the attempt to solve a basic design problem faced by all intelligent systems that are capable of autonomously adapting in real-time to unexpected changes in their world. This design problem is called the stability-plasticity dilemma and a theory called Adaptive Resonance Theory, or ART, is being developed that suggests a solution to this problem.

Springer Series in Synergetics Vol. 42: **Neural and Synergetic Computers**
Editor: H. Haken   ©Springer-Verlag Berlin Heidelberg 1988

The stability-plasticity dilemma asks how a learning system can be designed to remain plastic, or adaptive, in response to significant events, yet also remain stable in response to irrelevant events? How does the system know how to switch between its stable and its plastic modes to achieve stability without rigidity and plasticity without chaos? In particular, how can it preserve its previously learned knowledge while continuing to learn new things; what prevents the new learning from washing away the memories of prior learning?

One of the key computational ideas that is rigorously demonstrated within Adaptive Resonance Theory is that top-down learned expectations focus attention upon bottom-up information in a way that protects previously learned memories from being washed away by new learning, and enables new learning to be automatically incorporated into the total knowledge base of the system in a globally self-consistent way.

The ART achitectures that are discussed herein are neural networks that self-organize stable recognition codes in real-time in response to arbitrary sequences of input patterns. Within such an ART architecture, the process of adaptive pattern recognition is a special case of the more general cognitive process of hypothesis discovery, testing, search, classification, and learning. This latter property opens the possibility of applying ART systems to more general problems of adaptively processing large abstract information sources and data bases. The present chapter outlines the main computational properties of these ART architectures. ART model derivation and analysis are described in more detail elsewhere [1-4].

3. Competitive Learning Models

ART models grew out of an analysis of a simpler type of adaptive pattern recognition network which is often called a *competitive learning* model. Competitive learning models were developed in the early 1970's through contributions of VON DER MALSBURG [5] and GROSSBERG, leading in 1976 to the description of these models in several forms in which they are used today [6]. Authors such as AMARI [7], COOPER [8], and KOHONEN [9] have further developed these models. GROSSBERG [4] has provided a historical discussion of the development of competitive learning models.

In a competitive learning model (Fig. 1), a stream of input patterns to a network F_1 can train the adaptive weights, or long term memory (LTM) traces, that multiply the signals in the pathways from F_1 to a coding level F_2. In the simplest such model, input patterns to F_1 are normalized before passing through the adaptive filter defined by the pathways from F_1 to F_2. Level F_2 is designed as a competitive network capable of choosing the node which receives the largest total input ("winner-take-all"). The winning population then triggers associative pattern learning within the vector of LTM traces which sent its inputs through the adaptive filter.

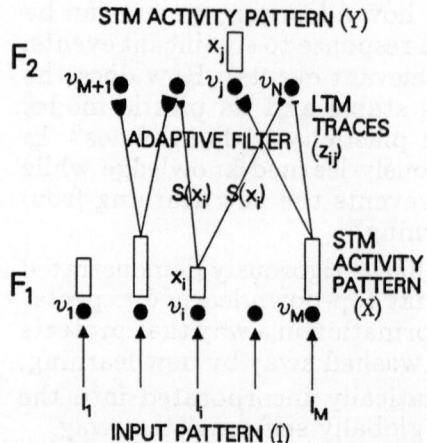

Figure 1. Stages of bottom-up activation: The input pattern I generates a pattern of STM activation $X = (x_1, x_2, \ldots, x_M)$ across F_1. Sufficiently active F_1 nodes emit bottom-up signals to F_2. This signal pattern S is multiplied, or gated, by long term memory (LTM) traces z_{ij} within the $F_1 \to F_2$ pathways. The LTM gated signals are summed before activating their target nodes in F_2. This LTM-gated and summed signal pattern T, where $T_j = \sum_i S_i z_{ij}$, generates a pattern of STM activation $Y = (x_{M+1}, \ldots, x_N)$ across F_2.

For example, as in Fig. 1, let I_i denote the input to the ith node v_i of F_1, $i = 1, 2, \ldots, M$; let x_i denote the activity, or short term memory (STM) trace, of v_i; let x_j denote the activity, or STM trace, of the jth node v_j of F_2, $j = M+1, \ldots, N$; and let z_{ij} denote the adaptive weight, or long term memory (LTM) trace, of the pathway from v_i to v_j. Then let

$$x_i = \frac{I_i}{\sum_{k=1}^{M} I_k} \tag{1}$$

be the normalized activity of v_i in response to the input pattern $I = (I_1, I_2, \ldots, I_M)$. For simplicity, let the output signal S_i of v_i equal x_i.

Let

$$T_j = \sum_{i=1}^{M} x_i z_{ij} \tag{2}$$

be the total signal received at v_j from F_1, let

$$x_j = \begin{cases} 1 & \text{if } T_j > \max(T_k : k \neq j) \\ 0 & \text{if } T_j < \max(T_k : k \neq j) \end{cases} \tag{3}$$

summarize the fact that the node x_j in F_2 which receives the largest signal is chosen for short-term memory storage, and let a differential equation

$$\frac{d}{dt}z_{ij} = \epsilon x_j(-z_{ij} + x_i) \tag{4}$$

specify that only the vector $Z_j = (z_{1j}, z_{2j}, \dots, z_{Mj})$ of adaptive weights which abut the winning node v_j is changed due to learning. Vector Z_j learns by reducing the error between itself and the normalized vector $X = (x_1, x_2, \dots, x_M)$ in the direction of steepest descent.

There are several equivalent ways to describe how such a system recognizes input patterns I that are presented to F_1. The winning node v_j in F_2 is said to code, classify, cluster, partition, compress, or orthogonalize these input patterns. In engineering, such a scheme is said to perform adaptive vector quantization. In cognitive psychology, it is said to perform categorical perception [3]. In categorical perception, input patterns are classified into mutually exclusive recognition categories which are separated by sharp categorical boundaries. A sudden switch in pattern classification can occur if an input pattern is deformed so much that it crosses one of these boundaries and thereby causes a different node v_j to win the competition within F_2. Categorical perception, in the strict sense of the word, occurs only if F_2 makes a choice. In more general competitive learning models, compressed but distributed recognition codes are generated by the model's coding level, or levels [3,4,9].

In response to certain input environments, a competitive learning model possesses very appealing properties. It has been mathematically proved [6] that, if not too many input patterns are presented to F_1, or if the input patterns form not too many clusters, relative to the number of coding nodes in F_2, then learning of the recognition code eventually stabilizes and the learning process elicits the best distribution of LTM traces that is consistent with the structure of the input environment.

Despite the demonstration of input environments that can be stably coded, it has also been shown, through explicit counterexamples [1,2,6], that a competitive learning model does not always learn a temporally stable code in response to an arbitrary input environment. In these counterexamples, as a list of input patterns perturbs level F_1 through time, the response of level F_2 to the *same* input pattern can be different on each successive presentation of that input pattern. Moreover, the F_2 response to a given input pattern might never settle down as learning proceeds.

Such unstable learning in response to a prescribed input is due to the learning that occurs in response to the other, intervening, inputs. In other words, the network's adaptability, or plasticity, enables prior learning to be washed away by more recent learning in response to a wide variety of input environments. In fact, there exist infinitely many input environments in which periodic presentation of just four input patterns can cause temporally unstable learning [1,2]. Learning can also become unstable due to simple changes in an input environment. Changes in the probabilities of inputs, or in the deterministic sequencing of inputs, can readily wash away prior learning. This instability problem is not, moreover, peculiar to competitive learning models. The problem is a basic one because it arises from a combination of the very features of an adaptive coding model that, on the surface, seem so desirable: its ability to learn from experience and its ability to code, compress, or categorize many patterns into a compact internal representation. Due to these properties, when a new input pattern I retrains a vector Z_j of LTM traces, the set of *all* input patterns coded by v_j changes too, because a change in Z_j in (2) can reverse the inequalities in (3) in response to many of the input patterns that were previously coded by v_j.

Learning systems which can become unstable in response to many input environments cannot safely be used in autonomous machines which may be unexpectedly confronted by one of these environments on the job. Adaptive Resonance Theory was introduced in 1976 to show how to embed a competitive learning model into a *self-regulating control structure* whose autonomous learning and recognition proceed stably and efficiently in response to an *arbitrary* sequence of input patterns.

4. Self-Stabilized Learning by an ART Architecture in an Arbitrary Input Environment

Figure 2 schematizes a typical example from a class of architectures called ART 1. It has been mathematically proved [1] that an ART 1 architecture is capable of stably learning a recognition code in response to an arbitrary sequence of *binary* input patterns until it utilizes its full memory capacity. Moreover, the adaptive weights, or LTM traces, of an ART 1 system oscillate at most once during learning in response to an arbitrary binary input sequence, yet do not get trapped in spurious memory states or local minima. After learning self-stabilizes, the input patterns directly activate the F_2 codes that represent them best.

As in a competitive learning model, an ART architecture encodes a new input pattern, in part, by changing the adaptive weights, or LTM traces, of a bottom-up adaptive filter. This filter is contained in the pathways leading from a feature representation field F_1 to a category representation field F_2. In an ART network, however, it is a second, top-down adaptive filter, contained in the pathways from F_2 to F_1, that leads to the crucial property of code self-stabilization. Such top-down adaptive signals play the role of learned expectations in an ART system. Before considering details about how the ART control structure automatically stabilizes the learning process, we will sketch how self-stabilization occurs in intuitive terms.

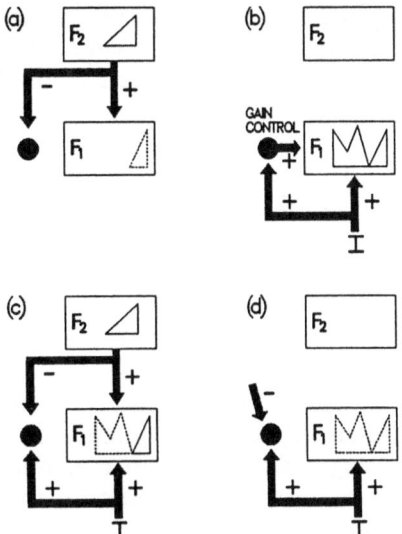

Figure 2. Matching by the 2/3 Rule: (a) A top-down expectation from F_2 inhibits the attentional gain control source as it subliminally primes target F_1 cells. Dotted outline depicts primed activation pattern. (b) Only F_1 cells that receive bottom-up inputs and gain control signals can become supraliminally active. (c) When a bottom-up input pattern and a top-down template are simultaneously active, only those F_1 cells that receive inputs from both sources can become supraliminally active. (d) Intermodality inhibition can shut off the F_1 gain control source and thereby prevent a bottom-up input from supraliminally activating F_1, as when attention shifts to a different input channel. Similarly, disinhibition of the F_1 gain control source in (a) may cause a top-down prime to become supraliminal, as during an internally willed fantasy.

Suppose that an input pattern I activates F_1. Let F_1 in turn, activate the code, or hypothesis, symbolized by the node v_{j_1} at F_2 which receives the largest total signal from F_1. Then F_2 quickly reads-out its learned top-down expectation to F_1, whereupon the bottom-up input pattern and top-down learned expectation are matched across F_1. If these patterns are badly matched, then a mismatch event takes place at F_1 which triggers a reset burst to F_2. This reset burst shuts off node v_{j_1} for the remainder of the coding cycle, and thereby deactivates the top-down expectation controlled by v_{j_1}. Then F_1 quickly reactivates essentially the same bottom-up signal pattern to F_2 as before. Level F_2 reinterprets this signal pattern, conditioned on the hypothesis that the earlier choice v_{j_1} was incorrect, and another node v_{j_2} is automatically chosen. The parallel search, or hypothesis testing, cycle of bottom-up adaptive filtering from F_1 to F_2, code (or hypothesis) selection at F_2, read-out of a top-down learned expectation from F_2 to F_1, matching at F_1, and code reset at F_2 now repeats itself automatically at a very fast rate until one of three possibilities occurs: a node v_{j_m} is chosen whose top-down expectation approximately matches input I, or a previously uncommitted F_2 node is selected, or the full capacity of the system is used and cannot accomodate input I. Until one of these outcomes prevails, essentially no learning occurs, because all the STM computations of the hypothesis testing cycle proceed so quickly that the more slowly varying LTM traces in the bottom-up and top-down adaptive filters cannot change in response to them. Significant learning occurs in response to an input pattern only after the hypothesis testing cycle that it generates comes to an end.

If the hypothesis testing cycle ends in an approximate match, then the bottom-up input pattern and the top-down expectation quickly deform the activity pattern $X = (x_1, x_2, \ldots, x_M)$ across F_1 into a net pattern that computes a fusion, or consensus, between the bottom-up and top-down information. This fused pattern represents the attentional focus of the system. When fusion occurs, the bottom-up and top-down signal patterns mutually reinforce each other via feedback and the system gets locked into a resonant state of STM activation. Only then can the LTM traces learn. What they learn is any new information about the input pattern that is represented within the fused activation pattern across F_1. The fact that learning occurs only in the resonant state suggested the name Adaptive Resonance Theory. Thus the system allows one of its prior learned codes to be altered only if an input pattern is sufficiently similar to what it already knows to risk a further refinement of its knowledge.

If the hypothesis testing cycle ends by selecting an uncommitted node at F_2, then the bottom-up and top-down adaptive filters that are linked to this node learn the F_1 activation pattern that is generated directly by the input. No top-down alteration of the F_1 activation pattern occurs in this case. If the full capacity has been exhausted and no adequate match exists, learning is automatically inhibited.

In summary, an ART network either refines its already learned codes based upon new information that can be safely accomodated into them via approximate matches, or selects new nodes for initiating learning of novel recognition categories, or defends its fully committed memory capacity against being washed away by the incessant flux of new input events.

5. Comparison with Alternative Learning Schemes

ART models are compared with other learning schemes in Table 1.

The most robust differences between ART and alternative learning schemes are that ART architectures are designed to learn quickly and stably in real-time in response to a possibly nonstationary world with an unlimited number of inputs until it utilizes its full memory capacity. Many alternative learning schemes become unstable unless they learn slowly in a controlled stationary environment with a carefully selected total number of inputs and do not use their full memory capacity [12]. For example, a learning system that is not self-stabilizing experiences a capacity catastrophe in response to an unlimited number of inputs: new learning washes away memories of prior learning if too many inputs perturb the system. To prevent this from happening, either the total number of input patterns that perturbs the system needs to be restricted, or the learning process itself must be shut off, before the capacity catastrophe occurs.

TABLE 1

ART Architecture	Alternative Learning Properties
Real-time (on-line) learning	Lab-time (off-line) learning
Nonstationary world	Stationary world
Self-organizing (unsupervised)	Teacher supplies correct answer (supervised)
Memory self-stabilizes in response to arbitrarily many inputs	Capacity catastrophe in response to arbitrarily many inputs
Effective use of full memory capacity	Can only use partial memory capacity
Maintain plasticity in an unexpected world	Externally shut off plasticity to prevent capacity catastrophe
Learn internal top-down expectations	Externally impose costs
Active attentional focus regulates learning	Passive learning
Slow or fast learning	Slow learning or oscillation catastrophe
Learn in approximate-match phase	Learn in mismatch phase
Use self-regulating hypothesis testing to globally reorganize the energy landscape	Use noise to perturb system out of local minima in a fixed energy landscape
Fast adaptive search for best match	Search tree
Rapid direct access to codes of familiar events	Recognition time increases with code complexity
Variable error criterion (vigilance parameter) sets coarseness of recognition code in response to environmental feedback	Fixed error criterion in response to environmental feedback
All properties scale to arbitrarily large system capacities	Key properties deteriorate as system capacity is increased

Shutting off the world is not possible in many real-time applications. In particular, how does such a system subsequently allow a familiar input to be processed and recognized, but block the processing of a novel input pattern before it can destabilize its prior learning? In the absence of a self-stabilization mechanism, an external teacher must act as the system's front end to independently recognize the inputs and make the decision. Shutting off learning at just the right time to prevent either a capacity catastrophe or a premature termination of learning from occurring would also require an external teacher. In either case the external teacher must be able to carry out the recognition tasks that the learning system was supposed to carry out. Hence non-self-stabilizing learning systems are not capable of functioning autonomously in ill-controlled environments.

In learning systems wherein an external teacher is needed to supply the correct representation to be learned, the learning process is often driven by mismatch between desired and actual outputs [10,11]. Such schemes must learn slowly, and in a stationary environment, or else risk unstable oscillations in response to the mismatches. They can also be destabilized if the external teaching signal is noisy, because such noise creates spurious mismatches. These learning models are also prone to getting trapped in local minima, or globally incorrect solutions. Models such as simulated annealing and the Boltzmann machine [10] use internal system noise to escape local minima and to thereby approach a more global minimum. An externally controlled (temperature) parameter regulates this process as it is made to converge ever more slowly to a critical value. In ART, by contrast, approximate matches, rather than mismatches, drive the learning process. Learning in the approximate-match mode enables rapid and stable learning to occur while buffering the system's memory against external noise. The hypothesis testing cycle replaces internal system noise as a scheme for discovering a globally correct solution, and does not utilize an externally controlled temperature parameter or teacher.

6. Attentional Priming and Prediction: Matching by the 2/3 Rule

One of the key constraints upon the design of the ART 1 architecture is its rule for matching a bottom-up input pattern with a top-down expectation at F_1. This rule is called the 2/3 Rule [1]. The 2/3 Rule is necessary to regulate both the hypothesis testing cycle and the self-stabilization of learning in an ART 1 system.

The 2/3 Rule reconciles two properties whose simplicity tends to conceal their fundamental nature: In response to an arbitrary bottom-up input pattern, F_1 nodes can be *supraliminally* activated; that is, activated enough to generate output signals to other parts of the network and thereby to initiate the hypothesis testing cycle. In response to an arbitrary top-down expectation, however, F_1 nodes are only *subliminally* activated; they sensitize, prepare, or attentionally *prime* F_1 for future input patterns that may or may not generate an approximate match with this expectation, but do not, in themselves, generate output signals. Such a subliminal reaction enables an ART system to anticipate future events and to thereby function as an "intentional" machine. In particular, if an attentional prime is locked into place by a high gain top-down signal source, then an ART system can automatically suppress all inputs that do not fall into a sought-after recognition category, yet amplify and hasten the processing of all inputs that do [3].

In order to implement the 2/3 Rule, it is necessary to assure that F_1 can distinguish between bottom-up and top-down signals, so that it can supraliminally react to the former and subliminally react to the latter. In ART 1, this distinction is carried out by a third F_1 input source, called an *attentional gain control* channel, that responds differently to bottom-up and top-down signals.

Figure 2 describes how this gain control source works. When it is activated, it excites each F_1 node equally. The 2/3 Rule says that at least 2 out of 3 input sources are needed to supraliminally activate an F_1 node: a bottom-up input, a top-down input, and a gain control input. In the top-down processing mode (Fig. 2a), each F_1 node receives a signal from at most one input source, hence is only subliminally activated. In the bottom-up processing mode (Fig. 2b), each active bottom-up pathway can turn on the gain control node, whose output, once on, is independent of the total number of active bottom-up pathways. Then all F_1 nodes receive at least a gain control input, but only those nodes that also receive a bottom-up input are supraliminally activated. When both bottom-up and top-down inputs reach F_1 (Fig. 2c), the gain control source is shut off, so that only those F_1 nodes which receive top-down confirmation of the bottom-up input are supraliminally activated. In this case, the 2/3 Rule maintains supraliminal activity only within the spatial intersection of the bottom-up input pattern and the top-down expectation. Consequently, if a bottom-up input pattern, as in Fig. 2b, causes the read-out of a badly matched top-down expectation, as in Fig. 2c, then the total number of supraliminally active F_1 nodes can suddenly decrease, and thereby cause a decrease in the total ouput signal emitted by F_1. This property is used heavily in controlling the hypothesis testing and self-stabilization processes, as we now show.

7. Automatic Control of Hypothesis Testing by Attentional-Orienting Interactions

An ART architecture automates its hypothesis testing cycle through interactions between an attentional subsystem and an orienting subsystem. These subsystems in the ART 1 architecture are schematized in Fig. 3.

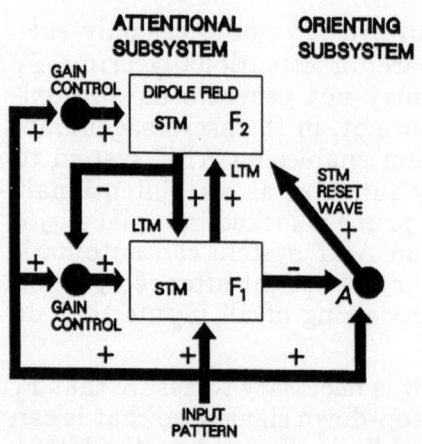

ATTENTIONAL SUBSYSTEM   ORIENTING SUBSYSTEM

GAIN CONTROL

DIPOLE FIELD

STM F_2

LTM

STM RESET WAVE

LTM

STM F_1

GAIN CONTROL

A

INPUT PATTERN

Figure 3. ART 1 system: Two successive stages, F_1 and F_2, of the attentional subsystem encode patterns of activation in short term memory (STM). Bottom-up and top-down pathways between F_1 and F_2 contain adaptive long term memory (LTM) traces which multiply the signals in these pathways. The remainder of the circuit modulates these STM and LTM processes. Modulation by gain control enables F_1 to distinguish between bottom-up input patterns and top-down priming, or expectation, patterns, as well as to match these bottom-up and top-down patterns by the 2/3 Rule. Gain control signals also enable F_2 to react supraliminally to signals from F_1 while an input pattern is on. The orienting subsystem generates a reset wave to F_2 when sufficiently large mismatches between bottom-up and top-down patterns occur at F_1. This reset wave selectively and enduringly inhibits previously active F_2 cells until the input is shut off.

The orienting subsystem A generates an output signal only when a mismatch occurs between bottom-up input pattern and top-down expectation at level F_1 of the attentional subsystem. Thus A functions like a novelty detector. The output signal from A is called an *STM reset wave* because it selectively inhibits the active node(s) at level F_2 of the attentional subsystem. The novelty detector A hereby disconfirms the F_2 hypothesis that led to the F_1 mismatch.

The 2/3 Rule controls the reset wave emitted by A as follows. When a bottom-up input pattern is presented, each of the active input pathways to F_1 also sends a signal to the orienting subsystem A, where all of these signals are added up. When the input pattern activates F_1, each of the activated F_1 nodes sends an inhibitory signal to A. The system is designed so that the total inhibitory signal is larger than the total excitatory signal. Thus in the bottom-up mode, there is a balance between active F_1 nodes and active input lines that prevents a reset wave from being triggered. (Note that level F_1 in ART 1 is not normalized as it was in (1) of the competitive learning model and in the

ART 2 systems discussed below. The decision of whether and how to normalize depends upon the design of the whole system.)

This balance is upset when a top-down expectation is read-out that mismatches the bottom-up input pattern at F_1. As in Fig. 2c, the total output from F_1 then decreases by an amount that grows with the severity of the mismatch. If the attenuation is sufficiently great, then inhibition from F_1 to A can no longer prevent A from emitting a reset wave. A parameter ρ called the *vigilance parameter* determines how large a mismatch will be tolerated before A emits a reset wave. High vigilance forces the system to search for new categories in response to small differences between input and expectation. Then the system learns to classify input patterns into a large number of fine categories. Low vigilance enables the system to tolerate large mismatches and to thus group together input patterns according to a coarse measure of mutual similarity. The vigilance parameter may be placed under external control, being increased, for example, when the network is "punished" for failing to distinguish two inputs that give rise to different consequences [1,3].

Figure 4 illustrates how these properties of the interaction between levels F_1, F_2, and A regulate the hypothesis testing cycle of the ART 1 system. In Fig. 4a, an input pattern I generates an STM activity pattern X across F_1. The input pattern I also excites the orienting subsystem A, but pattern X at F_1 inhibits A before it can generate an output signal. Activity pattern X also elicits an output pattern S which activates the bottom-up adaptive filter $T = ZS$, where Z is the matrix of bottom-up LTM traces. As a result an STM pattern Y becomes active at F_2. In Fig. 4b, pattern Y generates a top-down output U through the adaptive filter $V = \hat{Z}U$, where $\hat{Z}$ is the matrix of top-down LTM traces. Vector V is the top-down expectation that is read into F_1. Expectation V mismatches input I, thereby significantly inhibiting STM activity across F_1. The amount by which activity in X is attenuated to generate the activity pattern X^* depends upon how much of the input pattern I is encoded within the expectation V, via the 2/3 Rule.

When a mismatch attenuates STM activity across F_1, the total size of the inhibitory signal from F_1 to A is also attenuated. If the attenuation is sufficiently great, inhibition from F_1 to A can no longer prevent the arousal source A from firing. Fig. 4c depicts how disinhibition of A releases an arousal burst to F_2 which equally, or nonspecifically, excites all the F_2 cells. The cell populations of F_2 react to such an arousal signal in a state-dependent fashion. In the special case that F_2 chooses a single population for STM storage, the arousal burst selectively inhibits, or resets, the active population in F_2. This inhibition is long-lasting.

In Fig. 4c, inhibition of Y leads to removal of the top-down expectation V, and thereby terminates the mismatch between I and V.

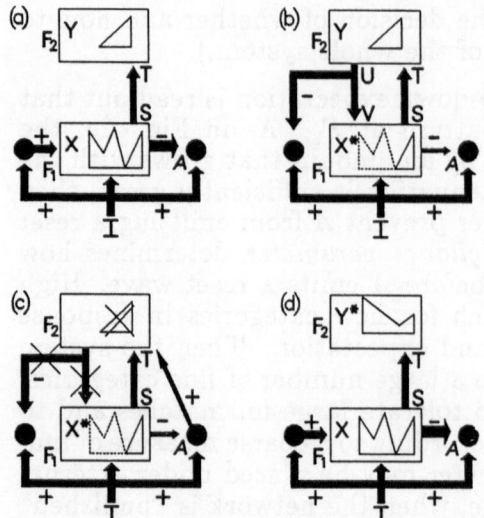

Figure 4. ART 1 hypothesis testing cycle: (a) The input pattern I generates the STM activity pattern X at F_1 as it activates A. Pattern X both inhibits A and generates the bottom-up signal pattern S. Signal pattern S is transformed via the adaptive filter into the input pattern $T = ZS$, which activates the compressed STM pattern Y across F_2. (b) Pattern Y generates the top-down signal pattern U which is transformed by the top-down adaptive filter $V = \hat{Z}U$ into the expectation pattern V. If V mismatches I at F_1, then a new STM activity pattern X^* is generated at F_1. The reduction in total STM activity which occurs when X is transformed into X^* causes a decrease in the total inhibition from F_1 to A. (c) Then the input-driven activation of A can release a nonspecific arousal wave to F_2, which resets the STM pattern Y at F_2. (d) After Y is inhibited, its top-down expectation is eliminated, and X can be reinstated at F_1. Now X once again generates input pattern T to F_2, but since Y remains inhibited T can activate a different STM pattern Y^* at F_2. If the top-down expectation due to Y^* also mismatches I at F_1, then the rapid search for an appropriate F_2 code continues.

Input pattern I can thus reinstate the original activity pattern X across F_1, which again generates the output pattern S from F_1 and the input pattern T to F_2. Due to the enduring inhibition at F_2, the input pattern T can no longer activate the original pattern Y at F_2. Level F_2 has been conditioned by the disconfirmation of the original hypothesis. A new pattern Y^* is thus generated at F_2 by I (Fig. 4d).

The new activity pattern Y^* reads-out a new top-down expectation V^*. If a mismatch again occurs at F_1, the orienting subsystem is again engaged, thereby leading to another arousal-mediated reset of

STM at F_2. In this way, a rapid series of STM matching and reset events may occur. Such an STM matching and reset series controls the system's hypothesis testing and search of LTM by sequentially engaging the novelty-sensitive orienting subsystem. Although STM is reset sequentially in time via this mismatch-mediated, self-terminating LTM search process, the mechanisms that control the LTM search are all parallel network interactions, rather than serial algorithms. Such a parallel search scheme continuously adjusts itself to the system's evolving LTM codes. The LTM code depends upon both the system's initial configuration and its unique learning history, and hence cannot be predicted *a priori* by a pre-wired search algorithm. Instead, the mismatch-mediated engagement of the orienting subsystem triggers a process of parallel self-adjusting search that tests only the hypotheses most likely to succeed, given the system's unique learning history.

The mismatch-mediated search of LTM ends when an STM pattern across F_2 reads-out a top-down expectation that approximately matches I, to the degree of accuracy required by the level of attentional vigilance, or that has not yet undergone any prior learning. In the former case, the accessed recognition code is refined based upon any novel information contained in the input I; that is, based upon the activity pattern resonating at F_1 that fuses together bottom-up and top-down information according to the 2/3 Rule. In the latter case, a new recognition category is then established as a new bottom-up code and top-down template are learned.

8. ART 2: Learning to Recognize an Analog World

Although self-organized recognition of binary patterns is useful in many applications, such as recognition of printed or written text, many other applications require the ability to categorize arbitrary sequences of analog (including binary) input patterns. A class of architectures, generically called ART 2, has been developed for this purpose [2]. Given the enhanced capabilities of ART 2 architectures, a sequence of arbitrary input patterns can be fed through an arbitrary preprocessor before the output patterns of the preprocessor are fed as inputs into an ART 2 system for automatic classification. Figure 5 illustrates how an ART 2 architecture has quickly learned to stably classify 50 analog input patterns, chosen to challenge the architecture in multiple ways, into 34 recognition categories after a single learning trial. Figure 6 illustrates how the same 50 input patterns have been quickly classified into 20 coarser categories after a single learning trial, using a smaller setting of the vigilance parameter.

ART 2 architectures can autonomously classify arbitrary sequences of analog input patterns into categories of arbitrary coarseness while suppressing arbitrary levels of noise. They accomplish this by modifying the ART 1 architecture to incorporate solutions of several additional design problems into their circuitry. In particular, level F_1 is split into

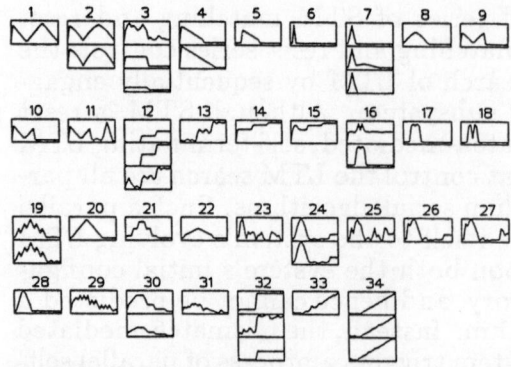

Figure 5. Category grouping by ART 2 of 50 analog input patterns into 34 recognition categories. Each input pattern I is depicted by a graph as a function of abscissa values i $(i = 1 \ldots M)$, with successive ordinate I_i values connected by straight lines. The category structure established upon one complete presentation of the 50 inputs remains stable thereafter if the same inputs are presented again.

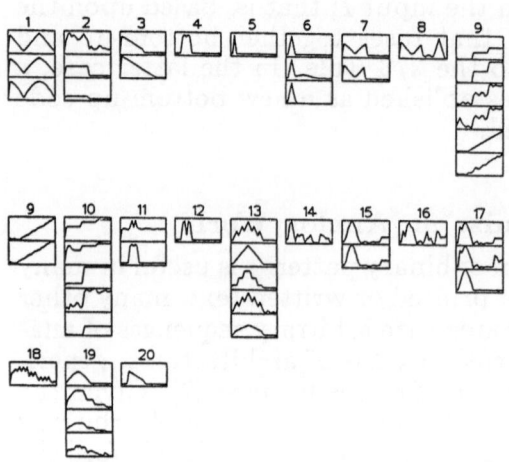

Figure 6. Lower vigilance implies coarser grouping. The same ART 2 system as used in Fig. 5 has here grouped the same 50 inputs into 20 recognition categories. Note, for example, that Categories 1 and 2 of Fig. 5 are joined in Category 1; Categories 19-22 are joined in Category 13; and Categories 8, 30, and 31 are joined in Category 19.

separate sublevels for receiving bottom-up input patterns, for receiving top-down expectations, and for matching the bottom-up and top-down data, as in Fig. 7. Three versions of the ART 2 architecture are now being applied to problems such as visual pattern recognition, speech perception, and radar classification.

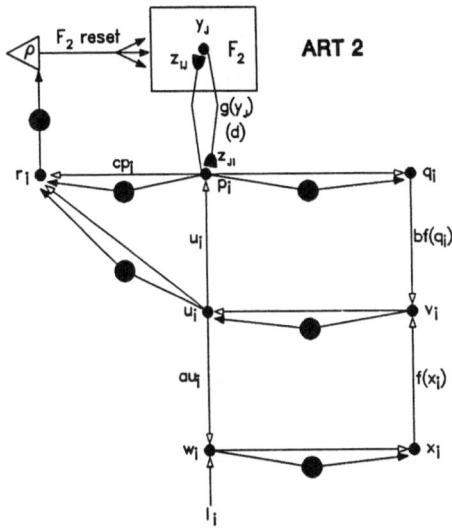

Figure 7. A typical ART 2 architecture. Open arrows indicate specific patterned inputs to target nodes. Filled arrows indicate nonspecific gain control inputs. The gain control nuclei (large filled circles) nonspecifically inhibit target nodes in proportion to the L_2-norm of STM activity in their source fields. As in ART 1, gain control (not shown) coordinates STM processing with input presentation rate.

In each ART 2 architecture, combinations of normalization, gain control, matching, and learning mechanisms are interwoven in generally similar ways. Although how this is done may be modified to some extent, in all of the ART 2 variations that we have discovered, F_1 needs to include different levels to receive and transform bottom-up input patterns and top-down expectation patterns, as well as an interfacing level of interneurons that matches the transformed bottom-up and top-down information and feeds the results back to the bottom and top F_1 levels. How the particular F_1 levels shown in Fig. 7 work will be described in Sections 10-12. Alternative ART 2 models are illustrated in CARPENTER AND GROSSBERG [2].

9. ART 2 Design Principles

We will now describe the main ART 2 design principles.

Stability-Plasticity Tradeoff: An ART 2 system needs to be able to learn a stable recognition code in response to an arbitrary sequence of analog input patterns. Since the plasticity of an ART system is maintained for all time, and since input presentation times can be of arbitrary duration, STM processing must be defined in such a way that a sustained new input pattern does not wash away previously learned information. Removal, or ablation, of one part of the F_1 internal feedback loop can lead to a type of instability in which a single input, embedded in a particular input sequence, can jump between categories indefinitely [2].

Search-Direct Access Tradeoff: An ART 2 system carries out a parallel search in order to regulate the selection of appropriate recognition codes during the learning process, yet automatically disengages the search process as an input pattern becomes familiar. Thereafter the familiar input pattern directly accesses its recognition code no matter how complex the total learned recognition structure may have become, much as we can rapidly recognize our parents at different stages of our life even though we may learn much more as we grow older.

Match-Reset Tradeoff: An ART 2 system needs to be able to resolve several potentially conflicting properties which can be formulated as variants of a design tradeoff between the requirements of sensitive matching and formation of new codes.

The system should on the one hand, be able to recognize and react to arbitrarily small differences between an active F_1 STM pattern and the LTM pattern being read-out from an established category. In particular, if vigilance is high, the F_1 STM pattern established by a bottom-up input exemplar should be nearly identical to the learned top-down $F_2 \rightarrow F_1$ expectation pattern in order for the exemplar to be accepted as a member of an established category. On the other hand, when an uncommitted F_2 node becomes active for the first time, it should be able to remain active, without being reset, so that it can encode its first input exemplar, even though in this case there is no top-down/bottom-up pattern match whatsoever. Section 14 shows how a combination of an appropriately chosen ART 2 reset rule and LTM initial values work together to satisfy both of these processing requirements. In fact, ART 2 parameters can be chosen to satisfy the more general property that learning increases the system's sensitivity to mismatches between bottom-up and top-down patterns.

STM Invariance under Read-out of Matched LTM: Further discussion of match-reset tradeoff clarifies why F_1 is composed of several internal processing levels. Suppose that before an uncommitted F_2 node is first activated, its top-down $F_2 \rightarrow F_1$ LTM traces are chosen equal to zero. On the node's first learning trial, its LTM traces will progressively learn the STM pattern that is generated by the top level of F_1. As noted above, such learning must not be allowed to cause a mismatch capable of resetting F_2, because the LTM traces have not previously learned any other pattern. This property is achieved by designing the bottom and middle levels of F_1 so that their STM activity patterns are not changed at all by the read-out of these LTM traces as they learn their first positive values.

More generally, F_1 is designed so that read-out by F_2 of a previously learned LTM pattern that matches perfectly the STM pattern at the top level of F_1 does not change the STM patterns circulating at the bottom and middle levels of F_1. Thus, in a perfect match situation, or in a situation where a zero-vector of LTM values learns a perfect match, the STM activity patterns at the bottom and middle F_1 levels are left invariant; hence, no reset occurs.

This invariance property enables the bottom and middle F_1 levels to nonlinearly transform the input pattern in a manner that remains stable during learning. In particular, the input pattern may be contrast enhanced while noise in the input is suppressed. If read-out of a top-down LTM pattern could change even the baseline of activation at the F_1 levels which execute this transformation, then the degree of contrast enhancement and noise suppression could be altered, thereby generating a new STM pattern for learning by the top-down LTM traces. The STM invariance property prevents read-out of a perfectly matched LTM pattern from causing reset by preventing any change whatsoever from occurring in the STM patterning at the lower F_1 levels.

Coexistence of LTM Read-out and STM Normalization: The STM invariance property leads to the use of multiple F_1 levels because the F_1 nodes at which top-down LTM read-out occurs receive an additional input when top-down signals are active than when they are not. The extra F_1 levels provide enough degrees of computational freedom to both read-out top-down LTM and normalize the total STM pattern at the top F_1 level before this normalized STM pattern can interact with the middle F_1 level at which top-down and bottom-up information are matched.

In a similar fashion, the bottom F_1 level enables an input pattern to be normalized before this normalized STM pattern can interact with the middle F_1 level. Thus separate bottom and top F_1 levels provide enough degrees of computational freedom to compensate for fluctuations in baseline activity levels. In the absence of such normalization, confusions between useful pattern differences and spurious baseline fluctuations could easily upset the matching process and cause spurious reset events to occur, thereby destabilizing the network's search and learning processes.

No LTM Recoding by Superset Inputs: Although read-out of a top-down LTM pattern that perfectly matches the STM pattern at F_1's top level never causes F_2 reset, even a very small mismatch in these patterns is sufficient to reset F_2 if the vigilance parameter is chosen sufficiently high. The middle F_1 level plays a key role in causing the attenuation of STM activity that causes such a reset event to occur.

An important example of such a reset-inducing mismatch occurs when one or more, but not all, of the top-down LTM traces equal zero or very small values and the corresponding F_1 nodes have positive STM activities. When this occurs, the STM activities of these F_1 nodes are suppressed. If the total STM suppression is large enough to reset F_2, then the network searches for a better match. If the total STM suppression is not large enough to reset F_2, then the top-down LTM traces of these nodes remain small during the ensuing learning trial, because they sample the small STM values that their own small LTM values have caused.

This property is a version of the 2/3 Rule that has been used to prove stability of learning by an ART 1 architecture in response to an arbitrary sequence of binary input patterns [1]. It also is necessary in order for ART 2 to achieve stable learning in response to an arbitrary sequence of analog input patterns (Section 17). In the jargon of ART 1, a *superset* bottom-up input pattern cannot recode a *subset* top-down expectation. In ART 1, this property was achieved by an attentional gain control channel (Fig. 3). In the versions of ART 2 developed so far, it is realized as part of F_1's internal levels. These design variations are still a subject of ongoing research.

Stable Choice until Reset: Match-reset trade-off also requires that only a reset event that is triggered by the orienting subsystem can cause a change in the chosen F_2 code. This property is imposed at any degree of mismatch between a top-down $F_2 \rightarrow F_1$ LTM pattern and the circulating F_1 STM pattern. Thus all the network's real-time pattern processing operations, including top-down $F_2 \rightarrow F_1$ feedback, the fast nonlinear feedback dynamics within F_1, and the slow LTM changes during learning must be organized to maintain the original $F_1 \rightarrow F_2$ category choice, unless F_2 is actively reset by the orienting subsystem.

Contrast Enhancement, Noise Suppression, and Mismatch Attenuation by Nonlinear Signal Functions: A given class of analog signals may be embedded in variable levels of background noise (Fig. 5). A combination of normalization and nonlinear feedback processes within F_1 determines a noise criterion and enables the system to separate signal from noise. In particular, these processes contrast enhance the F_1 STM pattern, and hence also the learned LTM patterns. The degree of contrast enhancement and noise suppression is determined by the degree of nonlinearity in the feedback signal functions at F_1.

A nonlinear signal function operating on the sum of normalized bottom-up and top-down signals also correlates these signals, just as squaring a sum $A + B$ of two L_2-normalized vectors generates $2(1 + A \cdot B)$. Nonlinear feedback signalling hereby helps to attenuate the total activation of F_1 in response to mismatched bottom-up input and top-down expectation patterns, as well as to contrast-enhance and noise-suppress bottom-up input patterns.

Rapid Self-Stabilization: A learning system that is unstable in general can be made more stable by making the learning rate so slow that LTM traces change little on a single input trial. In this case, many learning trials are needed to encode a fixed set of inputs. Learning in an ART system needs to be slow relative to the STM processing rate

(Section 10), but no restrictions are placed on absolute rates. Thus ART 2 is capable of stable learning in the "fast learning" case, in which LTM traces change so quickly that they can approach new equilibrium values on every trial. The ART 2 simulations in this article were all carried out under fast learning conditions, and rapid code self-stabilization occurs in each case. Self-stabilization is also sped up by the action of the orienting subsystem, but can also occur rapidly even without it [2].

Normalization: Several different schemes may be used to normalize activation patterns across F_1. In this article, we used nonspecific inhibitory interneurons (schematized by large black disks in Fig. 7). Each such normalizer uses $O(M)$ connections where M is the number of nodes to be normalized. Alternatively, a shunting on-center off-surround network could be used as a normalizer, but such a network uses $O(M^2)$ connections.

Local Computations: ART 2 system STM and LTM computations use only information available locally and in real-time. There are no assumptions of weight transport, as in back propagation, nor of an *a priori* input probability distribution, as in simulated annealing. Moreover, all ART 2 local equations have a simple form (Sections 10-13). It is the architecture as a whole that endows the model with its desirable emergent computational properties.

10. ART 2 STM Equations: F_1

The potential, or STM activity, V_i of the ith node at any one of the F_1 processing stages obeys a membrane equation [12] of the form

$$\epsilon \frac{d}{dt} V_i = -AV_i + (1 - BV_i)J_i^+ - (C + DV_i)J_i^- \qquad (5)$$

($i = 1 \ldots M$). Term J_i^+ is the total excitatory input to the ith node and J_i^- is the total inhibitory input. In the absence of all inputs, V_i decays to 0. The dimensionless parameter ϵ represents the ratio between the STM relaxation time and the LTM relaxation time. With the LTM rate $O(1)$, then

$$0 < \epsilon \ll 1. \qquad (6)$$

Also, $B \equiv 0$ and $C \equiv 0$ in the F_1 equations of the ART 2 example in Fig. 7. Thus the STM equations, in the singular form as $\epsilon \to 0$, reduce to

$$V_i = \frac{J_i^+}{A + DJ_i^-}. \qquad (7)$$

In this form, the dimensionless equations (8)-(13) characterize the STM activities, p_i, q_i, u_i, v_i, w_i, and x_i, computed at F_1:

$$p_i = u_i + \sum_j g(y_j) z_{ji} \tag{8}$$

$$q_i = \frac{p_i}{e + \| p \|} \tag{9}$$

$$u_i = \frac{v_i}{e + \| v \|} \tag{10}$$

$$v_i = f(x_i) + bf(q_i) \tag{11}$$

$$w_i = I_i + au_i \tag{12}$$

$$x_i = \frac{w_i}{e + \| w \|} \tag{13}$$

where $\| V \|$ denotes the L_2-norm of a vector V, y_j is the STM activity of the jth F_2 node, and z_{ji} is the LTM weight in the pathway from the jth F_2 node to the ith F_1 node. The nonlinear signal function f in (11) is typically of the form

$$f(x) = \begin{cases} \frac{2\theta x^2}{(x^2 + \theta^2)} & \text{if } 0 \le x \le \theta \\ x & \text{if } x \ge \theta, \end{cases} \tag{14}$$

which is continuously differentiable, or

$$f(x) = \begin{cases} 0 & \text{if } 0 \le x < \theta \\ x & \text{if } x \ge \theta, \end{cases} \tag{15}$$

which is piecewise linear. The graph of function $f(x)$ in equation (14) may also be shifted to the right, making $f(x) = 0$ for small x, as in (15). Since the variables x_i and q_i are always between 0 and 1 (equations (9) and (13)), the function values $f(x_i)$ and $f(q_i)$ also stay between 0 and 1. Alternatively, the signal function $f(x)$ could also be chosen to saturate at high x values. This would have the effect of flattening pattern details like those in category 17 of Fig. 5, sitting on the top of an activity peak.

11. ART 2 STM Equations: F_2

The category representation field F_2 is the same in ART 2 as in ART 1. The key properties of F_2 are contrast enhancement of the filtered $F_1 \rightarrow F_2$ input pattern, and reset, or enduring inhibition, of active F_2 nodes whenever a pattern mismatch at F_1 is large enough to activate the orienting subsystem.

Contrast enhancement is carried out by competition within F_2. Choice is the extreme case of contrast enhancememt. F_2 makes a choice when the node receiving the largest total input quenches activity in all other nodes. In other words, let T_j be the summed filtered $F_1 \to F_2$ input to the jth F_2 node:

$$T_j = \sum_i p_i z_{ij} \tag{16}$$

$(j = M + 1 \ldots N)$. Then F_2 is said to make a choice if the Jth F_2 node becomes maximally active, while all other nodes are inhibited, when

$$T_J = \max\{T_j : j = M + 1 \ldots N\}. \tag{17}$$

F_2 reset may be carried out in several ways, one being use of a *gated dipole field* network in F_2. When a nonspecific arousal input reaches an F_2 gated dipole field, nodes are inhibited or reset (Section 12) in proportion to their former STM activity levels. Moreover this inhibition endures until the bottom-up input to F_1 shuts off. Such a nonspecific arousal wave reaches F_2, via the orienting subsystem, when a sufficiently large mismatch occurs at F_1.

When F_2 makes a choice, the main elements of the gated dipole field dynamics may be characterized as

$$g(y_J) = \begin{cases} d & \text{if } T_J = \max\{T_j : \text{the } j^{th} \ F_2 \text{ node has not} \\ & \text{been reset on the current trial}\} \\ 0 & \text{otherwise.} \end{cases} \tag{18}$$

Equation (18) implies that (8) reduces to

$$p_i = \begin{cases} u_i & \text{if } F_2 \text{ is inactive} \\ u_i + d z_{Ji} & \text{if the } J^{th} \ F_2 \text{ node is active.} \end{cases} \tag{19}$$

12. ART 2 LTM Equations

The top-down and bottom-up LTM trace equations for ART 2 are given by

$$Top - down \ (F_2 \to F_1) : \frac{d}{dt} z_{ji} = g(y_j)[p_i - z_{ji}] \tag{20}$$

$$Bottom - up \ (F_1 \to F_2) : \frac{d}{dt} z_{ij} = g(y_j)[p_i - z_{ij}]. \tag{21}$$

If F_2 makes a choice, (18)-(21) imply that, if the Jth F_2 node is active, then

$$\frac{d}{dt} z_{Ji} = d[p_i - z_{Ji}] = d(1 - d)\left[\frac{u_i}{1 - d} - z_{Ji}\right] \tag{22}$$

and

$$\frac{d}{dt}z_{iJ} = d[p_i - z_{iJ}] = d(1-d)\left[\frac{u_i}{1-d} - z_{iJ}\right],$$ (23)

with $0 < d < 1$. For all $j \neq J$, $dz_{ji}/dt = 0$ and $dz_{ij}/dt = 0$. Sections 14 and 16 give admissible bounds on the initial values of the LTM traces.

13. ART 2 Reset Equations: The Orienting Subsystem

Since a binary pattern match may be computed by counting matched bits, ART 1 architectures do not require patterned information in the orienting subsystem (Fig. 3). In contrast, computation of an analog pattern match does require patterned information. The degree of match between an STM pattern at F_1 and an active LTM pattern is determined by the vector $r = (r_1 \ldots r_M)$, where for the ART 2 architecture of Fig. 7,

$$r_i = \frac{u_i + cp_i}{e + \parallel u \parallel + \parallel cp \parallel}.$$ (24)

The orienting subsystem is assumed to reset F_2 whenever an input pattern is active and

$$\frac{\rho}{e + \parallel r \parallel} > 1,$$ (25)

where the vigilance parameter ρ is set between 0 and 1.

For simplicity, we will henceforth consider an ART 2 system in which F_2 makes a choice and in which e is set equal to 0. Thus $\parallel x \parallel = \parallel u \parallel = \parallel q \parallel = 1$. Simulations use the piecewise linear signal function f in (15).

14. The Match-Reset Tradeoff: Choice of Top-down Initial LTM Values

Vector r gives rise to all the properties required to satisfy the match-reset tradeoff described in Section 9. Note first that, when the Jth F_2 node is active, (24) implies that

$$\parallel r \parallel = \frac{[\, 1 + 2 \parallel cp \parallel \cos(u,p) + \parallel cp \parallel^2 \,]^{\frac{1}{2}}}{1 + \parallel cp \parallel},$$ (26)

where $\cos(u,p)$ denotes the cosine of the angle between the vector u and the vector p. Also, by (19), the vector p equals the sum $u + dz_J$, where $z_J \equiv (z_{J1} \ldots z_{JM})$ denotes the top-down vector of LTM traces projecting from the Jth F_2 node. Since $\parallel u \parallel = 1$, the geometry of the vector sum $p = u + dz_J$ implies that

$$\parallel p \parallel \cos(u,p) = 1 + \parallel dz_J \parallel \cos(u,z_J).$$ (27)

Also,
$$\| p \| = [\, 1 + 2 \,\| dz_J \| \cos(u, z_J) + \| dz_J \|^2 \,]^{\frac{1}{2}}. \tag{28}$$
Equations (26)-(28) imply that

$$\| r \| = \frac{[\, (1 + o)^2 + 2(1 + c) \,\| cd\, z_J \| \cos(u, z_J) + \| cd\, z_J \|^2 \,]^{\frac{1}{2}}}{1 + [\, c^2 + 2c \,\| cd\, z_J \| \cos(u, z_J) + \| cd\, z_J \|^2 \,]^{\frac{1}{2}}}. \tag{29}$$

Both numerator and denominator equal $1 + c + \| cd\, z_J \|$ when $\cos(u, z_J) = 1$. Thus $\| r \| = 1$ when the STM pattern u exactly matches the LTM pattern z_J, up to a constant multiple.

Figure 8 graphs $\| r \|$ as a function of $\| cdz_J \|$ for various values of $\cos(u, z_J)$. The Jth F_2 node remains active only if $\rho \leq \| r \|$. Since $\rho < 1$, Fig. 8 shows that this will occur either if $\cos(u, z_J)$ is close to 1 or if $\| z_J \|$ is close to 0. That is, no reset occurs if the STM vector u is nearly parallel to the LTM vector z_J or if the top-down LTM traces z_{Ji} are all small. By (22), z_J becomes parallel to u during learning, thus inhibiting reset. Reset must also be inhibited, however, while a new category is being established. Figure 8 shows that this can be accomplished by making all $\| z_j \|$ small before any learning occurs; in particular, we let the top-down initial LTM values satisfy

$$z_{ji}(0) = 0, \tag{30}$$

for $i = 1 \ldots M$ and $j = M + 1 \ldots N$.

Condition (30) ensures that no reset occurs when an uncommitted F_2 node first becomes active. Hence learning can begin. Moreover, the learning rule (22) and the LTM initial value rule (30) together imply that z_J remains parallel to u as learning proceeds, so $\| r(t) \| \equiv 1$. Thus no reset ever occurs during a trial in which an uncommitted F_2 node is first activated.

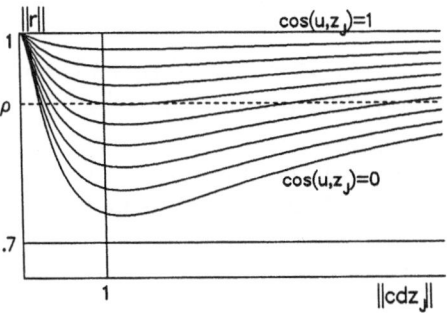

Figure 8. Graph of $\| r \|$ as a function of $\| cd\, z_J \|$ for values of $\cos(u, z_J)$ between 0 and 1 and for $c = .1$ and $d = .9$. F_2 reset occurs whenever $\| r \|$ falls below the vigilance parameter ρ.

15. Learning Increases Mismatch Sensitivity and Confirms Category Choice

Figure 8 suggests how to implement the property that learning increases sensitivity to mismatches between bottom-up and top-down patterns. Fig. 8 indicates that, for fixed $\cos(u, z_J)$, $\| r \|$ is a decreasing function of $\| cd\ z_J \|$ for $\| cd\ z_J \| \leq 1$. In fact, in the limit as $c \to 0$, the minimum of each curve approaches the line $\| cd\ z_J \| = 1$. By (22) and (30), $\| z_J \| < 1/(1-d)$ and $\| z_J \| \to 1/(1-d)$ during learning. Therefore implementation of the property that learning increases mismatch sensitivity translates into the parameter constraint

$$\frac{cd}{1-d} \leq 1. \tag{31}$$

The closer the ratio $cd/(1-d)$ is chosen to 1 the more sensitive the system is to mismatches, all other things being equal.

Parameter constraint (31) helps to ensure that learning on a given trial confirms the intial category choice on that trial. To see this note that, if an established category is chosen, $\| z_J \|$ is close to $1/(1-d)$ at the beginning and end of a "fast learning" trial. However $\| z_J \|$ typically decreases and then increases during a learning trial. Therefore if $cd/(1-d)$ were greater than 1, the reset inequality (25) could be satisfied while $\| z_J \|$ was decreasing. Thus, without (31), it would be difficult to rule out the possibility of unexpected F_2 reset in the middle of a learning trial.

16. Choosing a New Category: Bottom-up LTM Initial Values

Section 14 discusses the fact that the top-down initial LTM values $z_{ji}(0)$ need to be chosen small, or else top-down LTM read-out by an uncommitted node could lead to immediate F_2 reset rather than learning of a new category. The bottom-up LTM initial values $z_{ij}(0)$ also need to be chosen small, but for different reasons.

Let $z^J \equiv (z_{1J} \ldots z_{MJ})$ denote the bottom-up vector of LTM traces that project to the Jth F_2 node. Equation (23) implies that $\| z^J \| \to 1/(1-d)$ during learning. If $\| z^J(0) \|$ were chosen greater than $1/(1-d)$, an input that first chose an uncommitted node could switch to other uncommitted nodes in the middle of a learning trial. It is thus necessary to require that

$$\| z^J(0) \| \leq \frac{1}{1-d}. \tag{32}$$

Inequality (32) implies that if each $z^J(0)$ is uniform, then each LTM trace must satisfy the constraint

$$z_{ij}(0) \leq \frac{1}{(1-d)\sqrt{M}} \qquad (33)$$

for $i = 1 \ldots M$ and $j = M+1 \ldots N$. Alternatively, random numbers or trained patterns could be taken as initial LTM values. If bottom-up input is the sole source of F_2 activation, at least some $z_{iJ}(0)$ values need to be chosen positive if the Jth F_2 node is ever to become active.

Choosing equality in (33) biases the ART 2 system as much as possible toward choosing uncommitted nodes. A typical input would search only those nodes with which it is fairly well matched, and then go directly to an uncommitted node. If no learned category representation forms a good match, an uncommitted node will be directly accessed. Setting the initial bottom-up LTM trace values as large as possible, therefore, helps to stabilize the ART 2 network by ensuring that the system will form a new category, rather than recode an established but badly mismatched one, when vigilance is too low to prevent recoding by active reset via the orienting subsystem. Thus examples of instability occur only when, in addition to the removal of the orienting subsystem and the internal feedback at F_1, the initial bottom-up LTM trace values are significantly less than the maximum allowed by condition (33).

17. The Stability-Plasticity Tradeoff

ART 2 design principles permit arbitrary sequences of patterns to be encoded during arbitrarily long input trials, and the ability of the LTM traces to learn does not decrease with time. Some internal mechanism must therefore buffer established ART category structures against ceaseless recoding by new input patterns. ART 1 architectures buffer category structures by means of the 2/3 Rule for pattern matching (Fig. 4). During matching, an F_1 node in ART 1 can remain active only if it receives significant inputs both bottom-up and top-down. ART 1 implements the 2/3 Rule using an inhibitory attentional gain control signal that is read out with the top-down LTM vector (Fig. 3).

ART 2 architectures implement a weak version of the 2/3 Rule in which, during matching, an F_1 node can remain active only if it receives significant top-down input. It is possible, however, for a node receiving large top-down input to remain stored in memory even if bottom-up input to that node is absent on a given trial. The corresponding feature, which had been encoded as significant by prior exemplars, would hence remain part of the category representation although unmatched in the active exemplar. It would, moreover, be partially restored in STM. During learning, the relative importance of that feature would decline, but it would not necessarily be eliminated. However, a feature consistently absent from most category exemplars would eventually be removed from the category's expectation pattern z_J. The ART 2 matching rule implies that the feature would then not be relearned; if present in a given exemplar, it would be treated as noise.

18. Invariant Visual Pattern Recognition

An application of ART 2 to invariant visual pattern recognition is being carried out in a collaboration between Boston University and M.I.T. Lincoln Laboratory. This application uses a three stage preprocessor. First, the image figure to be recognized is detached from the image background using laser radar sensors. This can be accomplished by intersecting the images formed by two laser sensors: the image formed by a range detector focussed at the distance of the figure, with the image formed by another laser detector that is capable of differentiating figure from background, such as a doppler image when the figure is moving or the intensity of laser return when the figure is stationary [13]. The second stage of the preprocessor contains a neural network, called a Boundary Contour System [3,4], that detects, sharpens, regularizes, and completes the boundaries within noisy images. The third stage of the preprocessor contains a Fourier-Mellin filter, whose output spectra are invariant under such image transformations as 2-D spatial translation, dilation, and rotation [14]. Thus the input patterns to ART 2 are the invariant spectra of completed boundary segmentations of laser radar sensors. By setting ART 2 parameters to suppress (up to) a prescribed level of input noise and to tolerate (up to) a prescribed level of input deformation, this system defines a compact circuit capable of autonomously learning to recognize visual targets that are deformed, rotated, dilated, and shifted. Although this preprocessor does not purport to provide a biological solution to the problem of invariant visual object recognition, it is known that the mammalian visual cortex does carry out computations that are analogous to aspects of the second and third stages of this preprocessor [3,4,15].

19. The Three R's: Recognition, Reinforcement, and Recall

Recognition is only one of several processes whereby an intelligent system can learn a correct solution to a problem. Reinforcement and recall are no less important in designing an autonomous intelligent system.

Reinforcement, notably reward and punishment, provides additional information in the form of environmental feedback based on the success or failure of actions triggered by a recognition event. Reward and punishment calibrate whether the action has or has not satisfied internal needs, which in the biological case include hunger, thirst, sex, and pain reduction, but may in machine applications include a wide variety of internal cost functions. Reinforcement can modify the formation of recognition codes and can shift attention to focus upon those codes whose activation promises to satisfy internal needs based upon past experience. For example, both green and yellow bananas may be recognized as part of a single recognition category until reinforcement signals, contingent upon eating these bananas, differentiates them into separate categories.

Recall can generate equivalent responses or actions to input events that are classified by different recognition codes. For example, printed and script letters may generate distinct recognition codes, yet can also elicit identical learned naming responses.

Our own research program at Boston University during the past two decades has been devoted to discovering and implementing models of self-organizing biological systems wherein all the ingredients of recognition, reinforcement, and recall are joined together into a single integrated circuit [3,4,16]. The system depicted in Fig. 9 provides a framework for implementing some of these circuit designs. In particular, as ART 2 self-organizes recognition categories in response to the preprocessed inputs, its categorical choices at the F_2 classifying level self-stabilize through time. In examples wherein F_2 makes a choice, it can be used as the first level of an ART 1 architecture, or yet another ART 2 architecture, should one prefer. Let us call the classifying level of this latter architecture F_3. Level F_3 can be used as a source of pre-wired priming inputs to F_2. Alternatively, as in Fig. 9, self-stabilizing choices by F_3 can quickly be learned in response to the choices made at F_2. Then F_3 can be used as a source of self-organized priming inputs to F_2, and a source of priming patterns can be associated with each of the F_3 choices via mechanisms of associative pattern learning [3]. After learning of these primes takes place, turning on a particular prime can activate a learned $F_3 \rightarrow F_2$ top-down expectation. Then F_2 can be supraliminally activated only by an input exemplar which is a member of the recognition category of the primed F_2 node. The architecture

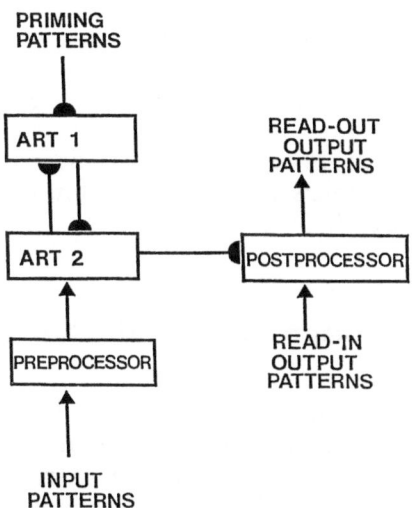

Figure 9. A self-organizing architecture for invariant pattern recognition and recall that can be expanded, as noted in the text, to include reinforcement mechanisms capable of focussing attention upon internally desired classes of external events.

ignores all but the primed set of input patterns. In other words, the prime causes the architecture to pay attention only to expected sources of input information. Due to the spatial invariance properties of the preprocessor, the expected input patterns can be translated, dilated, or rotated in 2-D without damaging recognition. Due to the similarity grouping properties of ART 2 at a fixed level of vigilance, suitable deformations of these input patterns, including deformations due to no more than anticipated levels of noise, can also be recognized.

The output pathways from level F_2 of ART 2 to the postprocessor can learn to recall any spatial pattern or spatiotemporal pattern of outputs by applying theorems about associative learning in a type of circuit called an avalanche [3]. In particular, distinct recognition categories can learn to generate identical recall responses. Thus the architecture as a whole can stably self-organize an invariant recognition code and an associative map to an arbitrary format of output patterns.

The interactions (priming $\rightarrow$ ART) and (ART $\rightarrow$ postprocessor) in Fig. 9 can be modified so that output patterns are read out only if the input patterns have yielded rewards in the past and if the machine's internal needs for these rewards have not yet been satisfied [3,4]. In this variation of the architecture, the priming patterns supply motivational signals for releasing outputs only if an input exemplar from an internally desired recognition category is detected. The total circuit forms a neural network architecture which can stably self-organize an invariant pattern recognition code in response to a sequence of analog or binary input patterns; be attentionally primed to ignore all but a designated category of input patterns; automatically shift its prime as it satisfies internal criteria in response to actions based upon the recognition of a previously primed category of input patterns; and learn to generate an arbitrary spatiotemporal output pattern in response to any input pattern exemplar of an activated recognition category. Such circuits, and their real-time adaptive autonomous descendents, may prove to be useful in some of the many applications where preprogrammed rule-based systems, and systems requiring external teachers not naturally found in the applications environments, fear to tread.

20. Self-Stabilization of Speech Perception and Production Codes: New Light on Motor Theory

The insights gleaned from the design of ART 2 have also begun to clarify how hierarchical learning systems with multiple ART levels can be designed. Fig. 10 schematizes a hierarchical ART system for learning to recognize and produce speech which self-stabilizes its learning in real-time without using a teacher. This ART architecture is being developed at Boston University by COHEN, GROSSBERG, STORK [17]. Top-down ART expectation mechanisms at several levels of this architecture help to self-stabilize learned codes and to self-organize the

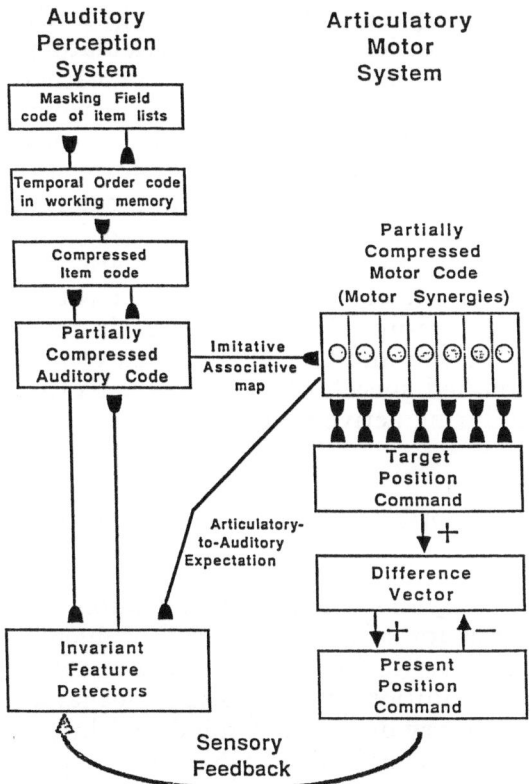

Auditory Perception System

Masking Field code of item lists

Temporal Order code in working memory

Compressed Item code

Partially Compressed Auditory Code

Invariant Feature Detectors

Imitative Associative map

Articulatory-to-Auditory Expectation

Articulatory Motor System

Partially Compressed Motor Code (Motor Synergies)

Target Position Command

Difference Vector

Present Position Command

Sensory Feedback

Figure 10. Schematic of some processing stages in an architecture for a self-organizing speech perception and production system. The left hand side of the figure depicts five stages of the auditory model, the right hand side four stages of the motor model. The pathways from the partially compressed auditory code to the motor system learn an imitative associative map which joins auditory feedback patterns to the motor commands that generated them. These motor commands are compressed via bottom-up and top-down adaptive filters within the motor system into motor synergies. The synergies read out top-down learned articulatory-to-auditory expectations, which select the motorically consistent auditory data for incorporation into the learned speech codes of the auditory system.

selection of invariant recognition properties. Of particular interest in this speech architecture is the role of top-down expectation signals from the architecture's articulatory, or motor, system to its auditory, or perception, system. These expectations help to explain classical results from motor theory, which state that speech is perceived in terms of how it would have been produced, even during passive listening.

The key insights of the motor theory take on new meaning in the theory through the self-stabilizing properties of top-down articulatory-to-auditory expectations. These expectations self-stabilize the learned imitative associative map that transforms the perceptual codes which represent heard speech into motor codes for generating spoken speech. In so doing, the articulatory-to-auditory expectations deform the bottom-up auditory STM patterns, via 2/3 Rule-like matching, into activation patterns which are consistent with invariant properties of the motor commands. These motorically-modified STM codes are then ecoded in long-term memory within a bottom-up adaptive filter within the auditory system itself. This bottom-up adaptive filter activates a partially compressed speech code at the auditory system's next processing level. The motorically-modified speech code is thus activated during passive listening as well as during active imitation.

21. Psychophysiological and Neurophysiological Predictions of ART

Although applications of ART to computer science depend upon the computational power of these systems for solving real-world problems, ART systems are also models of the biological processes whose analysis led to their discovery. In fact, in addition to suggesting mechanistic explanations of many interdisciplinary data about mind and brain, the theory has also made a number of predictions which have since been partially supported by experiments. For example, in 1976, it was predicted that both norepinephrine (NE) mechanisms and attentional mechanisms modulate the adaptive development of thalamocortical visual feature detectors [18]. In 1976 and 1978, KASAMATSU and PETTIGREW [19,20] described NE modulation of feature detector development. SINGER [21] reported attentional modulation in 1982. In 1978, a word length effect in word recognition paradigms was predicted [22]. In 1982 and 1983, SAMUEL, VAN SANTEN, JOHNSTON [23,24] reported a word length effect in word superiority experiments. In 1978 and 1980, a hippocampal generator of the P300 event-related potential was predicted [22,25]. In 1980, HALGREN and his colleagues reported the existence of a hippocampal P300 generator in humans [26]. The existence and correlations between other event-related potentials, such as processing negativity (PN), early positive wave (P120), N200, and P300 were also predicted in these theoretical articles. These predictions and supportive data are described in several recent books [3,4]. Additional experimental support for these predicted correlations has recently been reported [27,28].

Thus ART systems provide a fertile ground for gaining a new understanding of biological intelligence while suggesting novel computational theories and real-time adaptive neural network architectures with promising properties for tackling some of the outstanding problems in computer science and technology today.

ACKNOWLEDGEMENTS

We wish to thank Cynthia Suchta and Carol Yanakakis for their valuable assistance in the preparation of the manuscript and illustrations.

This research was supported in part by the Air Force Office of Scientific Research (AFOSR F49620-86-C-0037 and AFOSR F49620-87-C-0018), the Army Research Office (ARO DAAG-29-85-K-0095), and the National Science Foundation (NSF DMS-86-11959 and IRI-84-17756).

REFERENCES

1. G.A. Carpenter, S. Grossberg: *Computer Vision, Graphics, and Image Processing*, **37**, 1987, p.54.

2. G.A. Carpenter, S. Grossberg: *Applied Optics*, **26**, 1987, p.4919.

3. S. Grossberg (Ed.): **The Adaptive Brain, Volumes I and II.** Amsterdam: Elsevier/North-Holland, 1987.

4. S. Grossberg (Ed.): **Neural Networks and Natural Intelligence.** Cambridge, MA: MIT Press, 1988.

5. C. von der Malsburg: *Kybernetik*, **14**, 1973, p.85.

6. S. Grossberg: *Biological Cybernetics*, **23**, 1976, p.121.

7. S. Amari, A. Takeuchi: *Biological Cybernetics*, **29**, 1978, p.127.

8. E.L. Bienenstock, L.N. Cooper, P.W. Munro: *Journal of Neuroscience*, **2**, 1982, p.32.

9. T. Kohonen: **Self-Organization and Associative Memory.** New York: Springer-Verlag, 1984.

10. D.H. Ackley, G.E. Hinton, T.J. Sejnowski: *Cognitive Science*, **9**, 1985, p.147.

11. D.E. Rumelhart, G.E. Hinton, R.J. Williams: In D.E. Rumelhart and J.L. McClelland (Eds.), **Parallel Distributed Processing.** Cambridge, MA: MIT Press, 1986.

12. A.L. Hodgkin, A.F. Huxley: *Journal of Physiology*, **117**, 1952, p.500.

13. A.B. Gschwendtner, R.C. Harney, R.J. Hull: In D.K. Killinger and A. Mooradian (Eds.), **Optical and Laser Remote Sensing.** New York: Springer-Verlag, 1983.

14. D. Casasent, D. Psaltis: *Applied Optics*, **15**, 1976, p.1793.

15. E.L. Schwartz: *Vision Research*, **20**, 1980, p.645.

16. G.A. Carpenter, S. Grossberg: *Computer*, **21**, 1988, p.77.

17. M. Cohen, S. Grossberg, D. Stork: In M. Caudill and C. Butler (Eds.), **Proceedings of the IEEE International Conference on Neural Networks, IV**, 1987, 443-454.

18. S. Grossberg: *Biological Cybernetics*, **23**, 1976, p.187.

19. J.D. Pettigrew, T. Kasamatsu: *Science*, **194**, 1976, p.206.

20. J.D. Pettigrew, T. Kasamatsu: *Nature*, **271**, 1978, p.761.

21. W. Singer: *Human Neurobiology*, **1**, 1982, p.41.

22. S. Grossberg: In R. Rosen and F. Snell (Eds.), **Progress in theoretical biology**, Vol. 5. New York: Academic Press, 1978, p.233.

23. A. Samuel, J.P.H. van Santen, J.C. Johnston: *Journal of Experimental Psychology: Human Perception and Performance*, 8, 1982, p.91.

24. A. Samuel, J.P.H. van Santen, J.C. Johnston: *Journal of Experimental Psychology: Human Perception and Performance*, **9**, 1983, p.321.

25. S. Grossberg: *Psychological Review*, **1**, 1980, p.1.

26. E. Halgren, N.K. Squires, C.L. Wilson, J.W. Rohrbaugh, T.L. Babb, P.H. Crandall: *Science*, **210**, p.803.

27. J.-P. Banquet, J. Baribeau-Braun, N. Leseure: In R. Karrer, J. Cohen, and P. Tueting (Eds.), **Brain and information: Event related potentials**. New York: New York Academy of Sciences, 1984.

28. J.-P. Banquet, S. Grossberg: *Applied Optics*, **26**, 1987, p.4931.

Part II

Neural Networks

Problems for Nets and Nets for Problems

E.R. Caianiello

Dipartimento di Fisica Teorica e S.M.S.A., Università degli Studi di Salerno,
I-84081 Baronissi (Salerno), Italy

INTRODUCTION

A neural net can only be an element, or "organ", of a larger machine,whose "anatomy" and "physiology" must serve some purpose, or purposes; the name "computer", no matter what qualifications are added, seems inadequate for such a machine, which may be required to perform tasks which have nothing to do with computation, or to deal with "ambiguity" rather than with "tertium non datur", like animals or quantum physics. My personal experience has always led me to confront problems whose very existence was previously unknown; however satisfactory some results may have been to me, each step has only increased my feeling of ignorance and strengthened my belief that the whole field is still in its infancy.

For this reason I shall abstain from general consideration and limit my exposition to two such situations which I have met, as examples and justification of my attitude.

1. A PROBLEM FOR NETS: C-CALCULUS

1.1 Definition

I wish to state forthwith that C-calculus is simpler in principle than ordinary arithmetic; various fields can be envisaged in which it might prove of use, e.g. integration theory or fuzzy sets (where it might provide a natural tool for numerical computation).

I begin by remembering a game with which it started. Take any integer positive numbers, and apply to them the rules of arithmetic, with the restrictions that only the direct operations, addition and multiplication, be allowed; the inverse ones, subtraction and division, forbidden. Define furthermore the sum and the product of any two digits as follows

$$a + b = max(a, b)$$

$$a \times b = min(a, b) \quad . \tag{1}$$

We may thus "multiply" any two such numbers, e.g. 736 and 491

$$736 \times 491 = 47461 = 491 \times 736.$$

We find that multiplication and addition thus defined are always commutative for any such "numbers".

It would be an easy matter to demonstrate that, provided the "single digit" operations (1) are meaningful, one can operate in the same way on objects (subtraction and division being of course barred) such as vectors, matrices, etc., obtaining additivity and commutativity whenever they hold in arithmetic.

These "numbers" or "strings" of digits, with the operations (1) , form clearly a commutative semi-ring. As in arithmetic, each "digit" plays two different roles: one intrinsic to it ("cardinality"), the other ("position") relative to the string in which it belongs. The next remark is that standard set theory treats only intrinsic properties

Springer Series in Synergetics Vol. 42: **Neural and Synergetic Computers**
Editor: H. Haken   ©Springer-Verlag Berlin Heidelberg 1988

of sets. If in (1) we interpret + as "union" ∨ and x as "intersection" ∧ , we can immediately transport all that was said thus far to "strings of sets", or "composite sets", "C-sets" for short.

Operating on C-sets as before, with ∨ and ∧ in place of + and × in (1) (a,b denote now the "simple" sets of which C-sets are strings, as the digits in the former example), one has "C-calculus": a commutative semi-ring which permits , from some given C-sets, the generation of any number of other C-sets. Inverse operations are neither possible nor required in this context: only direct ones are permissible; one may perhaps see, though, advantages in being able to express in this way long lists of specifications in terms of a few basic ones.

An example of C-operation of a special relevance for our present purpose is the following. Consider a segment S partitioned in segments $a_1, a_2, a_3, \ldots, a_k$; this partition is $A = a_1 a_2 \ldots a_k$.

Consider now the same segment partitioned in a different way $B = b_1 b_2 \ldots b_l$.

Consider now A and B as C-sets: the elements of each partition or string, are "simple" sets; C-multiplication of A and B gives

$$A \times B = B \times A = a_1 a_2 \ldots a_k \times b_1 b_2 \ldots b_l = C = c_1 c_2 \ldots c_p$$

and it can be verified immediately that the simple sets of the product are obtained, in order, by joining on the segments the terminal points of both partitions A and B. The C-product of two partitions gives thus the refinement of one by the other: C-calculus is the natural way of composing partitions, or coverings. In fact, the same property holds true in any number of dimensions [4]. This is the key property of the C-calculus as regards its application to pattern recognition.

1.2 Application to Pattern Recognition

Granted a priori that a major crime of pattern recognition is the preliminary reduction of a (say) 2-dimensional image into pixels, and that we must so proceed because we are much less bright than a fly or a frog, we find that a rather peculiar situation then arises. Parcelling a picture into pixels (with tones of grey, or colour) is, logically, a parallel process, out of which we can gather more information, the finer the grid whose windows generate "homogenized pixels" (from each pixel only averages are taken) . Suppose now that the same grid is rigidly shifted, over the picture, by a fraction of its window size; we may proceed as before, and obtain some other amount of parallel information. The first question arises: can we use both pieces of information, the one from the first and the one from the second grid partitioning, to get better, more detailed information on the picture? Since we are taking only averages from each pixel, each time, the answer is no (unless, of course, we perform some mathematical acrobatics): one of the two readings has to be thrown away.

It would be nicer, one might feel, if there were a way of performing readings from the grid such as to permit the combination in a natural way of the readings of both grids to obtain more refined information on the picture (as might have been gathered by using a finer grid to begin with). If one can handle this situation, conceivably it may then be possible to use several times (serially) a single coarser grid, out of which is read the (parallel) information obtained by shifting the whole grid by one step, and so on. C-calculus can answer this question. The "reading" from a grid (of a suitable sort) becomes per se a C-set; two C-sets from different positions of the grid can be C-multiplied: this will give finer information, and so on.

Under "suitable" circumstances (to be defined explicitly in the following) this procedure can be carried through to the extreme limit of perfect recontruction of the original picture (as digitized at the finest possible level; e.g. with a $2^{10} \times 2^{10}$ grid for the original, it may be reconstructed by covering it stepwise with, say, a $2^3 \times 2^3$ grid). During this process, many things which one does with specific techniques such as contour extraction, contrast enhancement, feature extraction, etc., can be performed by interpolating in it steps which "answer" questions of this sort and become part of the algorithm. But the application of C-calculus will often fail; the original image may not be thus reconstructed. There is an element to be considered, which was before ignored with the adjective "suitable": the size of the window. It is a feature of our approach that the critical size, below which total reconstruction of the picture is impossible, is determined by the structure of the picture itself, and is not a matter of guesswork or trial and error.

One can arrange readings, and ways of analyzing them, from grids having sizes appropriate to constitute filters that see some wanted features and are blind to others. Typically, consider a saucer on a chessboard: things can be arranged so as to see only the saucer or only the chessboard (with a hole); or a specific component of a texture,ignoring all others, or to suppress some noise etc. Such filtering does not smear out or enhance; it gives at worst an indented contour to the saucer, as is natural when working with grids.

With one-dimensional patterns the procedure can be visualized immediately. We restrict our attention to "grids", i.e. to partitions of the abscissa x into segments, for simplicity, of equal length, the "windows" . Consider a graph whose ordinate y denotes greyness (or intensity of sound, or local pitch...). Discretize now the x coordinate with a grid G, of window w; change the graph into a sequence of rectangles, by replacing the portion of graph corresponding to a given window w_h by the rectangle which projects upon w into w_h and upon y into the segment having as upper and lower extrema the maximum M_h and the minimum m_h reached by the graph within w_h (no matter where, or how many times). We thus change the graph into a string of rectangles.

The ordered sequence of all these rectangles is the C-set determined from the graph by partitioning the x-axis with the given grid. We may now proceed as before, after shifting the grid by a step $1 < w$. Denoting with indices 1 and 2 the two C-sets thus obtained, we have, with an obvious notation,

$$\begin{cases} C_1 = R_{1,1}, \ldots, R_{1,k} \\ C_2 = R_{2,1}, \ldots, R_{2,k} \\ R_{i,h} = (w_{i,h}; m_{i,h}, M_{i,h}) \end{cases}$$

We may now define the product of two simple sets $R_{1,h}$ and $R_{2,l}$ as the intersection of the rectangles just defined, i.e.

$$R_{1,h} \times R_{2,l} = \begin{cases} \Phi & iff \quad W_{1,h} \wedge W_{2,l} = \Phi \\ (W_{1,h} \wedge W_{2,l}), & max(m_{1,h}, m_{2,l}), \quad min(M_{1,h}, M_{2,l}) \end{cases} .$$

In other words: instead of attaching to each window w one value, say the average of y in it, we take two, m_h and M_h; the difference $M_h - m_h = \Delta_h$ is known as the dynamic of the graph in w_h.

This modification is sufficient to carry out our proposed programme, because now it is evident that $C_1 \times C_2$ represents a finer partition of some strip within which the graph-line is contained.

Consider now any rectangle of the C-set $C_1 \times C_2$; its base is $w_{1,h} \wedge w_{2,h}$, the height is

$$y_h = min(M_{1,h}, M_{2,h}) - max(m_{1,h}, m_{2,h}).$$

It is thus evident that the dynamic of the graph in it is

$$\Delta(w_{1,h} \wedge w_{2,h}) \leq y_h,$$

which reduces to

$$\Delta(w_{1,h} \wedge w_{2,h}) = y_h$$

if the graph is monotonic in $w_{1,h} \vee w_{2,h}$.

This remark is essential in order to study under which conditions iterated C-multiplication of C-sets obtained by shifting a given grid will reproduce the given graph to maximum permissible accuracy (that of the original graph, which was supposed digitized at some finer level).

The criterion of convergence to be satisfied for total reconstruction of the original graph, or parts of it, must clearly be the following: convergence is achieved wherever one obtains, at the hth iteration, $m_{i,h} = M_{i,h}$ over the corresponding w_i's. One may substitute this criterion by the weaker one (especially if the ordinate is not discretized) that $M_{i,h} - m_{i,h} < \varepsilon$, prefixed, as small as convenient.

For equally spaced grids (it would be only a matter of convenience to relax or change this condition at any desired step of the procedure), there is a very simple formula which determines whether overall convergence is guaranteed: this will be the case if, and only if

$$W \le \frac{D}{2} + 1 \quad , \tag{2}$$

where D denotes the smallest distance between a minimum and a maximum of the graph. The same formula applies also in two (or more) dimensions if we now read W to mean the side of the square window, and D the euclidean distance in the plane between such extrema.

Of special interest is the case in which (2) is violated. Our procedure will then not reconstruct the original pattern, but produce a new pattern, which suppresses all those details of the original one which could be retrieved only by respecting (2). In other words, the procedure will act now as a filter. A trivial, intuitive example will convince us of this fact. Suppose that we wish to study in this way the above-mentioned chessboard. Operating with a square grid whose window is smaller than the case of the chessboard, we will readily reconstruct the chessboard. If, however, the window is larger than the case, no matter how we move the grid we shall always find $m = 0$, $M = 1$ (say) in any window: convergence is impossible. It is then a trivial matter to arrange things so that in the first case our procedure reconstructs the chessboard, in the second it yields a total blank: the chessboard is filtered away. If we have a saucer on the chessboard, we shall be able to retrieve only the saucer, obliterating the chessboard background. Likewise, one can proceed with textures : it is possible to see an object ignoring a textural background, or vice versa, to extract only some relevant textural elements.

Neural nets appear to be the natural tool for the implementation of C-calculus, which has been thus far only utilized through computer simulation.

2. NETS AND CELLULAR AUTOMATA FOR THE INVERSE PROBLEM

2.1 Notation

The "neuron" is a binary decision element, whose states can be better described as $x = (0,1)$, or $\xi=(-1,1)$, according to the specific purpose; of course

$$x = \frac{1+\xi}{2}.$$

The net has N neurons, whose interconnections determine its structure. We are not concerned here with specific structures; the NE (Neuronic Equations) [1] describe thus a general net as if it were a physical medium of which the NE describe the laws. Denote with

$$\xi \equiv \underline{\xi} \equiv \{\xi^1, \xi^2, \ldots, \xi^N\}; \quad \xi^h = \pm 1$$

variables, vectors, or one-column matrices, whose components have values as specified. Let $F(\xi)$ be any real functions subject only to the condition

$$F(\xi) \neq 0;$$

for any choice of variables ξ^h. This requirement (which is not in fact a restriction) will simplify remarkably our discussion. Call

$$\sigma[F] \equiv sgn[F] = \begin{cases} 1 & for\, F > 0 \\ -1 & for\, F < 0 \end{cases}.$$

Define

$$\langle F(\xi) \rangle = \frac{1}{2^N} \sum_{(\xi^1 = \pm 1, \ldots, \xi^N = \pm 1)} F(\xi^1, \ldots, \xi^N).$$

The tensor powers of ξ have 2^N components:

$$\eta^\alpha = \begin{cases} 1 \equiv \xi^0 \\ \vdots \\ \xi^h \qquad\qquad h = 1, \ldots, N \\ \vdots \\ \xi^{h_1} \ldots \xi^{h_r} \\ \vdots \qquad\qquad \alpha \equiv h_1, \ldots, h_r \\ \xi^1 \ldots \xi^N \end{cases}$$

$\eta \equiv \underline{\eta} \equiv \{\eta^\alpha\}$ is thus a vector in 2^N dimensions, $\eta^\alpha = \pm 1$; the α-ordering of the indices $0, 1, \ldots, 2^N - 1$ may be arranged to suit particular needs; we choose here

$$\eta = \begin{pmatrix} 1 \\ \xi^N \end{pmatrix} \times \begin{pmatrix} 1 \\ \xi^{N-1} \end{pmatrix} \times \cdots \times \begin{pmatrix} 1 \\ \xi^1 \end{pmatrix} = \begin{pmatrix} 1 \\ \xi^1 \\ \xi^2 \\ \xi^1 \xi^2 \\ \xi^3 \\ \vdots \\ \xi^1 \ldots \xi^N \end{pmatrix}.$$

Then, all the properties of boolean functions needed here can be readily derived from the evident ones

$$(\sigma[F])^2 = +1; \quad \sigma(\sigma[F]) = \sigma[F]; \quad \sigma[FG] = \sigma[F]\sigma[G].$$

In particular, one has the η-expansion

$$\sigma[F(\xi)] = \sum_{\alpha=0}^{2^N-1} f_\alpha \eta^\alpha \equiv f^T \eta,$$

where

$$f_\alpha = \langle \eta^\alpha \sigma[F(\xi)] \rangle = \langle \sigma[\eta^\alpha F(\xi)] \rangle.$$

2.2 Direct and Inverse Problem

It is convenient to work directly with tensorial expansions. If each neuron h of the net has as excitation function the real function $f^{(h)} = f^{(h)}(\xi^1, \ldots, \xi^N)$ we can write the NE for a general net as

$$\xi^h_{m+1} = \sigma[f^h(\xi^1_m, \ldots, \xi^N_m)] = \sum_\alpha f^h_\alpha \eta^\alpha_m = f^{hT}\eta_m.$$

We consider now the normalized ξ-state matrix of the net $\Phi_{(N)}$; with $N=3$, e.g.,

$$\Phi_{(3)} = 2^{-3/2} \begin{pmatrix} 1 & -1 & 1 & -1 & 1 & -1 & 1 & -1 \\ 1 & 1 & -1 & -1 & 1 & 1 & -1 & -1 \\ 1 & 1 & 1 & 1 & -1 & -1 & -1 & -1 \end{pmatrix}.$$

We can augment the $N \times 2^N$ ξ-matrix $\Phi_{(N)}$ to the $2^N \times 2^N$ η-state matrix, from

$$\eta = \begin{pmatrix} 1 \\ \xi_N \end{pmatrix} \times \ldots \times \begin{pmatrix} 1 \\ \xi_1 \end{pmatrix},$$

as follows:

$$\Phi_{(N)} = \begin{pmatrix} \frac{1}{2} & \frac{1}{2} \\ \frac{1}{2} & -\frac{1}{2} \end{pmatrix} \times \ldots \times \begin{pmatrix} \frac{1}{2} & \frac{1}{2} \\ \frac{1}{2} & -\frac{1}{2} \end{pmatrix}.$$

(N times)

$\Phi_{(N)} \equiv \Phi$ is an Hermite matrix such that

$$\Phi = \Phi^T; \qquad \Phi^2 = 1; \qquad \Phi = \Phi^{-1}$$

$$\det(\Phi_{(N)}) = (-1)^N .$$

We can thus also augment the N ξ-state NE to the 2^N η-state form

$$\eta_{m+1} = F\eta_m,$$

in which F is a $2^N \times 2^N$ matrix.

We obtain thus the central result that passage from ξ- to η-space linearizes the NE. Thus

$$\eta_m = F\eta_{m-1} = \ldots = F^m \eta_0.$$

That passage to functional space should linearize the NE is of course not surprising; the relevant feature is that 2^N is (of course) finite, and from now on standard matrix algebra can be used. Clearly, $F\Phi = \Phi P$, P a permutation.

We show next that NE exhibit normal modes, just as linear ones, though more complex than the simple periodic sequences typical of linearity in N-space; they intertwine into "reverberations" [1] since they stem from linearity in 2^N-space. Their interpretation is in principle the same as that expressed by Eigen and Schuster [5] for "quasispecies"in their classic discussion of hypercycles (for which NE might be an apt tool).

Let the matrix Δ, $\det(\Delta) \neq 0$, diagonalize F :

$$F\Delta = \Delta\Lambda, \qquad \Lambda \text{ diagonal.}$$

Then

$$P\Phi\Delta = \Phi\Delta\Lambda,$$

i.e. $\Phi\Lambda$ diagonalizes P.

Since P is a permutation matrix, its characteristic polynomial (as that of F) is necessarily of type

$$\lambda^\alpha \prod_b (\lambda^b - 1)^{c_b} = 0,$$

with

$$a + \sum_b bc_b = 2^N, \qquad c_b \geq 0.$$

Thus

$b = 1$ implies c_1 invariant states
$\lambda = 0$ implies transients

$b > 1$ implies c_b cycles of period b, corresponding to $\lambda_h = e^{\frac{2\pi hi}{b}}$
(b=1 can of course be regarded as a cycle of period 1).
If we set

$$\eta_m = \Delta\chi_m$$

the NE read

$$\Delta\chi_{m+1} = F\Delta\chi_m = \Delta\Lambda\chi$$

so that

$$\chi_{m+1} = \Lambda\chi_m$$

or

$$\chi_{\alpha,m+1} = \lambda_\alpha\chi_{\alpha,m}$$

express the wanted normal modes.
In conclusion:

The standard (or direct) problem (given the net or cellular automata), find its evolution) is solved by

$$P = \Phi F\Phi.$$

The inverse problem (given the wanted sequences of states, find the net or C.A. that performs it) is solved by

$$F = \Phi P\Phi.$$

The latter amounts to solving exactly the equations, and leads to many other results which are or will be reported elsewhere.

1. E. R. Caianiello: Journ. Theor. Biol. $\underline{1}$, 209 (1961)
2. E. R. Caianiello: In Brain Theory , ed. by G. Palm and A. Aertsen (Springer–Verlag 1986)
3. E. R. Caianiello, M. Marinaro: In Computer Simulation in Brain Science (Cambridge University Press 1986)

4. E. R. Caianiello: In Topics in the General Theory of Structures, ed. by E. R. Caianiello, M. A. Aizermann p.163 (Reidel Pub. Co. 1987) and references cited therein
5. M. Eigen, P. Schuster: The hypercycle - principle of natural self-organization (Springer–Verlag 1979)

Associative Memory and Its Statistical Neurodynamical Analysis

Shun-ichi Amari

University of Tokyo, Faculty of Engineering, Bunkyo-ku, Tokyo 113, Japan

1. Introduction

A neural network is a complex system consisting of a large number of mutually connected elements having a simple input-output relation. Its behavior is highly non-linear, so that it is in general difficult to analyze its dynamical behavior of information processing. In order to elucidate a typical behavior, we study a network whose connection weights or synaptic efficacies of connections are randomly generated and then fixed. Given a probability law of connection weights, we have an ensemble of randomly generated networks. Statistical neurodynamics provides a theoretical method to search for macroscopic behaviors which are shared by all typical random networks in the ensemble, i.e. those networks generated by the same probability law.

Amari [1], [2] studied the macroscopic dynamical behaviors of randomly generated neural networks, and then proposed a mathematical method of statistical neurodynamics [3]. Amari et al. [4] proved some fundamental lemmas of statistical neurodynamics. The statistical neurodynamical method is shown to be useful for analyzing behaviors of associative memory models (Amari and Maginu [5]). The present paper proposes various versions of associative memory models, and gives their dynamical behaviors and capacities by the statistical neurodynamical method.

Correlation type associative memory models were proposed by Nakano [6], Kohonen [7], and Anderson [8], independently. A mathematical analysis of their dynamical behaviors was given by Amari [9] (see also Amari [10]). Since Hopfield [11] pointed out the spin-glass analogy, there have appeared a number of theoretical works (e.g., Amit et al. [12], Meir and Domany [13], McEliece et al. [14], Amari and Maginu [5]), which used the statistical neurodynamical method in some sense.

There are a number of versions of the associative memory model. They are, for example, a cross-correlation associative memory, an autocorrelation associative memory, a sequence recalling associative memory model [9], a bilateral associative memory (BAM) [15], etc. We analyze the dynamical behaviors and the memory capacities of these models by the statistical neurodynamical method. Their behaviors are different. For example, the

Springer Series in Synergetics Vol. 42: **Neural and Synergetic Computers**
Editor: H. Haken   ©Springer-Verlag Berlin Heidelberg 1988

memory capacity of an autocorrelation sequence generator is about twice that of an autocorrelation associative memory. We show where this difference comes from.

2. Various types of associative memory models

A neural element, which we treat here, has the following simple input-output relation : It receives n input signals $x_1, \cdots x_n$ and emits one output y, where

$$y = sgn \left(\sum w_i x_i - h \right).$$

Here, w_i are called the connection weights or synaptic efficacies, sgn is the signature function taking a value $+1$ or -1 depending on the sign of its operand. Signals x_i and y also take on values $+1$ or -1. In the present paper, w_i's are assumed to be randomly generated in some manner, and we put $h = 0$ for the sake of simplicity. We show various types of associative memory models in the following.

1) Cross-correlation associative memory.

Let us consider a network consisting of n neurons. It receives a vector input signal $x = (x_1, x_2, \cdots, x_n)$ and emits a vector output signal $y = (y_1, y_2, \cdots, y_n)$, where y_i is the output of the i-th neuron. Let w_{ij} be the synaptic efficacy of the j-th component of a signal x entering into the i-th neuron. See Fig. 1. Then, the input-output behavior of the network is written, in the component form, as

$$y_i = sgn (\Sigma w_{ij} x_j). \tag{1}$$

We may write (1) as

$$y = Tx = sgn (Wx), \tag{2}$$

where T is the non-linear operator determined by the synaptic efficacy matrix $W = (w_{ij})$.

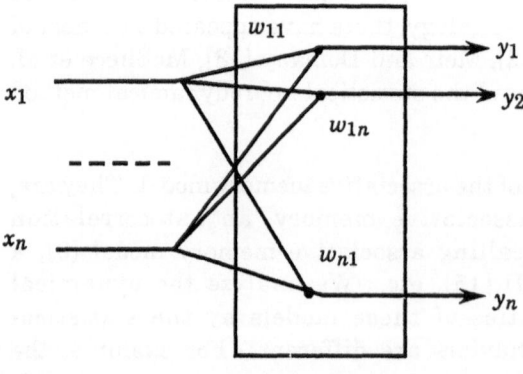

Fig. 1. Cross-associative memory

When m pairs of vectors $(s^1, q^1), \cdots, (s^m, q^m)$ are given, a cross-associative memory is required to emit q^μ when s^μ is input ($\mu = 1, 2, \cdots, m$). In other words, the input-output relation of the net is required to satisfy $Ts^\mu = q^\mu$ for all $\mu = 1, \cdots, m$. This implies that the q^μ is recalled from s^μ. Moreover, it is expected to have a noise reduction property such that, when input x is a noisy version of s^μ, the output Tx is equal to q^μ or much closer to q^μ. We use the inner product

$$a = \frac{1}{n} s^\mu \cdot x$$

to evaluate the similarity of x to s^μ. In the correlation associative memory, synaptic efficacy matrix W is put equal to

$$w_{ij} = \frac{1}{n} \sum_{\mu=1}^{m} q_i^\mu s_j^\mu \quad , \tag{3}$$

depending on the pairs (s^μ, q^μ) to be memorized. In the vector-matrix notion, this is written as

$$W = \frac{1}{n} \sum_{\mu=1}^{m} q^\mu (s^\mu)' \quad , \tag{4}$$

where q^μ and s^μ are assumed to be column vectors, and (s^μ)' denotes the transposition of s^μ.

We further assume that the components of the vectors q^μ and s^μ are randomly and independently determined such that they are $+1$ and -1 with probability 0.5 each. Then W is a randomly generated matrix. We search for the macroscopic property which holds for almost all W determined in this way, as the number n of the neurons becomes infinitely large.

2) *Cascade associative memory.*

Let us consider a cascaded series of cross-correlation associative memory models, $N_1, N_2, \cdots N_l, \cdots$ (Fig. 2).

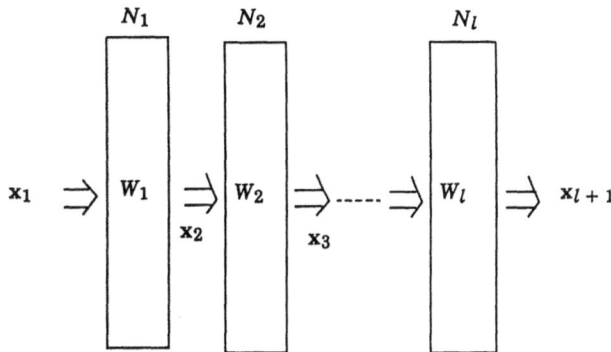

Fig. 2. Cascade associative memory

The cascaded associative memory was studied by Meir and Domany [13]. Let N_l be a network which receives input $\mathbf{x}^l$ and emits output $\mathbf{x}^{l+1}$,

$$\mathbf{x}^{l+1} = T_l \mathbf{x}^l, \qquad (5)$$

where T_l is the non-linear transformation. It is written as

$$T_l \mathbf{x} = sgn(W_l \mathbf{x}), \qquad (6)$$

where W_l is the synaptic efficacy matrix of N_l. A cascade associative system is a concatenation of networks $N_1, N_2, \cdots, N_k, \cdots$, such that an output $\mathbf{x}^l$ of the l-th network N_l becomes an input of the $(l+1)$st network N_{l+1}.

Let $S^1 = \{\mathbf{s}^1_1, \cdots, \mathbf{s}^1_{k+1}, \cdots\}, S^2 = \{\mathbf{s}^2_1, \cdots, \mathbf{s}^2_{k+1}, \cdots\}, S^m = \{\mathbf{s}^m_1, \cdots, \mathbf{s}^m_{k+1}, \cdots\}$ be m sequences of vectors. When $\mathbf{s}^\mu_{l+1} = T_l \mathbf{s}^\mu_l$ holds for $\mu = 1, \cdots, m$, the system recalls sequence S^μ by emitting $\{\mathbf{s}^\mu_{l+1}\}$ from N_l, when $\mathbf{s}^\mu_1$ is input to the first net N_1. When the input $\mathbf{x}_1$ to N_1 is close to $\mathbf{s}^\mu_1$, it is expected that $\mathbf{x}_{l+1} = T_l \mathbf{x}_l$ becomes closer and closer to $\mathbf{s}^\mu_{l+1}$. This is the noise reduction property, which we study in terms of the similarity or direction cosine $a^\mu_l = (1/n) \mathbf{s}^\mu_l \cdot \mathbf{x}_l$.

We assume that the matrix W_l is determined by

$$W_l = \frac{1}{n} \sum_{\mu=1}^{m} \mathbf{s}^\mu_{l+1} (\mathbf{s}^\mu_l)', \qquad (7)$$

where each component of $\mathbf{s}^\mu_l$ is determined randomly and independently as before. Without loss of generality, we assume that $\mathbf{x}_1$ is close to $\mathbf{s}^1_1$, and search for the dynamical relation of $a_l = (1/n) \mathbf{s}^1_l \cdot \mathbf{x}_l, l = 1, 2, \cdots$.

3) Cyclic associative memory and BAM.

A cyclic associative memory is obtained by adding a feedback connection from the output of a cascaded associative memory network to its input (Fig. 3). A cyclic associative memory is calld a k-AM, when it consists of k networks connected in the form of a ring. When a signal is transformed through component networks N_l sequentially one at a time, we have the following equation $\mathbf{x}^{(t)}_{l+1} = T_l \mathbf{x}^{(t)}_l$,

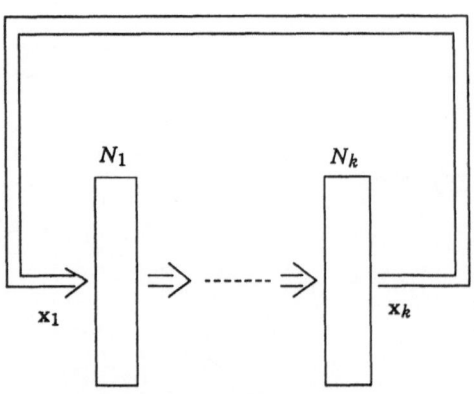

Fig. 3. Cyclic associative memory

where t is the number of times the circuit is circled and l is calculated with modulo k such that $x^{(t)}_{k+1}$ is put equal to $x^{(t+1)}_1$.

Let $S^\mu = \{s^\mu_1, \cdots, s^\mu_k\}$, $\mu = 1, \cdots, m$, be m sequences of vectors of length k. It is expected that $s^\mu_{l+1} = T_l s^\mu_1$ holds for all $\mu = 1, \cdots, m, l = 1, \cdots, k$ ($k + 1 = 1$), where the connection matrix W_l of N_l is given by

$$W_l = \frac{1}{n} \sum_{\mu=1}^{m} s^\mu_{l+1} (s^\mu_l)' \quad .$$

It is also expected that a k-AM has a good noise reduction property that, given an input x_1, the signal x_l converges to s^1_l provided x_1 is close to s^1_1, by circulating the network.

When $k = 2$, we have a 2-AM which is called a BAM (bilateral associative memory) (Kosko [15] , Okajima et al. [16]). As we see in the following, the behavior of a BAM is slightly different from other k-AM ($k \geq 3$).

4) *Autoassociative memory.*

Let us consider a network in which the output is fed back to its input (Fig. 4). This network can be regarded as a 1-AM. When its connections matrix is W, its dynamical behavior is given by

$$x^{t+1} = T x^t = sgn(W x^t) ,$$

where x^t is the output (the state vector) of the network at time t.

Given m patterns $s^1, \cdots s^m$, we have an autoassociative memory, when we put

$$W = \frac{1}{n} \sum_{\mu=1}^{m} s^\mu (s^\mu)' \quad .$$

It is expected that all s^μ are equilibria of the dynamics, satisfying $s^\mu = T s^\mu$ for all μ. The dynamics of this model is interesting. Its equilibrium behavior was

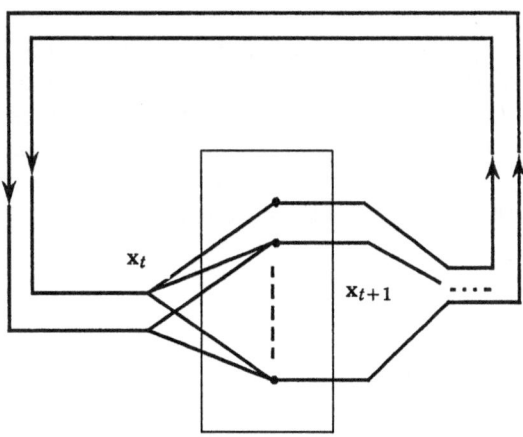

Fig. 4. Autoassociative memory

analyzed by Amit et al. [12], and its dynamical behavior was analyzed by Amari and Maginu [5].

5) *Associative sequence generator.*

Let us consider a sequence $S = \{s^1, s^2, \cdots, s^m\}$ of patterns. Let us also consider a network with recurrent connections (1-AM) whose synaptic efficacy matrix is given by

$$W = \frac{1}{n} \sum_{l=1}^{m} s^{l+1}(s^l)'\ ,$$

where $s^{m+1} = s^1$. It is expected that the network recalls the sequence S, $Ts^l = s^{l+1}$. Moreover, given an initial state x^1 which is within a neighborhood of some s^μ, the network recalls the sequence more and more precisely through dynamical state transitions. This is the noise reduction property. Such a model was proposed by Amari [9] and he gave a mathematical analysis of its stability, see also Cottrell [17].

Let S_μ be m sequences of length k_μ, $\mu = 1, 2, \cdots m$,

$$S_\mu = \{s_\mu^1, s_\mu^2, \cdots, s_\mu^{k_\mu}\}.$$

When W is put equal to

$$W = \frac{1}{n} \sum_{\mu=1}^{m} \sum_{l=1}^{k_\mu} s_\mu^{l+1}(s_\mu^l)'\ ,$$

given any s_μ^l or x in its neighborhood as the initial state, the model is expected to recall the remaining sequence by the dynamics,

$$s_\mu^{l+1} = T\, s_\mu^l\ .$$

We assume here that $k_\mu \geqq 3$.

3. Noise reduction property of cross associative memory

We study in the beginning the noise reduction property of the simplest cross-associative memory. Without loss of generality, we study the property of recalling q^1 from a noisy version of s^1. Given an input x whose similarity to s^1 is measured by the direction cosine to s^1, $a = a_1(x) = (1/n)\, s^1 \cdot x$, we search for the direction cosine a' or the similarity of the output $y = Tx$ to q^1, $a' = (1/n)\, q^1 \cdot y$. To this end, we calculate the i-th component of the output y as follows:

$$y_i = sgn\left(\sum w_{ij} x_j\right) = q_i^1\, sgn\, (a + q_i^1\, N_i) \tag{8}$$

where

$$N_i = \frac{1}{n} \sum_{\mu=2}^{m} \sum_{j=1}^{n} q_i^\mu s_j^\mu x_j \tag{9}$$

90

The term N_i represents the interference or crosstalk from the superposed pattern pairs in W other than (s^1, q^1). When $N_i = 0$ and $a > 0$, we have exact recalling $y = q^1$.

How large is the noise term N_i? Since the vectors (s^μ, q^μ) are generated randomly, we can evaluate it by using probability theory. The term N_i is a sum of $n(m - 1)$ random variables $q^\mu_i s^\mu_j x_j$ divided by n, so that the central limit theorem guarantees that N_i is normally distributed with mean 0 and variance $\sigma^2 = (m - 1)/n \doteq r$, where $r = m/n$ is the ratio of the number of the memorized pattern pairs to the number of neurons.

The probability that the output y_i is erroneous is given by

$$P = Prob\{a + q^1_i N_i < 0\} = Prob\{N_i < -a\} = \Phi(\frac{a}{\sigma}) \quad , \quad (10)$$

where

$$\Phi(u) = \int_u^\infty \frac{1}{\sqrt{2\pi}} \exp\{-\frac{t^2}{2}\} dt \quad . \quad (11)$$

The direction cosine a' of the output is given by $a' = (1/n) \sum y_i q^1_i$. Since the probability of $y_i = q^1_i$ is $1 - \Phi$ and the probability of y_i not eaqual to q^1_i is Φ, by the law of large numbers, a' converge to $a' = 1 - 2\Phi(a/\sigma) = F(a/\sigma)$, where

$$F(u) = \int_{-u}^u \frac{1}{\sqrt{2\pi}} \exp\{-\frac{t^2}{2}\} dt \quad . \quad (12)$$

This gives the answer to our simple problem.

Theorem 1. The noise reduction property of a cross associative memory is given by

$$a' = F(\frac{a}{\sqrt{r}}) \quad . \quad (13)$$

This is the well-known result (see, e.g., Amari [10], Kinzel [18]). The problem is whether this result can be generalized to be applicable to the dynamics of autoassociative memory or many other versions of correlation type associative memory models.

4. Dynamics of cascade associative memory, k-AM, and sequence generator

We now search for the noise reduction property of a cascade associative memory. Since this is a concatenation of cross-associative memory models, one may think that the overall noise reduction property is given also by concatenation,

$$a_{l+1} = F(a_l/\sqrt{r}) \, , \, l = 1, 2, \cdots$$

where

$$a_l = (1/n) s^1_l \cdot x^l$$

is the similarity of the output of the l-th layer to its expected output s^1_l. However, this is not the case. As is seen from (7), the connection matrices W_{l-1} and W_l of two successive layers are not stochastically independent. This is because they depend partly on the same random vectors s^μ_l. Therefore, the probability distribution of the crosstalk term $N_{l,i}$ of the l-th layer is not so simple as before.

The i-th component of x_l is written as

$$x_{l,i} = sgn(\frac{1}{n} s^1_{l,i} s^1_{l-1} \cdot x_{l-1} + N_{l,i}) = s^1_{l,i} \, sgn(a_{l-1} + s^1_{l,i} \, N_{l,i}) \quad ,$$

where

$$N_{l,i} = \frac{1}{n} \sum_{\mu=2}^{m} s^\mu_{l,i} s^\mu_{l-1} \cdot x_{l-1} \tag{14}$$

$$= \frac{1}{n} \sum_{\mu} \sum_{j} Z^\mu_{lij} \quad ,$$

with

$$Z^\mu_{lij} = s^\mu_{l,i} s^\mu_{l-1,j} x^{l-1}_j \quad .$$

We assume that $N_{l,i}$ is normally distributed with mean μ_l and variances σ^2_l. Although $N_{l,i}$ is a sum of Z^μ_{lij}, its variance is not simply equal to r. This is because

$$x^{l-1}_j = s^1_{l-1,j} \, sgn(a_{l-2} + s^1_{l-1,j} \, N_{l-1,j})$$

depends on s^μ_{l-1}, so that Z^μ_{lij} are not independent.

Since x^{l-1} does not depend on s^μ_l, we have

$$E[\, Z^\mu_{lij}\,] = 0 \quad .$$

This shows that

$$\mu_l = 0 \quad ,$$

so that $N_{l,i}$ is distributed with mean 0.

In order to calculate the variance

$$\sigma^2_l = E[(\sum Z^\mu_{lij})^2] \quad ,$$

we note that

1) $$E[(Z^\mu_{lij})^2] = 1 \quad .$$

2) $$E[Z^\mu_{lij} \, Z^{\mu'}_{lik}] = E[(s^\mu_{l+1,i} s^{\mu'}_{l+1,i}) s^\mu_{l,j} s^{\mu'}_{l,k} \, x^l_j x^l_k] = 0 \quad ,$$

provided $\mu \neq \mu'$. The correlational terms are, therefore, given by

$$A = E[Z^\mu_{lij} Z^\mu_{lij'}\,], \, j \neq j' \quad .$$

This term can be rewritten as

$$A = E[\, s s' \, sgn(a + sM + N) \, sgn(a + s'M + N')]$$

where

$$s = s_{l,j}^{\mu}, \quad s' = s_{l,j'}^{\mu}, \quad a = a_{l-1} \quad ,$$

and

$$M = \frac{1}{n} s_{l-1}^{\mu} \cdot x^{l-1}$$

represents the common factor in $N_{l-1,j}$ and $N_{l-1,j'}$. This is a small random variable subject to $N(0, \sigma^2_{l-1} / m)$, which are independent of the random variables N, N' subject to $N(0, \sigma^2_{l-1})$, which are the remaining terms in $N_{l-1,j}$ and $N_{l-1,j'}$, respectively. Since M, N and N' do not depend on s and s', we have

$$A = \frac{1}{2} E[sgn(a + M + N) sgn(a + M + N')] - \frac{1}{2} E[sgn(a + M + N) sgn(a - M + N')]$$

because $Prob\{s\,s' = 1\} = Prob\{s\,s' = -1\} = \frac{1}{2}$. By taking the expectation with respect to N and N', where M is fixed, i.e., by taking the conditional expection, we have

$$A = \frac{1}{2} E[F(\frac{a+M}{\sigma})F(\frac{a+M}{\sigma}) - F(\frac{a+M}{\sigma})F(\frac{a-M}{\sigma})] \quad ,$$

where $\sigma = \sigma_{l-1}$. Since M is a random variable of order $1/\sqrt{m}$, we expand the above A in the Taylor series of M and take the expectation with repeat to M. We then have

$$A = \frac{4}{m} \{p(\frac{a}{\sigma})\}^2 \quad ,$$

where

$$p(v) = \frac{1}{\sqrt{2\pi}} exp\{-\frac{v^2}{2}\} \quad .$$

This proves that

$$\sigma_l^2 = r + 4\{p(\overline{a_l})\}^2 \quad ,$$

where

$$\overline{a_l} = \frac{a_l}{\sigma_l} \quad .$$

The noise reduction property of a cascade network is then given by the following theorem, which was first obtained by Meir and Domany [13] by a different method.

Theorem 2. The direction cosine changes as

$$a_{l+1} = F(a_l/\sigma_l) \quad , \tag{15}$$

$$\sigma^{l+1} = r + 2\{p(a_l/\sigma_l)\}^2 \tag{16}$$

$$p(u) = \frac{1}{\sqrt{2\pi}} exp\{-\frac{u^2}{2}\}$$

in the process of recalling S^{μ} in a cascade associative memory system.

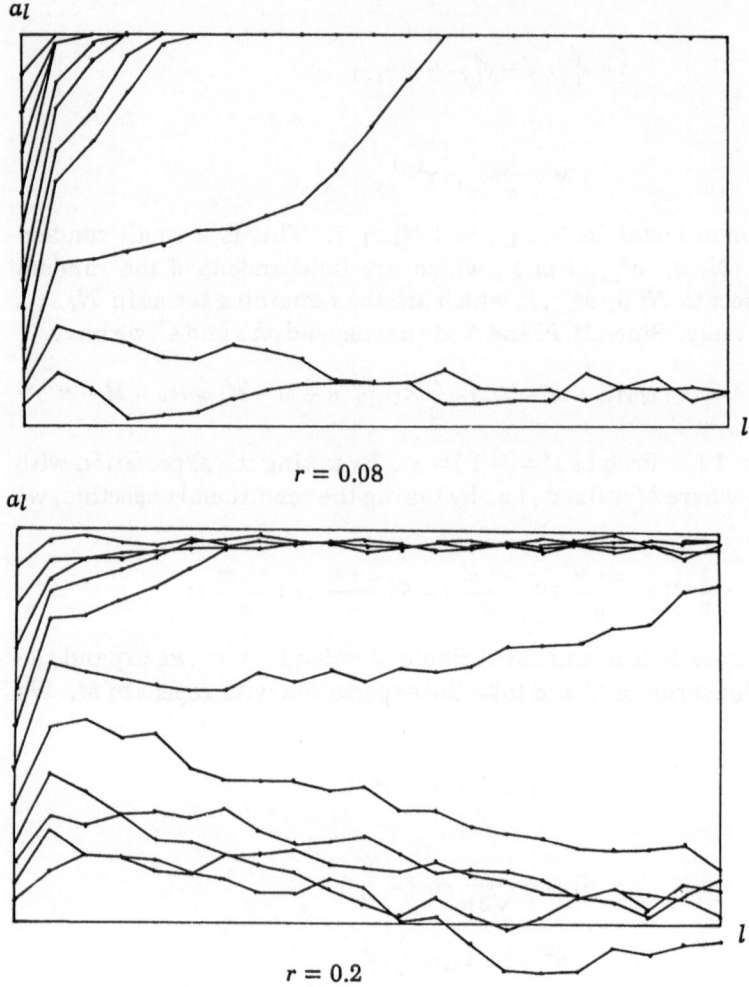

Fig. 5. Dynamical behavior of cascade networks

The system of equations (15), (16) demonstrate interesting properties :
There is a threshold r_c such that, when the pattern ratio $r = m / n$ is larger than r_c,
a_l never converges to 1 as l becomes large, even if a_1 is very large or even equal to
1. In this case, any sequence S^μ cannot be recalled. On the other hand, when r is
smaller than r_c, a_l converges to 1 as l becomes large, provided a_1 is larger than a
threshold a_c. Hence, we may call r_c the capacity of the system. When $r < r_c$,
there exists $a_c(r)$ such that, when $a_1 < a_c$, a_l converges to 1, but it otherwise does
not converge to 1. Therefore, a_c denotes the size of the basin of attractor from
which a good recall is possible. The two characteristic quantities r_c and $a_c(r)$ can
be calculated from Theorem 2. See Meir and Domany [13].

When the recalling process fails to recall S^μ, it sometimes occurs that a_l once
increases and becomes larger than the threshold $a_c(r)$. However, it decreases

a_l

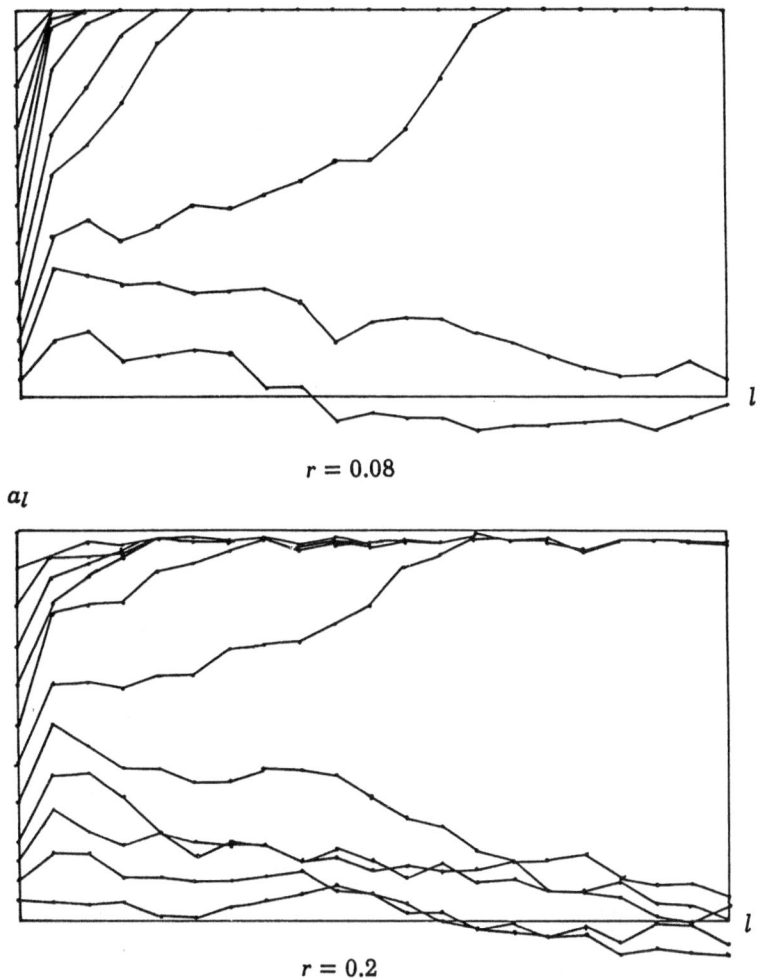

$r = 0.08$

a_l

$r = 0.2$

Fig. 6. Dynamical behavior of sequence generators

again. This interesting phemomenon is found also in the dynamics of auto-correlation associative memory (Amari and Maginu [5]). All of these characteristics are common to the auto-correlation associative memory analyzed by Amari and Maginu [5], although the dynamical equations are different.

The computer simulation of the noise reduction behavior of a cascaded network is shown in Fig. 5a where $r = 0.08$ and in Fig. 5b where $r = 0.2$. The theory is in good agreement.

In the case of a cyclic associative memory (k-AM), the noise reduction property would be a little different. In the case of a cascaded memory, in

$$x^{l+1} = sgn(W_l \, sgn(W_{l-1}x_{l-1})) \, ,$$

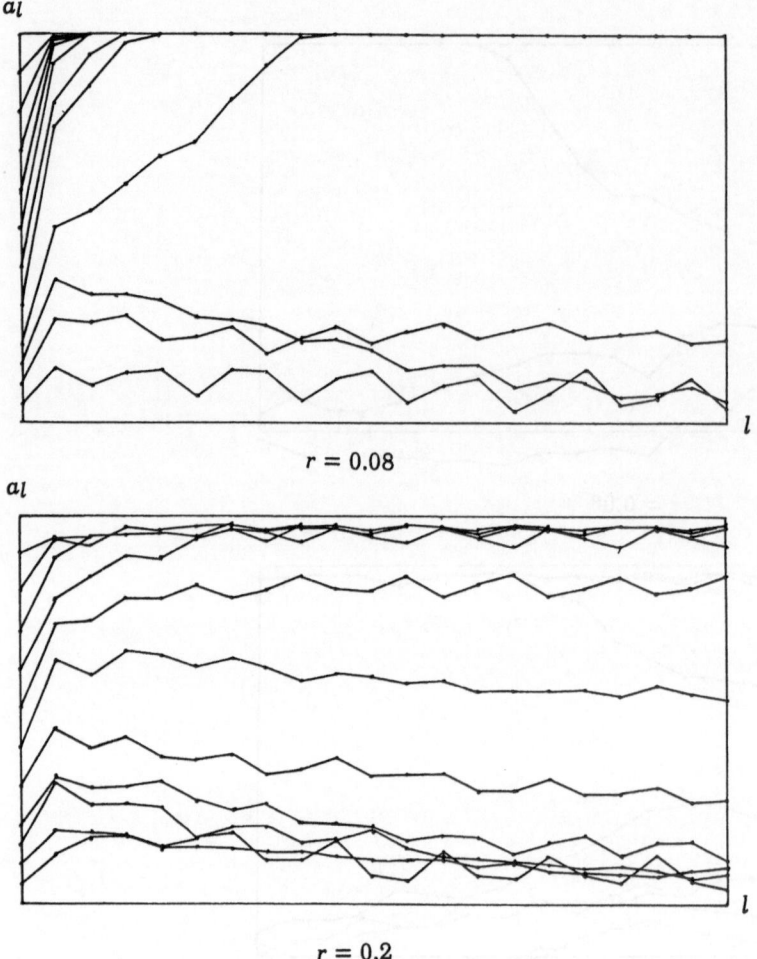

a_l

$r = 0.08$

a_l

$r = 0.2$

Fig. 7. Dynamical behavior of 3-AM

W_l is independent of x_{l-1}. Therefore, the direct correlation between W_l and x_l emerges through that of W_l and W_{l-1}. This is not true in the case of a k-AM, because x_{l-1} has some correlations with W_l. This correlation, however, is very diffuse, so that it might be neglected as n tends to infinity. Therefore, we use the following assumption, which was also used in Amari and Maginu [5].

Assumption. The direct correlations of W_l and x_{l-1} can be neglected in the calculation of correlations of W_l and x_l.

If this assumption is true, it is easy to show that the noise reduction property of a k-AM ($k \geqq 3$), or of an associative sequence generator is the same as a cascade associative memory. However, a 2-AM (BAM) and an autoassociative memory have different noise reduction properties.

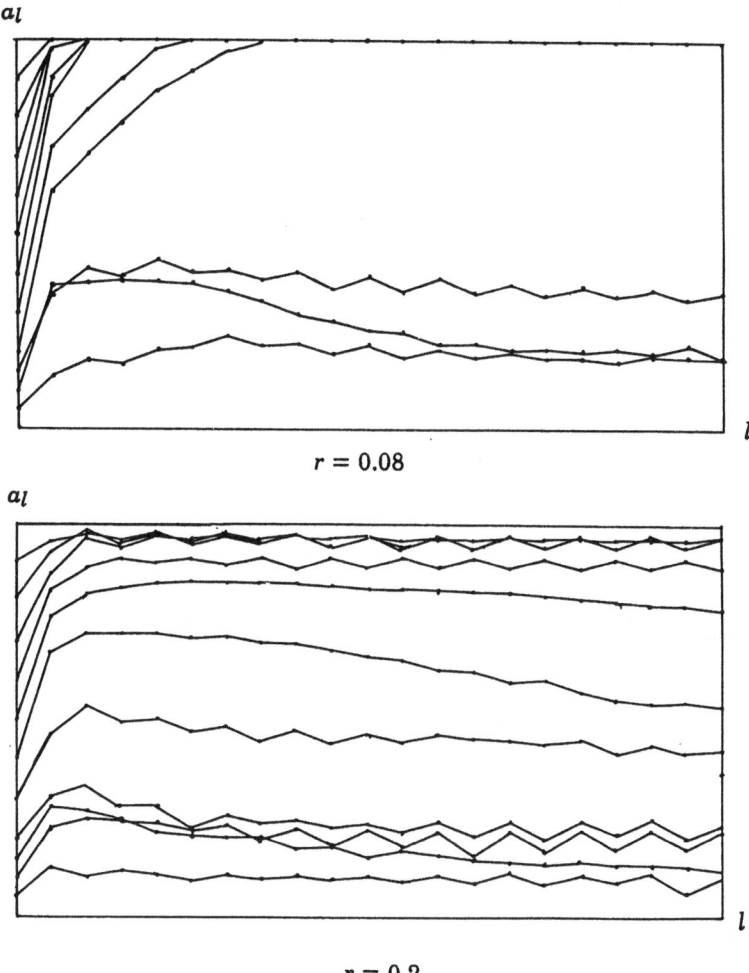

$r = 0.08$

$r = 0.2$

Fig. 8. Dynamical behavior of BAM

We show a computer simulation of the behavior of a sequence generator (Fig. 6), of a 3-AM (Fig. 7), and of a BAM (Fig. 8). We can see that the simulation is also in good agreement with the theory even in the case of a BAM. However, the behavior of an autoassociative memory is different (Fig. 9).

It should be emphasized that the memory capacity of a sequence generator is 0.27, which is twice   that of an autoassociative memory, although both models use networks of the same recurrent architecture.

Conclusions

We have shown theoretical derivations of dynamical behaviors of recalling processes of various types of associative memory models, by using the statistical neurodynamical method.

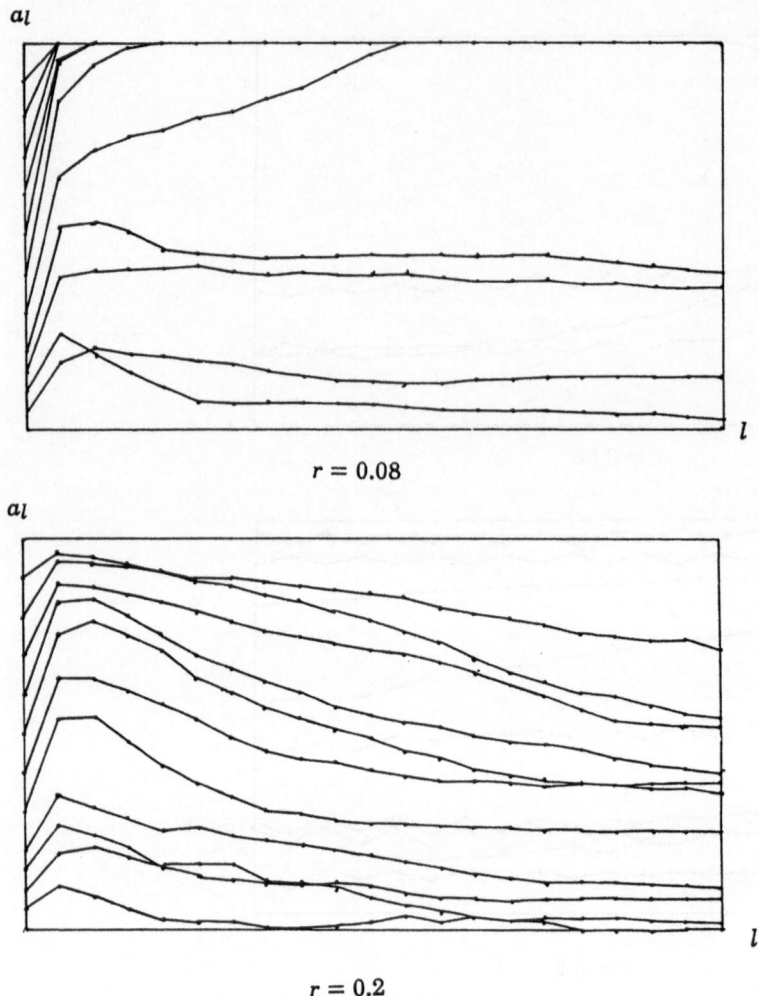

a_l

$r = 0.08$

a_l

$r = 0.2$

Fig. 9. Dynamical behavior of autoassociative memory

References

1. Amari, S. [1971] : "Characteristics of randomly connected threshold element networks and network systems", Proc. IEEE, 59, 35-47.

2. Amari, S. [1972a] : "Characteristics of random nets of analog neuron-like elements", IEEE Trans., SMC-2, 643-657.

3. Amari, S. [1974] : "A method of statistical neurodynamics", Kybernetik, 14, 201-215.

4. Amari, S. Yoshida, K., and Kanatani, K. [1977] : "A mathematical foundation for statistical neurodynamics", SIAM J. App. Math., 33, 95-126.

5. Amari, S. and Maginu, K. [1988] : "Statistical neurodynamics of associative memory", Neural Networks, 1, 63-73

6. Nakano, K. [1972] : "Association--- a model of associative memory", IEEE Trans., SMC-2, 381-388

7. Kohonen, T. [1972] : "Correlation matrix memories", IEEE Trans., C-21, 353-359.

8. Anderson, J. A. [1972] : "A simple neural network generating interactive memory", Math. Biosciences, 14, 197-220.

9. Amari, S. [1972] : "Learning patterns and pattern sequences by self-organizing nets of threshold elements", IEEE Trans., C-21, 1197-1206.

10. Amari, S. [1977] : "Neural theory of association and concept-formation", Biol. Cybernetics, 26, 175-185.

11. Hopfield, J. J.[1982] : "Neural networks and physical systems with emergent collective computational abilities", P. Nat. Acad. Sci. U.S.A., vol. 79, 2445-2458.

12. Amit, D. J., Gutfreund, H. and Sompolinsky, H. [1986] : "Spin-glass models of neural networks", Phys. Rev., A2, 1007-1018.

13. Meir, R. and Domany, E. [1987] : "Exact solution of a layered neural network memory", Phy. Rev. Lett., 59, 359-362.

14. McEliece, R. J. et. al. [1987] : "The Capacity of the Hopfield associative memory", IEEE Trans., Inf. Theory, IT-33, 461-482.

15. Kosko, B. [1987] : "Adaptive bilateral associative memory", Appl. Oct., 26, 4947-4960.

16. Okajima, K., Tanaka, S. and Fujiwara, S. [1977] : "A heteroassociative memory network wirh feedback connection", Proc. 1st Int. Conf. Neural Networks, II-711-II-718.

17. Cottrell, M. [1988] : "Stability and attractivity in associative memory networks", Biol. Cybern, 58, 129-139.

18. Kinzel, W.[1985] : "Learning and pattern recognition in spin glass models", Z. Physik, B60, 205-213.

Local Synaptic Rules with Maximal Information Storage Capacity

G. Palm

C. u. O. Vogt-Institut für Hirnforschung, Universität Düsseldorf,
D-4000 Düsseldorf, Fed. Rep. of Germany

1. Introduction

A mathematical model of a natural process is always a kind of carica-
ture. Today the physicists' most popular model for the process of
neural activation in the brain is the so-called spin-glass model - a
mathematical model that is derived from theoretical investigations on
Ising spins and has a certain formal similarity to mathematical
models of brain dynamics. This type of model is certainly a long way
from the reality of magnetic materials, but it is still further
remote from real brain dynamics. On the other hand, it has often
proved to be beneficial for a theoretical insight to step back from
too close a look at reality and to work with caricatures. It is in
this spirit that we want to compare and analyse a family of mathematical neural network models that investigate learning and synaptic
plasticity (e.g. [1],[3],[4]).

In all these models the values for neural axonal activity are
assumed to be discrete (one value for 'active', one for 'inactive').
So the axonal activity state can be described by a vector $(a_1,..,a_n)$,
where n is the total number of neurons in the network. The synaptic
connectivity matrix c_{ij} relates this axonal activity from the axons
to the dendrites and gives rise to a <u>dendritic potential</u> by linear
combination: in the j-th neuron

(1) $d_j = \sum_i a_i \cdot c_{ij}.$

If this potential d_j is high enough, then in the next time-step the
corresponding axonal output activity value a_j is set to 'active'.
In the various models there are considerable variations in this
procedure, for example the decision may be stochastic or deterministic, it may be synchroneous for all neurons or asynchroneous, etc.

In all models the connectivity matrix c_{ij} is formed by a <u>local
learning rule</u> in the following sense:

Springer Series in Synergetics Vol. 42: **Neural and Synergetic Computers**
Editor: H. Haken   ©Springer-Verlag Berlin Heidelberg 1988

(2) $$c_{ij} = f(\sum_{k=1}^{M} R(a^k_i, b^k_j)),$$

where f is a non-decreasing and R an arbitrary function. In many cases f is just the identity.

The essential meaning of this is the following. There are M pairs of vectors $(a^k, b^k)_{k=1,..,M}$ of axonal activity values that are considered as <u>patterns</u> to be learned. The strength of the connection c_{ij} from neuron i to neuron j is assumed to be built in a additive way from a number of incremental changes R (a^k_i, b^k_j) that depend only on the <u>locally</u> <u>available</u> values of the various patterns (a^k, b^k) i.e. on the i-th component of the 'input' pattern a^k and the j-th component of the corresponding 'output' pattern b^k.

In this paper I shall try to compare the effectivity for information storage of various local rules R(x,y) for synaptic changes. The purpose of the network to serve as an (associative) memory can be further specified in two directions:

- Auto-association, where $a^k = b^k$, and information is retrieved from the memory by iterative pattern completion or pattern correction.

- Hetero-association, where $a^k \neq b^k$, and information is retrieved by observing the network's response to an input pattern a'. If a' is sufficiently close to one of the stored patterns a^l, the network's response should be b^l.

2. Optimization of the learning rule.

The essential problem in both cases is a simple decision problem that has to be solved by every single neuron in the network.

Given a certain input activity to the neuron (e.g. neuron no. j) and its afferent connectivities $(c_{ij})_{i=1,..,n}$ the neuron forms its dendritic potential d_j and has to use this variable to decide whether it should be active or not. In other words: If the input activity is sufficiently similar to one of the stored vectors a^l, the neuron j should decide from the size of d_j whether it belongs to the corresponding pattern b^l or not. This decision can be made by means of a threshold criterion on d_j. In fact, if

(3) $$c_{ij} = \sum_k R(a^k_i, b^k_j) \qquad \text{and} \qquad R(x,y) = x \cdot y$$

(which is a special case of (2)), then

$$(4) \qquad d_j = \sum_i a_i\, c_{ij} = \underbrace{\sum_{k \neq 1} \sum_i a_i\, R(a^k{}_i, b^k{}_j)}_{N} + \underbrace{a_i\, R(a^1{}_i, b^1{}_j)}_{S}.$$

Here S can be regarded as the signal for the decision problem, since b_j is the desired response, and S is indeed a measure for the similarity between (a_i) and $(a_i{}^1)$. On the other hand, N has to be regarded as noise. If the values a_i and b_j are randomly generated with a certain probability p of being a 1, then N is a nearly binomically distributed random variable, and can - for sufficiently large network-size n - be assumed to be almost normally distributed. Thus we can invoke conventional signal detection theory to find the proper detection threshold θ for the neuron and the corresponding error probabilities in terms of the signal-to-noise ratio, i.e. the signal difference for the on-signal and the off-signal divided by the standard deviation of the noise.

A local learning rule $R(x,y)$ is described by 4 numbers as shown in Table 1.

Table 1

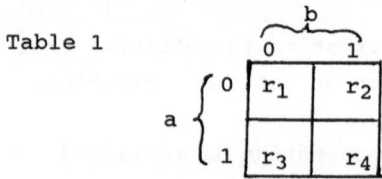

In the following we will simply write such a rule as a row vector, i.e. instead of Table 1 we write (r_1, r_2, r_3, r_4).
Since the signal-to-noise ratio is the same for a rule $R(x,y)$ and any rule $a + b \cdot R(x,y)$ we can normalize a general rule to a description by only two parameters, for example as $(a,b,0,1)$.

For such a rule one can easily compute the signal-to-noise ratio in (4). The formula will involve the axonal activity values a_i of the neurons. According to the most common models they are either 1 and 0 or 1 and -1. To accommodate both cases we will now assume them to be 1 and -c, with an arbitrary $c \geq 0$.

Incidentally, the use of $a_i = -1$ in (1) is again unrealistic for real neurons, since it means that no activity on an axon has a synaptic effect on the postsynaptic dendrite! This interpretation cannot be easily avoided by rewriting the variables, since a change of variables from $a_i \in \{-1,1\}$ to $a'_i = a_i + 1 \in \{0,2\}$ would result in

$$(5) \qquad \sum_i a'_i\, c_{ij} = \sum_i a_i\, c_{ij} + \sum_i c_{ij} \,.$$

For this transformation one has to keep track of the total
strengths of all afferent synapses $\sum_i c_{ij}$ for every neuron j. It
seems unlikely that each neuron should somehow be informed about this
quantity - or even be able to keep it constant.

In this framework we can now optimize the signal-to-noise ratio in
terms of the two parameters a and b defining the learning rule, while
keeping constant the other paramters describing the network, namely
c, the probability of activation p, M and n. The result of this
optimization is given in Theorem 1.

Theorem 1: The optimal local synaptic rule is $(cp,-cq,-p,q)$, where
          $q = 1-p$, and its signal-to-noise ratio is $x = (p \cdot q \cdot \alpha)^{-1/2}$,
          where $\alpha = M/n$.

Note that the signal-to-noise ratio is independent of the parameter c. Note also that for $c = 1$ and $p = 1/2$ we obtain Hopfield's rule
and that for $c = 0$ and $p \longrightarrow 0$ we obtain Hebb's rule. For these two
important special cases, the signal-to-noise ratios are given in
Theorem 2.

Theorem 2: For the Hopfield rule $Ho = (1,-1,-1, 1)$ and $c = 1$ the
          signal-to-noise ratio is $x = (2pq(1-2pq) \cdot \alpha)^{-1/2}$, where
          $q = 1-p$. For the Hebb rule $He = (0,0,0,1)$ and $c = 0$ the
          signal-to-noise ratio is $x = (p(1-p^2) \cdot \alpha)^{-1/2}$.

As the next step we invoke a fidelity criterion on the signal-to-
noise ratio x and the resulting error probabilities e_1 and e_2. For
simplicity we may chose the detection threshold θ in such a way that
$e_1 = e_2 = e$; this results (with a small asymptotically vanishing
deviation) in $\theta = x/2$ and - assuming normal distribution, which is
possible for large n - in

(6) $e = N(-x/2)$.

For fixed n we may now require that $e \leq \varepsilon$ for a chosen error bound
$\varepsilon > 0$. Asymptotically we may require that e.n remains bounded or even
goes to zero for $n \rightarrow \infty$. A requirement of this kind will be called a
high fidelity criterion. Observe that this criterion can only be ful-
filled if $x \rightarrow \infty$ as $n \rightarrow \infty$. By Theorem 1 this implies that $p \cdot \alpha \rightarrow 0$.
So if one does not want to store only a vanishing number of patterns
per neuron, one has to work with $p \rightarrow 0$, i.e. with extremely sparse
patterns. For constant p and asymptotically large n one has to accept
a much weaker error criterion, called a low fidelity criterion,

namely $e < \mathcal{E}$ or equivalently $x > \mathfrak{z}$ for fixed $\mathcal{E}$ or $\mathfrak{z}$ as $n \longrightarrow \infty$. Such a criterion obviously implies that the expected number of retrieval-errors grows proportionally to n.

Such requirements lead to a restriction on M when c, p and n are kept constant. This is quite obvious, because the errors will increase if one tries to store more patterns in the same matrix. Furthermore, it is well known that the increase in error probability becomes quite catastrophic at a certain critical value of M [1]. For the case of auto-association, i.e. iterative pattern restoration this value is known to be around $M = 0.14 \cdot n$.

In the high fidelity case these considerations lead to the following asymptotic choice of M.

<u>Proposition 1</u>: One can achieve asymptotically vanishing error probabilities for $x^2 = -8 \cdot \ln p$ and asymptotically vanishing p. This leads to the corresponding value of M according to the rule chosen (see Theorems 1 and 2).

3. Information storage capacity

It is the purpose of this section to optimize the only remaining parameter p. This cannot be done in terms of the signal-to-noise ratio anymore, because a decrease of p leads on the one hand to an increase of the signal-to-noise ratio x, and on the other hand to an decrease of the information content of one pattern. In fact, this information content is

(7) $i(p) = -p \log_2(p) - (1-p) \log_2(1-p)$.

For the optimization of p we therefore need a criterion which takes into account not only the total number M of patterns that can be stored within a certain error-criterion [$e < \mathcal{E}$], but also their information content I.

The natural measure for this optimization is the information-storage capacity I. Formally, this capacity can be defined as the channel capacity of the channel depicted in Fig. 1.

Clearly, the evaluation of this storage capacity I requires a specification of the storage and retrieval procedures, and these are obviously different in different models. We consider two cases as indicated in the introduction.

104

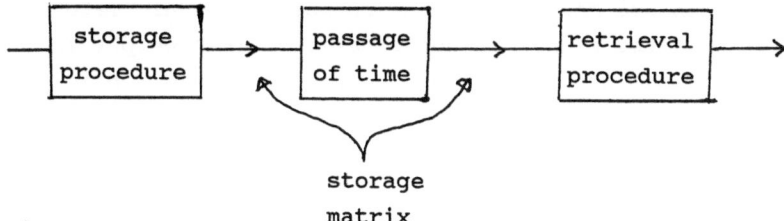

Fig. 1

(i) Hetero-association:

The matrix $C = (c_{ij})$ is formed as in (2). Retrieval is done by means of $x \mapsto x \cdot C$ and threshold detection.

(ii) Auto-association:

The matrix (c_{ij}) is again formed as in (2), but $x = y$. Retrieval is done by means of an iteration of the mapping $x \mapsto \text{sign}(x \cdot C)$ towards a fixed point.

For the second case it is much harder to calculate the storage capacity I in a rigorous way. I have carried out a number of approximate calculations which lead to the following conjecture.

Conjecture: For every local synaptic rule, the optimal storage capacity for auto-association is (depending on the choice of the retrieval dynamics) at most half the optimal capacity for hetero-association.

This conjecture is intuitively plausible because auto-association leads to a symmetric matrix where only half of the entries contain the total information. Since auto-association is most interesting for the usual spin-glass models and since these work with a fixed p, usually with $p = 1/2$, I have also analysed this case for a low fidelity criterion that corresponds to the familiar requirement that $M < 0.14 \cdot n$.

In the following I want to concentrate on the case (i) of hetero-association, since the non-dynamical situation is much easier to analyse (and I will use the above error criterion in the low fidelity case).

The storage capacity can be achieved when the set of the M patterns to be stored is generated stochastically independently. This means that we may assume that each x_i^k and each y_i^k is a two valued random variable with a probability p of being 1, and all these variables are independent. Now the storage capacity I can be calculated as

(8) $I = M \cdot n \, (i(p) - i(e)).$

For large n we can use the approximate normality to calculate e according to (6) from the signal-to-noise ratio x. Then we put in Proposition 1 for M and optimize for p. The results are given in the following theorems and figures. In Figs. 2, 3 and 4, where the relative storage capacity I/n^2 is plotted against p, I have used a mixture of a low-fidelity criterion for large p which amounts to $M/n < 0.14$, and the high fidelity criterion of Proposition 1 for small p, namely

(9) $x^2 = 8 \cdot (d \cdot (2p)^{1/d} - \ln p)$ with d = 2.8783.

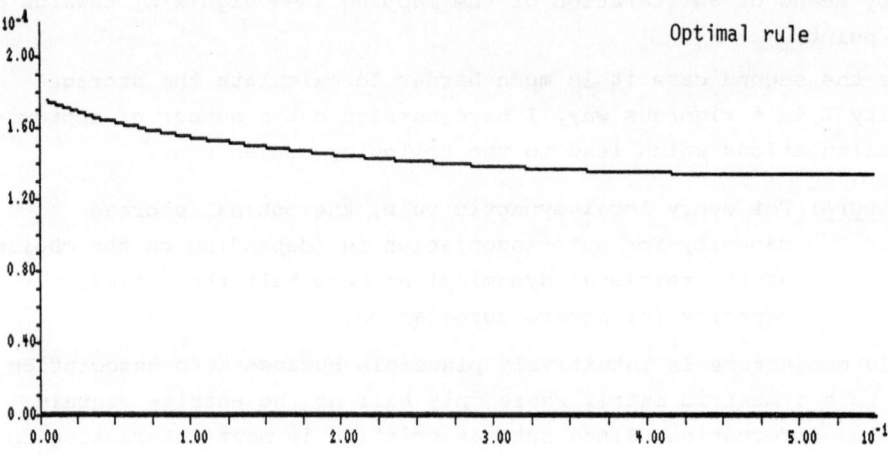

Fig. 2

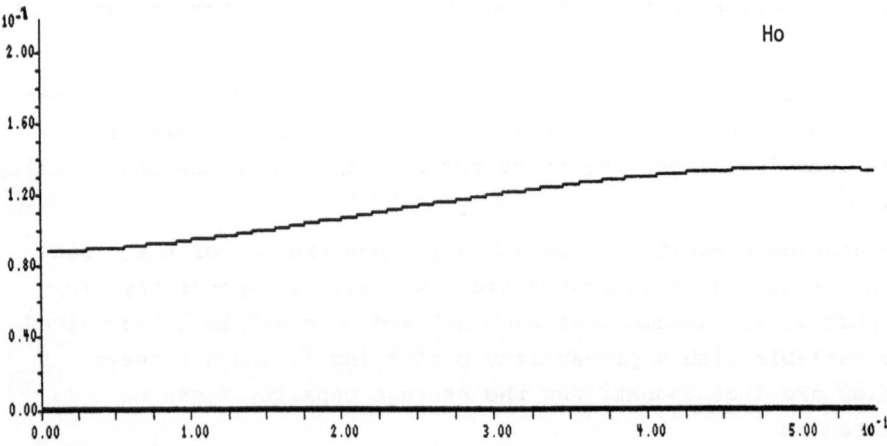

Fig. 3

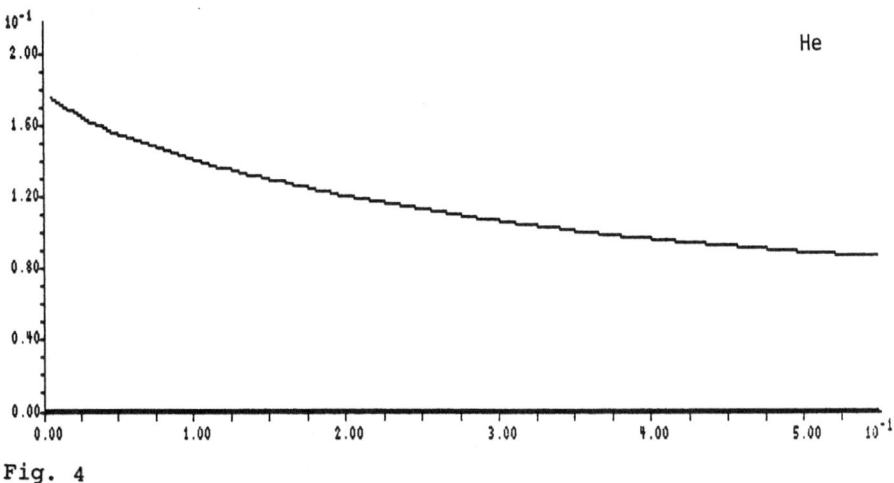

Fig. 4

Theorem 3: (i) For the optimal local rule (Th. 1) the capacity I(n)
           of a network of size n is under the hifi condition
           (Prop. 1) asymptotically given by $I(n)/n^2 \longrightarrow 1/(8\cdot\ln 2)$,
           (ii) for the Hebb rule He it is asymtotically the same,
           (iii) for the Hopfield rule Ho, $I(n)/n^2 \longrightarrow 1/(16\cdot\ln 2)$.

4. The effects of clipping

Up to now we have considered only a linear superposition of synaptic
changes according to (3).

    In this section we want to consider the effects of clipping, i.e.
of equation (2), where f is a two-valued function. The discussion
will be restricted to the two most common cases, i.e. clipped
Hopfield, where the rule is Ho and the clipping function is
f(x) = sign(x), and clipped Hebb, where the rule is He and the
clipping function is f(x) = min (x,1).

    For the clipped Hopfield rule the calculations can be done exactly
along the lines of the preceding sections, taking into account the
slightly lower signal-to-noise ratio. One result on the clipped
Hopfield rule (p = 1/2) is given in Theorem 4.

Theorem 4: For the clipped Hopfield rule the signal-to-noise ratio
           (compare Theorem 1) is $x = (\pi\cdot pq(1-2pq)\cdot\alpha)^{-1/2}$, where
           q = 1-p, and the storage capacity I(n) is in the hifi case
           asymptotically given by $I(n)/n^2 \longrightarrow 1/(8\pi\ln 2)$.

The clipped Hebb rule gives rise to a new possibility, namely to a different choice of the detection threshold θ. Indeed, if the input vector x has k ones and we choose $\theta = k$, then one of the error probabilities e_1 and e_2 is still zero! Observe that in all other cases $\theta = k/2$ would have been a judicious choice for the detection threshold with about equal nonzero error probabilities e_1 and e_2.

Making use of this different choice of detection threshold we can arrive at lower error probabilities and higher information content in spite of a lower signal-to-noise ratio. This can be seen from Theorem 5, which summarises the results. In the calculations for Theorem 5 one cannot use the normal distribution anymore. Instead one has to calculate the error probability e_2 directly (as in [3]). This can approximately be done from the density q of zeroes in the storage matrix after storage of the M patterns as follows:

(10) $q = (1-p^2)^M;$ $e = (1-q)^{pn}$.

The result of these calculations is shown in Fig. 5 for n = 1000. The asymptotic result was already known ([3],[6]) and is restated in the last theorem.

<u>Theorem 5</u>: For the clipped Hebb rule (and c = 0) the signal-to-noise ratio is $x^2 = p \cdot n \cdot q/(1-q)$ and the optimal storage capacity I(n) is in the hifi case asymptotically given by $I(n)/n^2 \longrightarrow \ln 2$.

clipped He

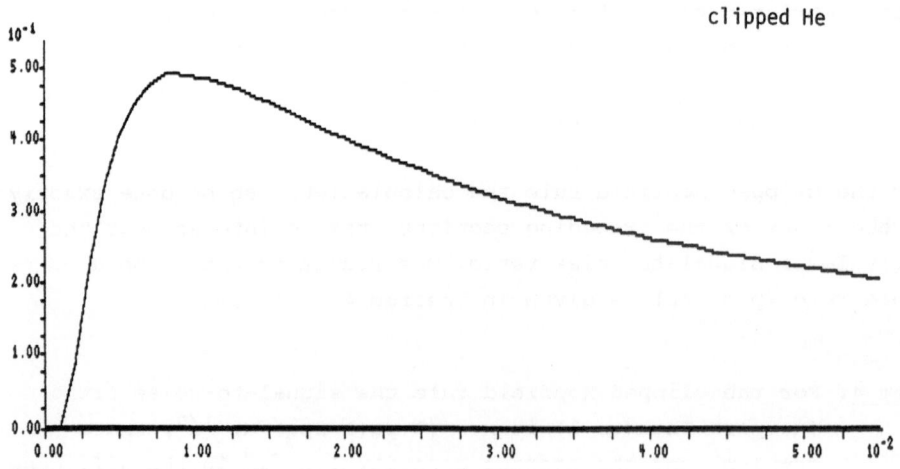

Fig. 5

This theorem shows a clear superiority for the clipped Hebb rule for associative information storage. This superiority is even more surprising if we consider the fact that it needs only 1 bit per synapse for its implementation. Thus a hardware implementation in VLSI of this special type of associative memory is highly recommended.

In fact we have already gained some experience with such a hardware implementation ([2],[5]), albeit not yet in VLSI. It turns out that our special purpose machine, now implementing a 3148 x 3148 associative matrix, can indeed be put to effective use in specific applications. Of course, the sparse coding has to be specially designed for each particular application - and we are presently gaining insight in such coding strategies. Let me finally show some statistical data on one such application, where the memory was used in a checker-playing program to memorize the evaluation of checker positions (Fig. 6). It shows that for a good coding the error probabilities may be considerably lower than for randomly generated patterns.

In summary, these considerations emphasize the importance of sparse coding of the information to be stored in an associative memory, especially if one wants to store a large number of patterns in a economical way and yet with high fidelity retrieval.

In addition, they show the superiority of the simple Hebb rule He for sparsely coded patterns, and in particular, when one makes use of clipping or some other simple algorithm that reduces the range of possible synaptic values. In fact, if one considers the ratio of effectively retrievable bits to necessary hardware bits for the implementations, then the clipped Hebb rule with its asymptotic efficiency of 69 % is clearly superior to all other local rules for associative memories.

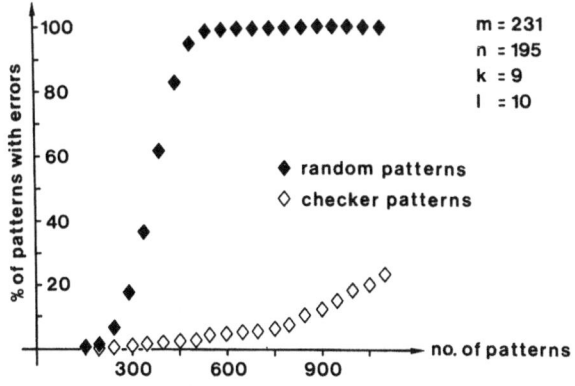

Fig. 6

References

1. J. J. Hopfield, PNAS USA <u>79</u>, 2554-2558 (1982)
2. M. Marcinowski, Diplomarbeit, Tübingen (1986)
3. G. Palm, Biol. Cybern. <u>36</u>, 19-31 (1980)
4. G. Palm, Science <u>235</u>, 1227-1228 (1987)
5. G. Palm and T. Bonhoeffer, Biol. Cybern. <u>51</u>, 201-204 (1984)
6. D. J. Willshaw, O. P. Buneman and H. C. Longuett-Higgins, Nature <u>222</u>, 960-961 (1969)

Hopfield Model, Boltzmann Machine, Multilayer Perceptron and Selected Applications

M. Weick

Corporate Research and Technology,
Corporate Laboratories for Information Technology,
Siemens AG, Otto-Hahn-Ring 6, D-8000 Munich 83, Fed. Rep. of Germany

1. Introduction

For problems that digital computers can't readily solve, we might soon find that artificial neural networks offer a practical alternative. Because neural networks are parallel computing architectures, they aren't hindered by the bottlenecks that can result when a serial processor attempts parallel processing. Neural networks are best at solving problems for which no algorithm exists or for which an algorithmic solution is too slow. Learning methods like learning from examples will never replace programming techniques which are currently used, but they will supplement these methods.

One characteristic feature of neural networks is the robustness. Unlike traditional computers, destruction of units at random in the net had no effect for instance upon correct classification of the tools of an engineer.

Our brain is not as good in arithmetic operations as a digital computer. But when it comes to operations such as association, categorization, feature extraction and generalization it outperforms even the most powerful of today's computers. Our brain processes information in parallel by means of massively interconnected networks of neurons. Each neuron independently evaluates its state and decides to change it or not depending on whether the sum of its synaptic inputs exceeds a given threshold or not. Each of the 10^{10} to 10^{11} neurons has 10^3 to 10^4 synaptic connections with neighboring neurons, which leads to 10^{13} to 10^{15} connections. Even when we assume the neuron to take only two states: firing or not firing, i. e., binary neurons, the total number of degrees of freedom of the brain is truly astronomical, reaching $2^{10^{15}}$.

For surveys about the different models and learning rules for artificial neural networks (ANNs) and a comprehensive compilation about the main concepts of neural net research and development see CARPENTER, GROSSBERG [6], GROSSBERG [14], KOHONEN [20], LIPPMAN [23], and McCLELLAND, RUMELHART [25, 32]. For a description of the state of the art in neural nets, readers are referred to the four-volume Proceedings of the IEEE First International Conference on Neural Networks, San Diego, 1987 [1].

2. Hopfield model

In 1982, HOPFIELD [17] introduced a fully interconnected network with symmetric connections, in which the processor elelements (neurons) update their state asynchronously. The assumption of symmetric connections is one of the most biologically unrealistic features of the Hopfield model. The Hopfield network is an example of a feedback system, whose input or initial state is for instance an incomplete address or picture. This network reaches a final stable state within a few feedback cycles. A final state is that item of the learned addresses or pictures which is most similar to the input.

By an analogy with the energy function of a physical system, he was able to show that this network displays collective computation properties.

The Hopfield energy function has the following form:

$$E = -\frac{1}{2} \sum_{i,j=1}^{N} T_{ij} \, s_i \, s_j + \sum_{i=1}^{N} I_i \, s_i \tag{2.01}$$

where

T_{ij} is the strength of connection between units i and j,

$s_i = \pm 1$ is the binary state value of unit i

and

I_i is a threshold.

Equation (2.01) shows, that the energy of a given state is a function of the T_{ij}. Therefore, information can be stored in the Hopfield net through proper selection of the T_{ij}. We assume a number of patterns $S^\eta = [s_1\eta, s_2\eta, ..., s_N\eta]$ to be stored. Hopfield gives the following Hebbian learning rule for selection of the T_{ij}:

$$T_{ij} = \sum_a (2 \, s_i^a - 1)(2 \, s_j^a - 1) \qquad \text{with} \, T_{ii} = 0. \tag{2.02}$$

This function has the effect of storing a 1 if both states are either on or off, and a -1 otherwise. The Hebbian rule produces local minima in the energy function for each of the stored patterns S^η, up to the storage capacity. Equation (2.02) isn't appropriate when internal (hidden) units are included, since their values for different patterns are unknown. Therefore, the Hopfield model can perform a content-addressable- or an auto-associative-memory function, which is characterized through the ability of recalling the storage contents by incomplete data. It's able to store about 0.15 x N different pictures, where N is the number of processing elements or units in the system. The extension from binary to real valued weights changes this capacity by a factor of π/2. For a detailed overview about the capacity of the Hopfield associative memory see McELIECE et al. [26]. The Hopfield network can also be used to classify binary patterns created from radar cross sections [11], from consonants and vowels extracted from spoken words [10], and from lines in an image [13].
The Hopfield model [18] is suitable for VLSI implementations because of the simple architecture and components such as operational amplifiers and resistors.
PSALTIS and FARHAT [30] discussed several optical implementations of the Hopfield model as associative memories. Such optical systems with bistable optical amplifiers and holograms as basic elements have the following attractive features:

• An optical neural net can have more than 10^6 parallel processing elements (units).

• Each of these processing elements can optically interconnected to the 10^6 other elements.

An electronic implementation of a Hebbian-type neural network with 32 neurons and a 1024-programmable binary connection matrix is described by MOOPENN et al. [28]. The chip is cascadable and, therefore, can be used as a basic building block for a large connection matrix up to 512 x 512 in size. The connection matrix consisted of CMOS analog switches each in series with a 10^6 Ω feedback resistance, while each of the threshold amplifiers consisted of a current summing amplifier followed by a voltage comparator.

GRAF et al. at Bell Labs [12] fabricated a CMOS chip with 54 simple units fully interconnected and programmable -1, 0, and +1 connection strengths. The chip can be used as a coprocessor to a workstation and operate as an associative memory and pattern classifier. The recognition rate for hand-written digits is in the order of 90 percent and can be improved by a higher resolution for the digits.

3. Boltzmann machine

The Hopfield model was designed originally as a content-addressable-memory, and it is apt to converge into one of the neighboring local minima of an energy function E. HINTON et al. [16] developed the Boltzmann machine, which can escape from the local minima controlling a temperature parameter T, because many applications require the global minima as solutions and a Hopfield model without noise can never escape from local minima of E, see HOPFIELD [18].

The Boltzmann machine is a neural network structure with a distributed knowledge representation and nondeterministic processor elements (units).

A first reference to the relationship between massively parallel systems and the Boltzmann distribution dates from John von Neumann:

All of this will lead to theories [of computation] which are much less rigidly of an all-or-none nature than past and present format logic. They will be of a much less combinatorial, and much more analytical, character. In fact, there are numerous indications to make us believe that this new system of formal logic will move closer to another discipline which has been little linked in the past with logic. This is thermodynamics, primarily in the form it was received from Boltzmann, and is that part of theoretical physics which comes nearest in some of its aspects to manipulating and measuring information.

John von Neumann, Collected Works, Vol. 5, p. 304

An energy function of the Boltzmann machine is defined as follows:

$$E = - \sum_{i < j} w_{ij}\, s_i\, s_j + \sum_{i} \Theta_i\, s_i \qquad (3.01)$$

where

w_{ij} is the strength of connection between units i and j,

s_i is 1 if unit i is on and 0 otherwise

and

Θ_i is a threshold.

$$\Delta E_k = \sum_{i} w_{ki}\, s_i - \Theta_k \qquad (3.02)$$

where

ΔE_k is the difference in the global energy between having unit k on instead of off.

The last term in equations (3.01) and (3.02) can be eliminated by introducing an unit s_{on}, which is permanently on with $\omega_{on_i} = \Theta_i$.
Now, equations (3.01) and (3.02) can be written as:

$$E = - \sum_{i < j} w_{ij} \, s_i \, s_j \tag{3.03}$$

$$\Delta E_k = \sum_i w_{ki} \, s_i \tag{3.04}$$

A state value of a unit is determined by the following thermodynamic probability function:

$$P_k = \frac{1}{(1 + e^{-\Delta E_k / T})} \tag{3.05}$$

where

p_k is the probability of setting $s_k = 1$ regardless of the previous state

and

T is a parameter which acts like the temperature of a physical system.

The calculation of a state's value by p_k means, that even when a global solution has been reached, there is still a small probability of changing the state.

The Boltzmann machine uses simulated annealing [19] to reach a global energy minimum since the relative probability of two global states α and β follows the Boltzmann distribution:

$$\frac{P_\alpha}{P_\beta} = e^{-(E_\alpha - E_\beta)/T} \tag{3.06}$$

where

P_α is the probability of being in the αth global state
and

E_α is the energy of that state.

$$\frac{\partial \ln P_\alpha}{\partial w_{ij}} = \frac{1}{T} \left[s_i^\alpha s_j^\alpha - p_{ij}' \right] \tag{3.07}$$

where

s_i^α is the state of the $i th$ unit in the αth global state, so

$s_i^\alpha s_j^\alpha = 1$ only if units i and j are both on in state α,

114

and

$\overset{\cdot}{P_{ij}}$ is the probability of finding two units i and j on at the same time when the system is at equilibrium.

$$G = \sum_{a} P(V_a) \, ln \, \frac{P(V_a)}{P'(V_a)} \qquad (3.08)$$

where

G is an information-theoretic measure of the discrepancy between the internal model and the environment,

$P(V_a)$ is the probability of the ath state of the visible units when their states are determined by the environment,
and

$P'(V_a)$ is the corresponding probability when only the input units clamped.

It is the aim of the learning algorithm to reduce G. Therefore, it is necessary to calculate $\partial G/\partial \omega_{ij}$ for each of the connections:

$$\frac{\partial G}{\partial w_{ij}} = -\frac{1}{T} \, (p_{ij} - p'_{ij}) \qquad (3.09)$$

where

p_{ij} is the probability of units i and j being on together when all the visible units are clamped,

and

p'_{ij} is the corresponding probability with only the input units clamped.

The learning algorithm is divided into two parts:

Part I: The input and output units are clamped to a special pattern that is desired to be learned while the whole network comes to a state of low energy by an annealing procedure.

Part II: The whole network comes to a state of low energy by simulated annealing but now only the input units are clamped.

The goal of the learning algorithm is to find weights such that the learned outputs as described in part II match the desired outputs in part I as nearly as possible. Therefore, after each pattern has been processed through parts I and II, the weights are updated (often by a fixed amount for all connections) according to

$$\Delta w_{ij} = \varepsilon \, (p_{ij} - p'_{ij}) \qquad (3.10)$$

where

ε scales the size of each weight change.

Equation (3.10) uses only locally available information. Each connection change depends only on the statistics which are collected in parts I and II about the two units connected via the connection.

For a comprehensive description of the Boltzmann machine, readers are referred to ACKLEY et al. [2]. DERTHICK [7] gives a detailed description about Boltzmann machine learning algorithms.

SEJNOWSKI et al. [33] described a Boltzmann machine learning algorithm which can discover mirror, rotational and translational symmetries in input patterns. The network consists of 10 x 10 input units, 12 hidden units and 3 output units which represent the horizontal, vertical and right diagonal mirror symmetries. For the 10 x 10 mirror symmetry problem with $\approx 10^{15}$ possible input patterns SEJNOWSKI et al. reached a performance of 90 % after 100000 presentations.
A Boltzmann machine simulation by PRAGER et al. [29] is able to distinguish between the 11 steady-state vowels in English with an accuracy of 85%. The net consists of 128 multi-valued or 2048 binary input units for the different frequencies, 40 hidden units and 8 output units, because each vowel is represented by an eight-bit code. The simulation time on a VAX 11/750 for a learning cycle is approximately 6 min.
ALSPECTOR and ALLEN [4, 5] have designed a 2µm CMOS VLSI implementation of a modified Boltzmann machine with a chip size of 6.5 x 6.5 mm^2. The units on this chip are implemented by amplifiers, the weights by variable resistors and the nondeterministic behavior of a Boltzmann machine by an electronic noise generator. The speedup factor for the learning mode of such a VLSI chip is in the order of one million compared with an equivalent simulation on a VAX 11/780 computer.

4. Multilayer perceptrons

The multilayer perceptron is an example for a feedforward neural network. The term "feedforward" implies that no processing output can be an input for a processing element on the same layer or a preceding layer, which means all connections are unidirectional. The input layer normally performs no function other than the buffering of the input signals. The next layer is named the hidden layer because its outputs are internal to the neural network. Typically only one hidden layer exists, but for complex mappings, additional layers might appear between the input and output layer.

During learning the learning algorithm modifies the connection weights to reduce the error at the output of the neural network. An example of a standard learning algorithm for multilayer perceptrons is the back-propagation algorithm of RUMELHART et al. [31]. This learning algorithm adjusts the weights connected to the output layer and then works backward toward the input layer, adjusting weights in each layer.

RUMELHART et al. remove the centralized nature of gradient-descent techniques by casting the gradient-descent expressions as weight-update rules that use information available locally to each unit plus a quantity that is back-propagated from other units. Let us consider a unit j whose output is connected to a number of other units with indices in the set IND(jk), as shown in Fig. 1.

Let the back-propagated value from unit k to unit j be $\delta_k \cdot \omega_{jk}$. The value δ_j for any unit j is

$$\delta_j = \begin{cases} (d_j - y_j)\, y_j\, (1 - y_j)\,, \, if\, j\, \varepsilon\, IND \\ \sum_{k\, \varepsilon\, IND(jk)} (\delta_k\, \omega_{jk})\, y_j\, (1 - y_j)\,, otherwise, \end{cases} \qquad (4.01)$$

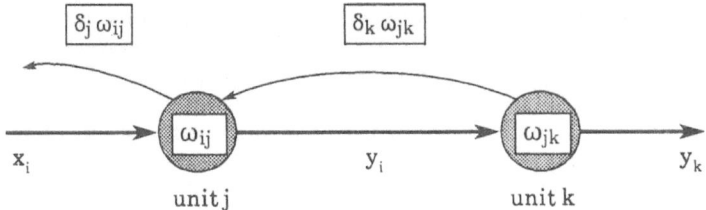

Figure 1:Back-Propagation of Gradient Information

where IND is the set of indices for the output units, and d_j is the desired output value for output-unit j.

This quantity is used to update the weights of unit j according to

$$\Delta\omega_{ij} = \rho\,\delta_j\,x_i, \tag{4.02}$$

before the unit back-propagates $\delta_j\,\omega_{ij}$ to unit i. Since this algorithm is deterministic, a random process is used to initialize the weights to small random values, otherwise the units in a hidden layer would evolve identical weights.

RUMELHART et al. add a modification to this update rule that they refer to as a momentum term for dealing with the problem of searching ravines:

$$\Delta\omega_{ij}[t+1] = \rho\,\delta_j\,x_i + \rho_m\,\Delta\omega_{ij}[t], \tag{4.03}$$

where $\rho_m > 0$ controls the amount of past weight changes that are added to the current weight change. This has the effect of increasing the step size when the gradient does not change sign, so a smaller value of ρ can be used to limit cross-ravine jumps. The momentum of each weight is distinct, enabling large changes in some weights while maintaining small changes in others. This has the desired effect for ravines that are oriented parallel to a dimension of weight space, i. e., a change in the magnitude of weight changes constrain the search to weight values near the floor of the ravine.

The back-propagation learning algorithm for the multilayer perceptron isn't suitable for a VLSI implementation because it requires non-local information for the adjusting weights that must be propagated down from higher layers. Therefore, weight updating isn't a parallel operation.

An extension of the back-propagation algorithm to non-feedforward nets has been done by ALMEIDA [3]. The learning rule is local, therefore the data needed for the update of each weight can be obtained from the two units that it connects, so that this modified algorithm is also suitable for a VLSI implementation.

KUNG and HWANG [21] provide an analytical solution based on the algebraic projection analysis to two of the main points in the back-propagation algorithm namely:

- the disrimination capability by a given number of hidden units

and

- the speed of convergence in learning.

The aim of this approach is an optimal number of hidden units.

SEJNOWSKI and ROSENBERG [34] wrote a program called NETtalk, which learns to pronounce written English text. For this it is necessary to perform the task of converting strings of letters to strings of phonemes. The throughput during learning is of about 2 letters per second on a VAX 11/780. They used a three-layer simulated multilayer perceptron. This system starts completely untrained, and through comparison of the words and speech patterns of a six-year old child learns both the pronunciation and speech control laws autonomously. The verbal output of the net starts out as a babble, gradually becoming quiet between words and finally becoming clearly understandable. The information in the neural net chosen for NETtalk is distributed such that no single unit or connection is essential. Therefore, the net is fault tolerant and it recovered from damage in a shorter time than it took to learn initially.

A three layer time delay neural net (TDNN) which is trained for the speaker-dependent recognition of phonemes is demonstrated by WAIBEL et al. [37]. The net can learn the temporal structure of acoustic events, the temporal relationships between such events, and is is translation invariant, that means, the features learned by the net are insensitive to shifts in time. They achieve a recognition rate of 98.5% for the phonemes "B", "D", and "G" in varying phonetic contexts and testing data from three speakers.

A three layer perceptron for the shape-from-shading problem is simulated by LEHKY and SEJNOWSKI [22]. It consists of an input layer with 122 units, similar to neurons found in the retina, an hidden layer with 27 units, and an output layer with 24 units, which were selective for both the magnitude and orientation of curvature. It is trained with 2000 images, synthesized by calculating light reflected from the paraboloid surface. The correlation between the correct and the actual outputs reached a plateau of 0.88 after 40000 presentations. It is now able to generalize well for images that were not part of the training set.

GALLANT [9] described a neural net as knowledge base in an expert system. He has written a neural net knowledge base generator that uses several learning algorithms and a stand-alone expert system inference engine that interprets such knowledge bases.

A multilayer optical neural net based on holographically interconnected units is presented by WAGNER and PSALTIS [36]. It allows a massively parallel implementation of the back-propagation algorithm.

Recently, Fujitsu demonstrated mobile robots controlled with neurocomputers. Each robot has visual, auditory and tactile sensors, four wheels and two motors, and a neural network simulator for a three-layered neural network.

5. Simulation of neural networks

It is difficult or impossible, except for simple cases, to know in advance how well a particular neural network will perform in a given application. Therefore, up to now, simulation is important for developing neural networks.

Most neural networks are currently being simulated and implemented on digital computers. The simulation speed of neural networks is typically measured in the number of simulated connections per second. A simulation of a neural net of 2000 units (computing the weighted sum of their inputs) on a SUN 3/260 for instance, each with 100 connections, performed 100 simulation steps in 83 seconds, which leads to $\approx$ 240000 connections per second [8].

Today two commercial IBM PC/AT compatible coprocessor boards suitable for neural network simulations exist. The ANZA-Plus™ from Hect-Nielsen Neurocomputer Corp., San Diego, supports up to 1 million processing elements and 1.5 million connections with a performance in the range of 1 to 10 million simulated connection updates per second, depending on whether the network is in a learning or a nonlearning mode. The Delta floating-point processor from Science Applications International Corp., San Diego, supports up to 1 million processing elements and connections with a performance in the same order as the ANZA-Plus™ board [15].

6   Conclusions

Neural networks aren't right for every application, but they give us another option for solving problems. Up to now there exist several factors that determine the success of the neural network approach.

One is the degree of parallelism. Before you can expect to take advantage of the high speeds possible with massively parallel implementations of neural networks, integrated circuits that pack more neural network building blocks into one chip are needed. Several research organizations are currently working on this problem [4, 5, 12, 24, 27, 35].

More research must be done to investigate the feasibility and capabilities of optical implementations of neural networks, because this kind of implementation is able to solve the problem of the number of connections in large neural networks.

Learning algorithms are another factor that will determine the ultimate success of neural networks, because the learning algorithms which are currently used can't learn well or quickly enough (for instance in real-time).

Other open research questions and problems are:

Are specific models more appropiate for given classes of computations than other models ?

Is it useful to combine supervised and unsupervised learning ?

How can a task be modularized to speed learning ?

What tasks can be learned in polynomial time ?

What hardware is best for supporting the particular neural network models ?

Existing networks may not be sufficiently general to reflect all learning situations.

The generalization and representability are sensitive to the architecture of the network.

References

1.   The Institute of Electrical and Electronics Engineers, Inc. (Ed.):
     Proceedings of the IEEE First International Conference on Neural Networks. San Diego, California, June 21-24, 1987. Vol. I-IV.

2.   David H. Ackley, Geoffrey E. Hinton, and Terrence J. Sejnowski:
     A Learning Algorithm for Boltzmann Machines.
     Cognitive Science 9, 141-169(1985).

3.   Luis B. Almeida:
     A learning rule for asynchronous perceptrons with feedback in a combinatorial environment.
     In: The Institute of Electrical and Electronics Engineers, Inc. (Ed.):
     Proceedings of the IEEE First International Conference on Neural Networks, San Diego, California, June 21-24,1987. Vol. II, pp. 609-618.

4.   Joshua Alspector and Robert B. Allen:
     A Neuromorphic VLSI Learning System.
     In: Paul Losleben (Ed.): Advanced Research in VLSI. Proceedings of the 1987 Stanford Conference. pp. 313-349. The MIT Press, Cambridge, MA, 1987.

5. Joshua Alspector, Robert B. Allen, Victor Hu, and Srinagesh Satyanarayana:
   Stochastic Learning Networks and their Electronic Implementation.
   Proceedings of the IEEE Conference on Neural Information Processing Systems -
   Natural and Synthetic-, Denver, Colorado, November 8-12, 1987.

6. Gail A. Carpenter and Stephen Grossberg:
   The ART of Adaptive Pattern Recognition by a Self-Organizing Neural Network.
   IEEE Computer 21, No. 3, 77-88(1988).

7. Mark Derthick:
   Variations on the Boltzmann Machine Learning Algorithm.
   Technical Report No. CMU-CS-84-120(1984). Carnegie-Mellon University,
   Pittsburgh, PA 15213, August, 1984.

8. Jerome A. Feldman, Mark A. Fanty, and Nigel H. Goddard:
   Computing with Structured Connectionist Networks.
   IEEE Computer 21, No. 3, 91-103(1988).

9. Stephen I. Gallant:
   Connectionist Expert Systems.
   Communications of the ACM 31, No. 2, 152-169(1988).

10. Bernard Gold:
    Hopfield Model Applied to Vowel and Consonant Discrimination.
    In: John S. Denker (Ed.):Neural Networks for Computing. In: Rita G. Lerner (Ser.
    Ed.): American Institute of Physics (AIP) Conference Proceedings 151. pp. 158-164.
    American Institute of Physics, New York, 1986.

11. R. Paul Gorman, Terrence J. Sejnowski:
    Analysis of Hidden Units in a Layered Network Trained to Classify Sonar Targets.
    Neural Networks 1, No. 1, 75-89(1988).

12. Hans Peter Graf, Lawrence D. Jackel, and Wayne E. Hubbard:
    VLSI Implementation of a Neural Network Model.
    IEEE Computer 21, No. 3, 41-49(1988).

13. P. M. Grant and J. P. Sage:
    A Comparison of Neural Network and Matched Filter Processing for Detecting Lines
    in Images.
    In: John S. Denker (Ed.):Neural Networks for Computing. In: Rita G. Lerner (Ser.
    Ed.): American Institute of Physics (AIP) Conference Proceedings 151. pp. 194-199.
    American Institute of Physics, New York, 1986.

14. Stephen Grossberg:
    Nonlinear Neural Networks: Principles, Mechanics, and Architectures.
    Neural Networks 1, No. 1, 17-61(1988).

15. Robert Hecht-Nielsen:
    Neurocomputing: picking the human brain.
    IEEE Spectrum 25, No. 3, 36-41(1988).

16. Geoffrey E. Hinton, Terrence J. Sejnowski, and David H. Ackley:
    Boltzmann Machines: Constraint Satisfaction Networks that Learn.
    Technical Report No. CMU-CS-84-119. Carnegie-Mellon University, Pittsburgh,
    PA 15213, May, 1984.

17. John J. Hopfield:
    Neural networks and physical systems with emergent collective computational abilities.
    Proceedings of the National Academy of Sciences of the United States of America **79**, 2554-2558(1982).

18. John J. Hopfield:
    Neurons with graded response have collected computational properties like those of two-state neurons.
    Proceedings of the National Academy of Sciences of the United States of America **81**, 3088-3092(1984).

19. Scott Kirkpatrick, C. D. Gelatt,Jr., and M. P. Vecchi:
    Optimization by Simulated Annealing.
    Science **220**, No. 4598, 671-680(1983).

20. Teuvo Kohonen:
    An Introduction to Neural Computing.
    Neural Networks 1, No. 1, 3-16(1988).

21. S.Y. Kung and J.N. Hwang:
    An Algebraic Projection Analysis for Optimal Hidden Units: Size and Learning Rates in Back-Propagation Learning.
    Proceedings of the IEEE Annual International Conference on Neural Networks, San Diego, California, July 24-27,1988.

22. Sidney R. Lehky and Terrence J. Sejnowski:
    Network model of shape-from-shading: neural function arises from both receptive and projective fields.
    Nature **333**, 452-454(1988).

23. Richard P. Lippmann:
    An Introduction to Computing with Neural Nets.
    IEEE ASSP Magazine , 4-22(1987).

24. Stuart Mackie, Hans Peter Graf, Daniel B. Schwartz, and John S. Denker:
    Microelectronic Implementations of Connectionist Neural Networks.
    Proceedings of the IEEE Conference on Neural Information Processing Systems - Natural and Synthetic-, Denver, Colorado, November 8-12, 1987.

25. James L. McClelland, David Rumelhart, and the PDP Research Group (Eds.):
    Parallel Distributed Processing. Explorations in the Microstructures of Cognition. Volume 2: Psychological and Biological Models.
    The MIT Press, Cambridge, Massachusetts, 1986.

26. Robert McEliece, Edward C. Posner, E. R. Rodemich, and Santosh S. Venkatesh:
    The Capacity of the Hopfield Associative Memory.
    IEEE Transactions on Information Theory **IT-33**, No. 4, 461-482(1987).

27. Carver A. Mead and M.A. Mahowald:
    A Silicon Model of Early Visual Processing.
    Neural Networks 1, No. 1, 91-97(1988).

28. A. Moopenn, John Lambe, and A. P. Thakoor:
    Electronic Implementation of Associative Memory Based on Neural Network Models.
    IEEE Transactions on Systems, Man, and Cybernetics **SMC-17**, No. 2, 325-331(1987).

29. R.W. Prager, T.D. Harrison, and F. Fallside:
    Boltzmann machines for speech recognition.
    Computer Speech and Language 1, 3-27(1986).

30. Demetri Psaltis and Nabil Farhat:
    Optical information processing based on an associative-memory model of neural nets with thresholding and feedback.
    Optics Letters **10**, No. 2, 98-100(1985).

31. David E. Rumelhart, Geoffrey E. Hinton, and Ronald J. Williams:
    Learning Internal Representations by Error Propagation. ICS Report 8506. September 1985. Institute for Cognitive Science. University of California, San Diego.

32. David Rumelhart, James L. McClelland, and the PDP Research Group (Eds.):
    Parallel Distributed Processing.Explorations in the Microstructures of Cognition. Volume 1: Foundations. The MIT Press, Cambridge, Massachusetts, 1986.

33. Terrence J. Sejnowski, Paul K. Kienker, and Geoffrey E. Hinton:
    Learning Symmetry Groups with Hidden Units: Beyond the Perceptron.
    Physica **22 D**, 260-275(1986).

34. Terrence J. Sejnowski, Charles R. Rosenberg:
    Parallel Networks that Learn to Pronounce English Text.
    Complex Systems 1, 145-168(1987).

35. A. P. Thakoor, A. Moopenn, John Lambe, and S. K. Khanna:
    Electronic hardware implementations of neural networks.
    Applied Optics **26**, No. 23, 5085-5092(1987).

36. Kelvin Wagner and Demetri Psaltis:
    Multilayer optical learning networks.
    Applied Optics **26**, No. 23, 5061-5076(1987).

37. A. Waibel, T. Hanazawa, Geoffrey Hinton, K. Shikano, and K. Lang:
    Phoneme Recognition Using Time-Delay Neural Networks.
    ATR Technical Report TR-I-0006, October, 1987.

Computation in Cortical Nets

W. von Seelen, H.A. Mallot, and F. Giannakopoulos

Institut für Zoologie III (Biophysik), Johannes Gutenberg-Universität,
D-6500 Mainz, Fed. Rep. of Germany

1. Introduction

1.1 The Purpose of Biological Information Processing

The evolution of information processing has led to a mutual adaptation of sense organs, brains, and patterns of behavior. In order to understand how animals deal with visual information, it is important to consider the sequence in which the competence for the various patterns of visually guided behavior occurred; a review of this issue is given by HORRIDGE /1/. Simple tasks, such as obstacle avoidance or exploration were solved prior to more complicated ones such as pattern recognition. Therefore, existing solutions to simple problems could serve as *preadaptations* on which the more sophisticated mechanisms for higher functions could improve. An example showing how preadaptations from a simple vision task can be used for a more complicated one has been presented by REICHARDT & POGGIO /2/ for a figure-ground discriminator based on simple motion detection. Another example is the development of elaborate eye-movements in mammals that reduces the relevance of position invariance in biological pattern recognition.

Interestingly, a similar approach has recently been proposed by BROOKS /3/ for the implementation of visually guided behavior in autonomous mobile robots. Simple tasks such as obstacle avoidance, exploration, or tracking are implemented first and can serve as building blocks for more advanced behavior.

In applying strategies of neural networks to information processing tasks, it therefore seems natural to start with the simplest possible patterns of visual behavior. Most applications mentioned in the sequel will deal with *early* vision. However, it is not implied that visual behavior necessarily requires middle and high level vision to follow this level of processing. Rather, we are interested in tasks that can be based on a simple level of processing alone.

1.2 Structured Neural Networks

On the level of global architecture as opposed to single units, realistic neural networks exhibit much more structure than is usually acknowledged. This structure is characteristic for the task the network has to deal with. In accordance with a basic idea of biological science, i.e. the correlation of structure and function, one could think of a correspondence between the structure of a neural network and the information processing strategy it pursues.

In both the mammalian visual cortex and the optic tectum of the vertebrates in general, the gray matter is organized as a cortex, i.e. in two-dimensionally extended layers of neurons with strong vertical connectivity. Over the past decade, many details of cortical organization in several sensory and motor systems have been elu-

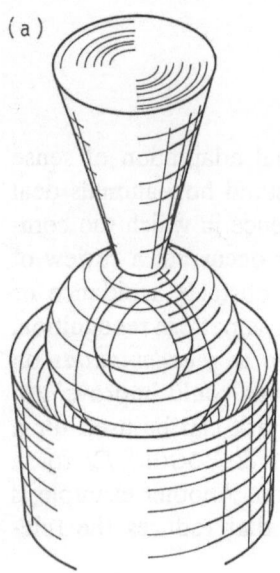

(a)

Fig. 1. Network model of intrinsic cortical organization based on the average anatomy. a. General shape of pre- and postsynaptic connectivity domains of a pyramidal cell. The large sphere and the cone represent basal and apical dendrites, respectively, i. e. regions of excitatory inputs. The inner sphere represents inhibitory inputs mediated by basket cells. The excitatory output fibers of the cell sparsely fill the lower cylinder. b. Block diagram of the complete model. The arrows represent two-dimensional distributions of excitation (solid lines) or inhibition (dashed lines). The spatial behavior is determined by the connectivity domains discussed above. The temporal behavior is due to synaptic delays and time constants as well as to propagation times

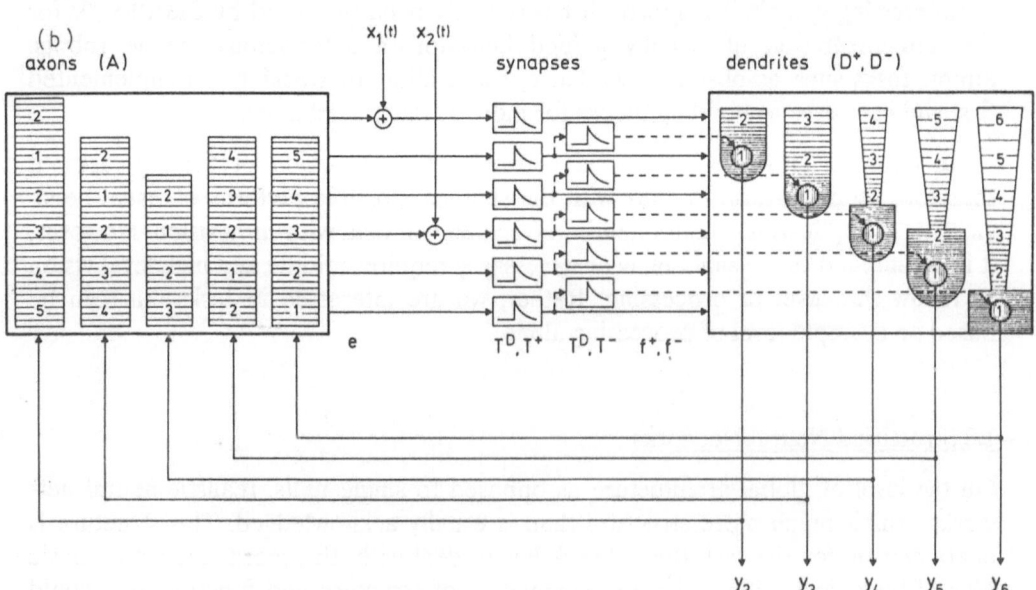

cidated. In earlier papers /4, 5/ we formulated a number of *structural principles*, the first two of which will be discussed here in more detail.

Average Anatomy: The fiber anatomy of cortical neurons is basically uniform within a cortical area. Average intrinsic coupling is characterized by layering, strong vertical connectivity, and positive feedback via the pyramidal cell axon collaterals (Fig. 1).

Retinotopic Mapping: Parallel projections between different regions of the brain usually preserve continuity but cause systematic distortions. By altering the effect of the uniform intrinsic operation in a systematic way, these distortions act as a pre-processing applied to the input.

Patchy Connectivity and **Columnar Organization:** Projections of different areas to a common target segregate into distinct patches or stripes (e.g. ocularity stripes in area V1). This too leads to new neighborhoods for intrinsic processing. Columnar or periodic organization of the intrinsic operation adds to the space-variance produced by different types of mapping.

Processing information in a two-dimensional cortex is natural in the visual system where two-dimensional images are the input. It seems, however, to be a useful strategy in other sensory systems as well. As an example, consider the auditory system, where a stimulus which by its physical nature is purely sequential is transformed into a spatio-temporal distribution of excitation by the basilar membrane. Most of the above features can be found in the auditory system as well.

In the next two sections of the paper, we discuss the information processing capabilities that can be inferred from these features of cortical organization. In Section 4, we present an example of how these capabilities can be used as a *neural instruction set* to implement a simple vision task.

2. Lamination and Intrinsic Feedback

Cortices are two-dimensionally extended, layered structures with strong vertical connectivity. We designed a continuous model of the network of pyramidal and basket cells on the basis of average dendritic and axonal domains (average anatomy; cf. Fig. 1). Due to the overall uniformity of the visual cortex, we can thus study the effects of dense two-dimensional connectivity and feedback. For a detailed description of the model and the properties of the linear version, see KRONE et al. /6/ and VON SEELEN et al. /4/.

The main results of the **linear version** of the network concern the spatio-temporal behavior induced by feedback (Fig. 2). In the open loop, spatial filtering is due to the axonal and dendritic domains depicted in Fig 1a. Temporal characteristics included in the model are synaptic delays and time constants as well as very short propagation times between the layers. In the closed loop, spatio-temporal behavior is no longer separable into two independent parts. This is a general property of dis-

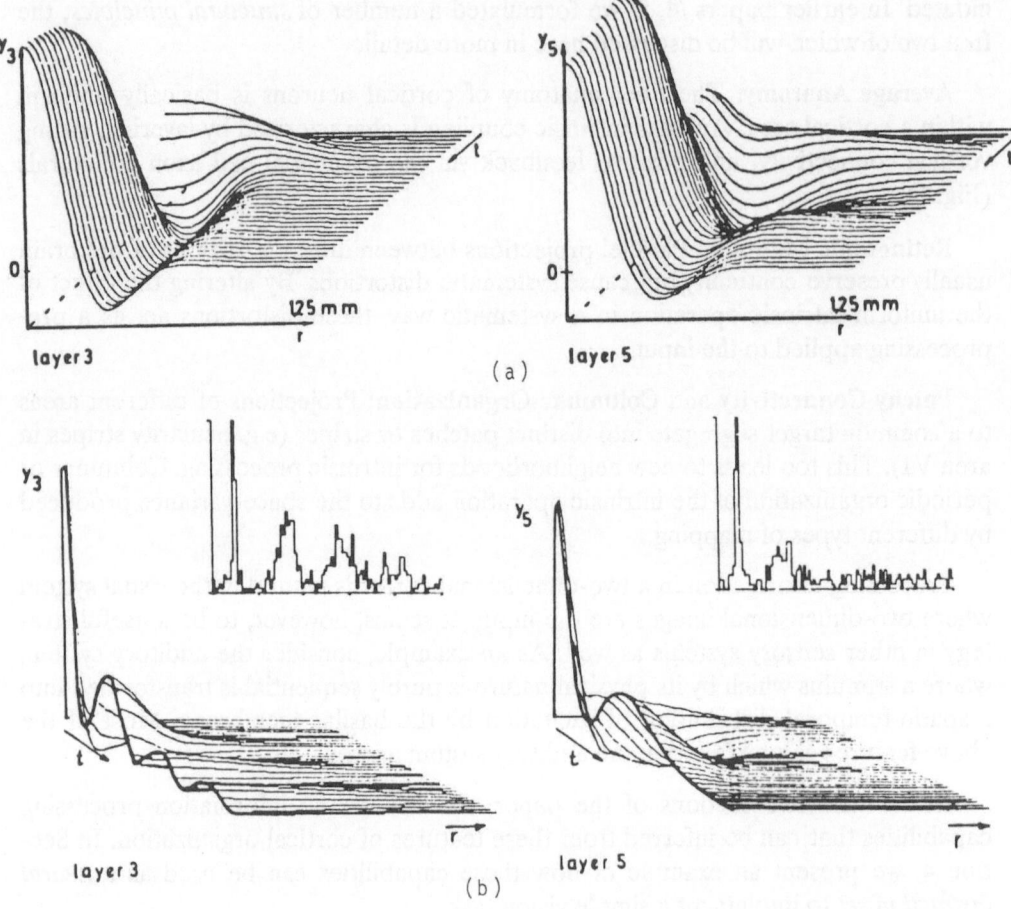

Fig. 2. Spatiotemporal activity profiles for layers 3 and 5 of the cortex model of Fig. 1. Point stimulation at position $r = 0$ in layer 4. The total simulated time is *100 ms*. a. Sustained stimulation (spatial impulse, temporal step response) b. Flashed (transient) stimulation (spatio-temporal impulse response). a. and b. show results for different network parameters. **Insets:** Post Stimulus Time Histograms recorded from the cat visual cortex (Area 17) after on-off stimulation. The results are directly comparable to the simulated responses at the site of stimulation *(r = 0)*.

tributed feedback systems /4/. The model predicts a number of features of spatio-temporal interaction in cortical receptive fields that could be confirmed in electro-physiological experiments on the cat's visual cortex by BEST & DINSE /7/:

Oscillations: After the stimulus onset, the excitation oscillates for about half a second with a frequency on the order of 10 to 20 Hz (cf. Fig. 2b).

Spatio-temporal receptive fields: The temporal structure of the stimulus affects the apparent spatial organization of the receptive field. The features detected by cortical neurons (if any) are therefore spatio-temporal in nature.

Dynamic specificities: Tuning curves for barlength (hypercomplexity) and orientation depend on the velocity or the duration of the stimulus, respectively.

The spatio-temporal distribution of excitation simulated in our model can be described as a damped oscillation in time modulating a spatial spread or contraction (Fig. 2). In Fig. 3, we summarize the basic types of spatio-temporal receptive field organization that have been described. While the occurrence of true spatio-temporal separability is doubtful, there is evidence for piecewise separability (retinal ganglion cells /8/) and motion-type nonseparability (psychophysical channels for motion detection /9/). The spreading behavior found in our model and the recordings of BEST & DINSE /7/ can be related to "looming detectors" (i.e. detectors for the divergence of a velocity vector field) in a certain spatio-temporal frequency channel as have been proposed by KOENDERINK & VAN DOORN /10/ for the analysis of optical flow. Psychophysical evidence for the existence of looming detectors in the human visual system has been presented by REGAN & BEVERLY /11/. Spatio-temporal nonseparability is advantageous for all types of motion perception, since the optimal filters for these tasks are spatio-temporally oriented /12, 13/.

In terms of pattern recognition or feature extraction, spatio-temporal filters should not simply be considered edge-detectors of a special kind. Temporal modulation of a stimulus critically influences the spatial features extracted. In other words, the primitives used by the visual system are elementary events rather than elements of a static sketch of the scene.

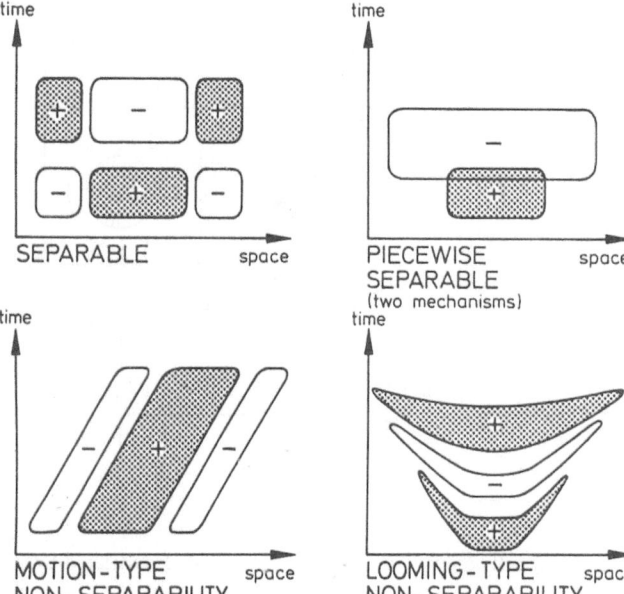

Fig. 3. Types of spatio-temporal receptive field organization. a. Separability. b. Piecewise separability, i.e., a small number of separable "mechanisms". c. Motion-type non-separability. d. Looming-type non-separability as found in our model and the recordings of /7/

A **nonlinear version** of the model (Fig. 4) was built by including a stationary non-linearity in the positive feedback path. For the sake of analytical treatability, the nonlinear model was restricted to two layers. In numerical studies, more layers can easily be included. We obtain the following equations:

$$\tau \dot{y}(t) = -y(t) + \quad d_1^+ * a_1^+ * f(y(t-T_1)) + d_2^+ * a_2^+ * f(y(t-T_2))$$
$$- \quad d_2^+ * a^- * d^- * v(t-T_2') + d_1^+ * E_1 + d_2^+ * E_2 ,$$
$$\tau \dot{v}(t) = -v(t) + \quad a_2^+ * f(y(t-T_2'')) + E_2 . \tag{1}$$

Here, * denotes the spatial convolution and τ is the time constant of synaptic trans-mission. $T_{1,2}$, T_2' and T_2'' account for the latencies of both the synapses and the dendritic and axonal conduction. For further notations, cf. Fig. 4. Details of the model are discussed in /4/.

While the spatio-temporal interactions of the linear model were found in this nonlinear version as well, a number of additional features occurred that extend the computational capabilities of the layered network. Characteristical simulations of (1) for four different sets of parameters are shown in Fig. 5. Analytically, a para-meter classification can be achieved for the simplified (space independent) system of Fig. 6:

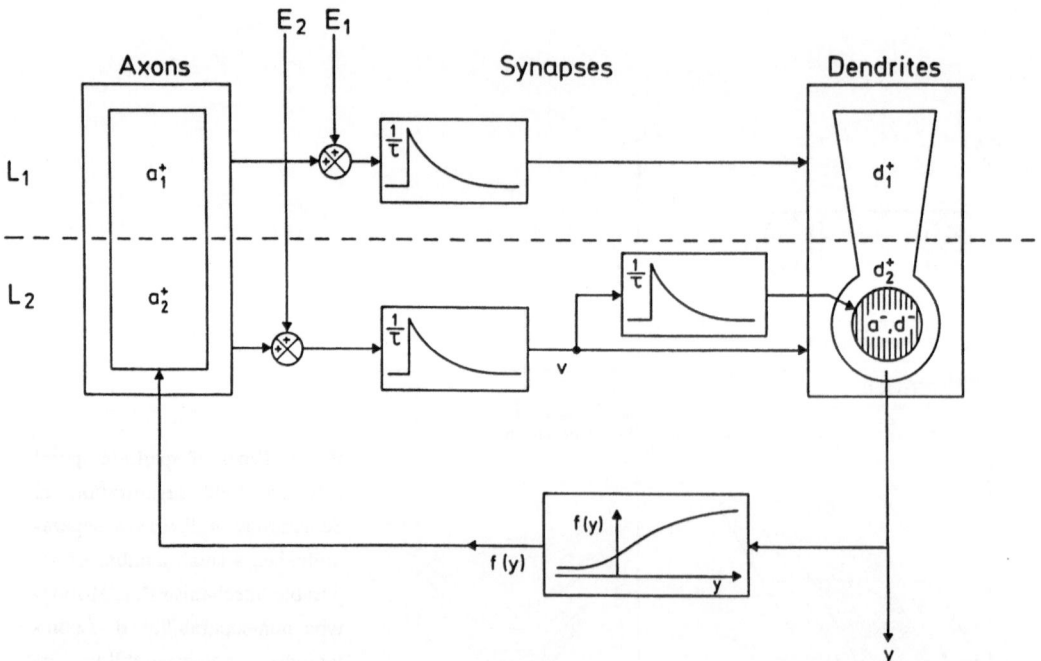

Fig. 4. Structure of the two-layered cortical network model. The range of connectivity is indicated by the regions labelled a^+ for axons of excitatory neurons, d^+ for their dendrites, and so forth

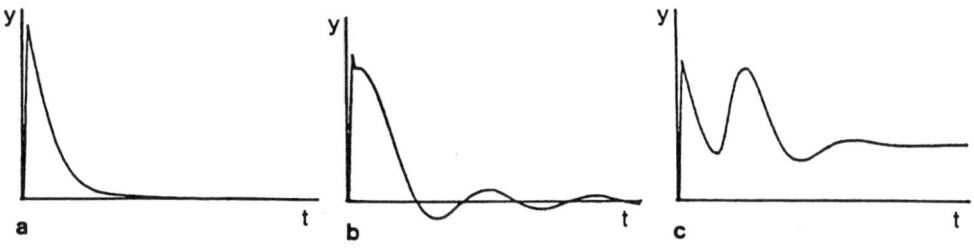

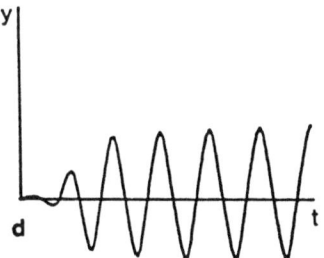

Fig. 5. Numerical solutions of (1) for four different parameter values. Only the temporal course at the site of stimulation is depicted. a. Without positive and negative feedback. b. Weak positive and strong negative feedback. c. Strong positive and weak negative feedback. d. Strong positive and negative feedback

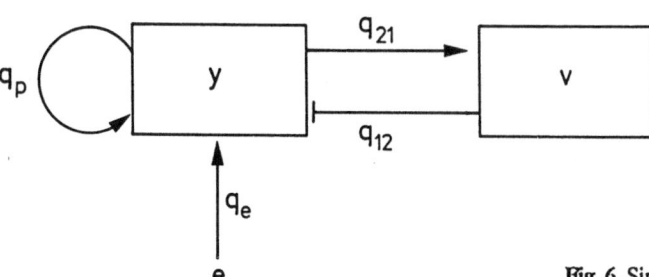

Fig. 6. Simplified system as described by (2)

$$\tau \dot{y}(t) = -y(t) + q_p f(y(t)) - q_{12} v(t) + q_e E_1$$
$$\tau \dot{v}(t) = -v(t) + q_{21} f(y(t)). \tag{2}$$

q_p, q_{12}, q_{21}, and q_e are non-negative constants derived from the axonal and dendritic gain factors. q_p and $q_n := q_{12} q_{21}$ describe the positive and the negative feedback, respectively. The parameter classification for this system appears in Fig. 7.

All solutions of (2) are special space-independent solutions of (1) without latencies and $E_2 = 0$. Space-independent solutions describe the dynamics of cell populations in local columns of cortical tissue /14,15/. The most important properties of the system (2) are:

Bifurcation and Hysteresis: For sustained stimulation, system configuration can change due to stimulus strength. We distinguish two cases: *Hopf Bifurcation:* For strong negative and positive feedback, the system undergoes a transition from stable equilibrium dynamics to limit cycle behavior. *Hysteresis:* If only positive feedback is

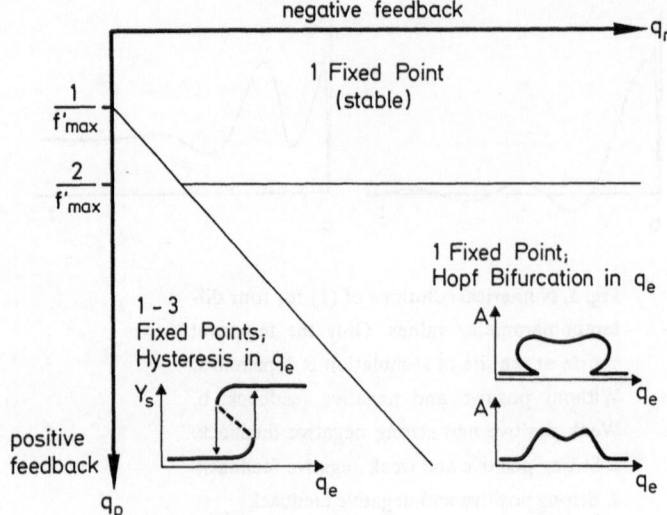

Fig. 7. Parameter classification for the network. While for these analytical results latencies and space-dependences were neglected, the results could be confirmed numerically for the complete model (cf. Fig. 5)

strong, changes of stimulus strength result in state transitions with hysteresis effects (cf. Fig. 7).

Nonlinear oscillations: For sinusoidal stimuli, the system operates as a nonlinear oscillator. Thus network excitation and external stimulus can be synchronized. Although rather complicated dynamics occurred, chaotic behavior could not be demonstrated (cf. ref. /4/).

Bistable flip-flop: Multiple stable equilibria can be interpreted as different processing modes for small inputs. Suitable stimuli can switch the network between two different processing modes (Fig. 8).

As was shown numerically, most of the listed properties occur in the space-dependent system (1) also if the latencies are small (cf. Fig. 5).

The models presented by WILSON & COWAN /14,15/ and COWAN & ERMEN-TROUT /16/ exhibit similar properties. Formal differences include the latencies and the strong asymmetry in our model as well as the confinement to just one non-linearity located in the positive feedback loop. Our model focuses on the special cytoarchitecture of the visual cortex (as summarized in /6/) that we consider crucial for the understanding of visual information processing in the brain.

We conclude that the proposed network is both realistic and treatable enough to model local neural operations in the neocortex. As compared to network models based on single neurons, it has the advantage of including *realistic cortex anatomy* and the effects of large numbers of neurons. The behavior found, i.e., spatio-temporal filters, bifurcations, switching, hysteresis, and synchronization, contributes to the *instruction set* that neural computation has to rely on.

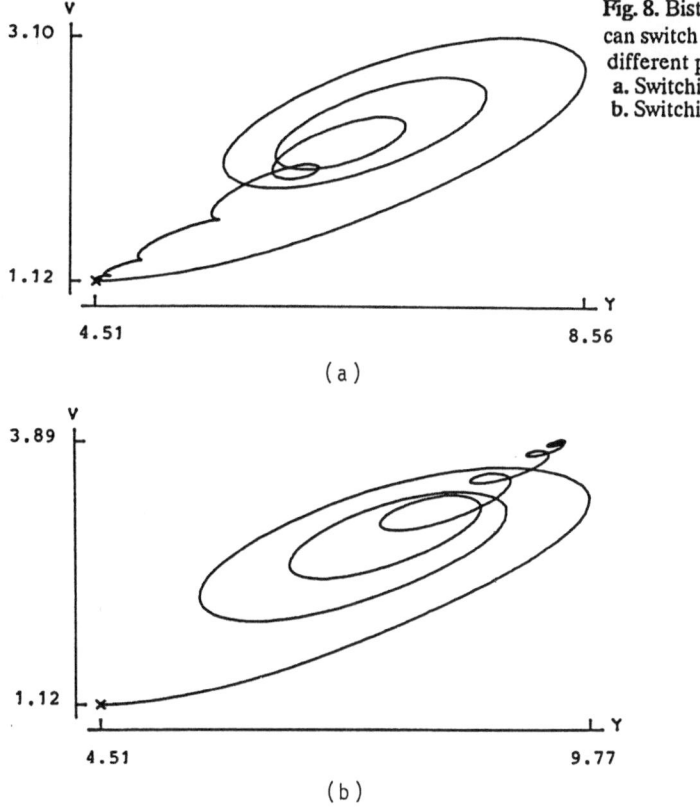

Fig. 8. Bistable flip-flop. Suitable stimuli can switch the network between two different processing modes.
a. Switching failed.
b. Switching occurred

(a)

(b)

3. Retinotopic Mapping

Receptotopic mapping is a general feature of almost all higher sensory centers. Coordinate transforms in massively parallel projections cannot be dealt with on the single cell level. In Fig. 9, topographic maps of three areas of the visual system of the cat are shown together with a model that describes all three maps by minor variations of the same mapping function. This function is a composition of four steps,

$$R : \mathbb{R}^2 \longrightarrow \mathbb{R}^2, \ R = P_2 \circ A_2 \circ P_1 \circ A_1 . \tag{3}$$

The mappings P_2 and P_1 combine to the complex power function, i.e. in real polar coordinates:

$$P_1(r, \beta) = (r^p, \beta)$$
$$P_2(r, \beta) = (r, p\beta); \ p > 0, \ r > 0, \ -pi < \beta < pi. \tag{4}$$

The mappings A_1 and A_2 are affine (shifted linear) transformations. If one of them contains a reflection at the vertical meridian, an area 18 type map with a branched representation of the horizontal meridian results. For details, see MALLOT /17/.

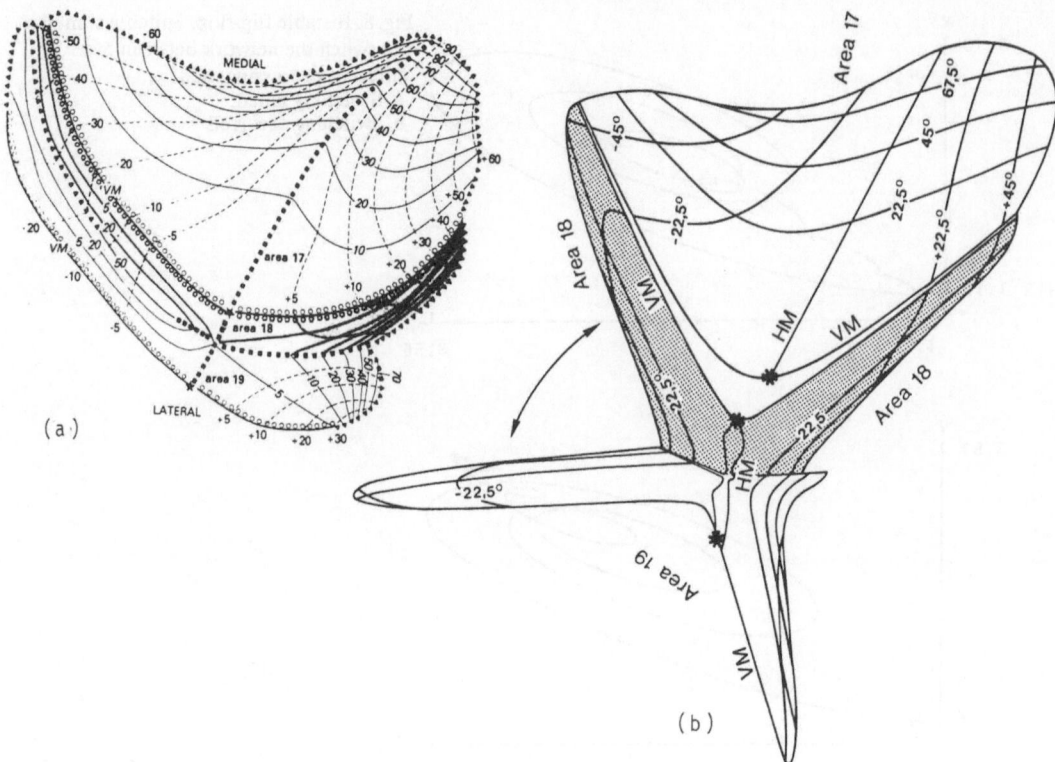

Fig. 9. a. Multiple representation of the visual field in the cat's cortical areas 17, 18, and 19 after /24/. **b.** Formalization by mathematical functions, in this case by a composition of a *complex power function* with certain affine transformations. The functions for all three areas are closely related /17/. The model fits the overall shape of the areas, the differences between upper and lower visual field in area 17, the magnification factors, the continuous bordering between areas, and the branching of the horizontal meridian in areas 18 and 19. The gap between the lower peripheries in areas 18 and 19 (arrow) can be closed in a three-dimensional representation. The resulting fold corresponds to the lateral sulcus of the cat brain

Unlike the mapping in area V1 in the monkey, topographic maps in the cat visual cortex are not conformal, nor can they be approximated by the complex logarithm as proposed by FISCHER /18/ and SCHWARTZ /19/. This is important since most applications of coordinate transforms in computational problems that have been proposed so far rely heavily on the special properties of the logarithmic map (e.g., /20-22/). This runs counter to the fact that in all investigated mammals, multiple retinotopic representations of the visual field with different maps are found. For example, in macaque monkeys where some 15 retinotopically organized areas can be distinguished /23/ it is only the area V1 map that the complex logarithm approximates. In the cat, logarithmic mapping is not found at all /24/.

The computational advantage of coordinate transforms lies in the fact that they prepare data from distant parts of the visual field for subsequent processing by just one space-invariant operation. Consider an intrinsic operation performed by a cortical area such as the spatio-temporal filtering described in Section 2. Retinotopic mapping changes the effective surrounds of the neurons in the target area and thereby influences the meaning of local intrinsic computations with respect to the original image. Consider a simple space-invariant difference-of-gaussians (DOG) filter in a topographically mapped area (Fig. 10a). Let R denote the mapping function, u,v the cortical and x,y the retinal coordinates, i.e. $R(x,y) = (u,v)$. Further let $k(u,v)$ denote the kernel of the intrinsic convolution, i.e. the DOG function in Fig. 10. The composition of mapping and convolution transforms a stimulus distribution $s(x,y)$ into a distribution of excitation $e(u,v)$ by the following equation:

$$e(u,v) = \int\int s(R^{-1}(u',v'))\, k((u',v')-(u,v))\, du'\, dv' \tag{5a}$$

$$= \int\int s(x,y)\, k\, (R(x,y)-(u,v))\, /J_R(x,y)/\, dx\, dy \tag{5b}$$

/17, 25/. Here, J_R denotes the determinant of the Jacobian of R. The resulting receptive field, i.e. the kernel of the linear operator composed of mapping and convolution ("*mapped filter*"), (5b), is depicted for two different maps in Fig. 10b, c. For an affine transformation (b), a simple distortion of the mask's range occurs. In general, however, the mask becomes asymmetric in a space-variant way. This is shown for the complex logarithm in Fig. 10c.

In Fig. 10d,e, the space-variance of the resulting operation is shown for the simulated map of the cat's area 17 by suitable contour lines of the corresponding kernels. In this case, radial orientations are imposed on isotropic cortical operators

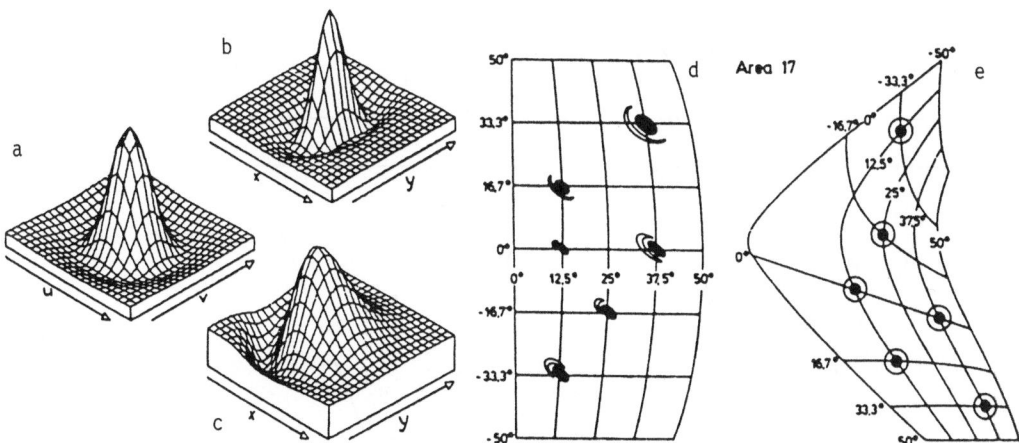

Fig. 10. Effect of retinotopic mapping on an intrinsic kernel (5b). a. - c. Effect on single kernel. a. DOG function b. resultant kernel for an affine coordinate transformation c. resultant kernel for the complex logarithmic mapping. d. - e. Space-variance of the resultant kernel for the simulated area 17 map (Fig. 9). Contour lines for 20% excitation and 80% inhibition. Kernel as in Fig. 10a

by retinotopic mapping. This is in spite of the fact that the complex power function used in this example does not map radial directions onto parallel ones. In ongoing work, we are investigating mappings and kernels that produce certain desired space-variances e.g. for applications in optical flow analysis.

In summary, retinotopic mapping appears as a tool to make the most out of the parallelism available. In modular architectures, just one convolution type operation may subserve many different tasks if the corresponding input is preprocessed by an appropriate mapping.

4. Application to Optical Flow Analysis

In this section, we briefly mention a computational application of a "retinotopic mapping" and a subsequent local operation for the problem of obstacle avoidance. The proposed mechanism leads to the detection of obstacles without any pattern recognition being involved.

Consider the optical flow generated by an observer moving between stationary obstacles in a plane (Fig 11a). Let us suppose for the moment that egomotion is known. For a given direction of view, the stimulus velocity determines the distance of the object from the observer, or, equivalently, its elevation above the ground

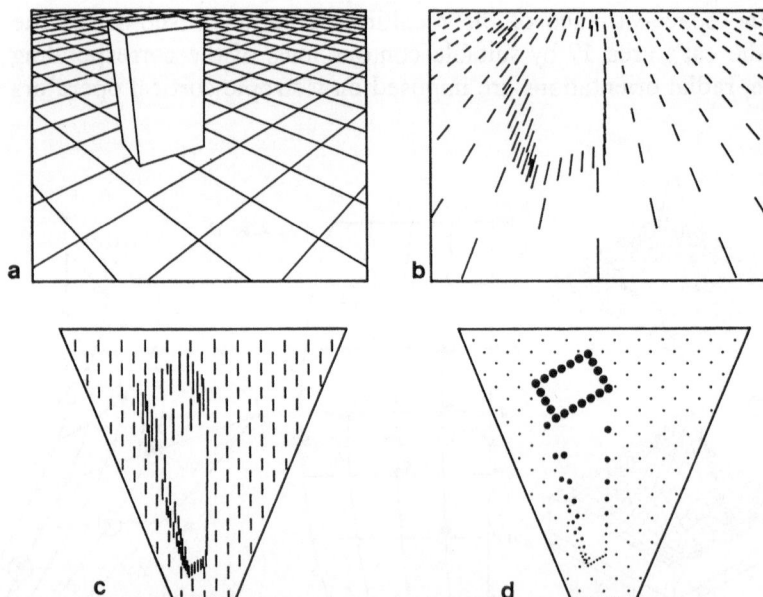

Fig. 11. Optical flow analysis by retinotopic mapping. By inverting the perspective projection of the floor into the image plane, the velocity vectors of elevated points can be detected. The method is robust against small changes in elevation (potholes or bumps), since the output scales with elevation. **a.** Scene with obstacle. **b.** Needle-plot of the velocity field. **c.** Mapped velocity field. **d.** Detected obstacle

(11b). If we want to extract this information with a simple space-invariant filter, we may apply a coordinate transform as is shown in Fig. 11c. Based on prior knowledge of the environment, the inverse of the perspectivic projection of the floor unto the image plane is selected. By this mapping, the velocity vectors assigned to points in the plane become equal and by means of a suitable filter and threshold, elevated points can be detected (Fig. 11d). Elevation above the ground plane is thus taken as an operational definition of an obstacle. Egomotion is the smallest velocity vector occurring in the mapped images (Fig 11c). Note that, at least for stationary objects, the aperture problem is avoided, since one component of the image flow suffices for obstacle detection. A more formal description of the algorithm and the inverse perspective mapping is given in references /5, 26/. In ongoing work, we combine this mapping principle with a parallel image flow algorithm, cf. LITTLE et al. /27/ and MALLOT et al. /28/.

5. Conclusions

Neural networks show obvious anatomical differences that can be studied directly. Especially in the field of neural computation, the direct investigation of existing structures with the methods of comparative biology may be more rewarding than the search for emerging structure in randomly wired nets.

We propose a basic neural network for visual information processing which contains the following elements: First, a space-invariant intrinsic operation with linear non-separable spatio-temporal interactions or non-linear characteristics and, second, a number of mappings and other connectivity schemes (such as patches) which recombine information into new neighborhoods suitable for being processed by the local intrinsic operation. In future research, we attempt to further validate this basic network in technical vision applications.

Acknowledgement. This work was supported by the *Deutsche Forschungsgemeinschaft* (Grant Se251/30-1) and by the *Stiftung Volkswagenwerk* (Grant I/62979). The simulations shown in Fig. 5 were kindly provided by Mr. K. BOHRER. We are grateful to Ms. K. REHBINDER and Ms. M. GROSZ who carefully prepared the figures.

References

1. G. A. Horridge: *Proc. Roy. Soc. Lond.* B **230**, 279 (1987)
2. W. Reichardt & T. Poggio: *Biol. Cybern.* **35**, 81 (1979)
3. R. A. Brooks: In *AI in the eighties and beyond*, ed. by W. E. L. Grimson and R. S. Patil (The MIT Press, Cambridge, MA, 1987).
4. W. von Seelen, H. A. Mallot, F. Giannakopoulos: *Biol. Cybern.* **56**, 37 (1987)
5. H. A. Mallot & W. von Seelen: In *Visuomotor Coordination: Amphibians, Comparisons, Models, and Robots,* ed. by J.-P. Ewert & M. Arbib (Plenum Press, New York 1988, in press)

6. G. Krone, H. A. Mallot, G. Palm, A. Schüz: *Proc. Roy. Soc. Lond.* B **226**, 421 (1986)
7. J. Best & H. R. O. Dinse: In *Proc. nEuro '88*, ed. by G. Dreyfus & L. Personnaz (Paris, 1988, in press)
8. S. Dawis, R. Shapley, E. Kaplan, D. Tranchina: *Vision Research* **24**, 549 (1984)
9. D. C. Burr, J. Ross, M. C. Morrone: *Proc. Roy. Soc. Lond.* B **227**, 249 (1986)
10. J. J. Koenderink & A. J. Van Doorn: *J. Opt. Soc. Am.* **66**, 717 (1976)
11. D. Regan & K. I. Beverly: *Vision Res.* **18**, 415 (1978)
12. A. Korn & W. von Seelen: *Kybernetik (Biol. Cybern.)* **10**, 64 (1972)
13. E. H. Adelson & J. R. Bergen: *J. Opt. Soc. Am.* A **2**, 322 (1985)
14. H. R. Wilson & J. D. Cowan: *Biophys. J.* **12**, 1 (1972)
15. H. R. Wilson & J. D. Cowan: *Kybernetik (Biol. Cybern.)* **13**, 81 (1973)
16. J. D. Cowan & G. B. Ermentrout: In *Studies in Mathematical Biology I*, ed. by S. A. Levin, (Math. Association of America 1978) p. 67
17. H. A. Mallot: *Biol. Cybern.* **52**, 45 (1985)
18. B. Fischer: *Vision Res.* **13**, 2113 (1973)
19. E. L. Schwartz: *Vision Res.* **20**, 645 (1980)
20. A. A. Sawchuk: *J. Opt. Soc. Am.* **64**, 138 (1974)
21. H. J. Reitboeck & J. Altmann: *Biol. Cybern.* **51**, 113 (1984)
22. R. Jain, S. L. Barlett, N. O'Brien: *IEEE Trans. Pattern Anal. Machine Intel. (PAMI)* **9**, 356 (1987)
23. D. Van Essen: In: (eds.) *Cerebral Cortex, Vol 3*, ed. by A. Peters & E. G. Jones, (Plenum Press, New York & London 1985) p 259
24. R. J. Tusa, A. C. Rosenquist, L. A. Palmer: *J. Comp. Neurol.* **185**, 657 (1979)
25. H. A. Mallot: *Trends in Neurosci.* **10**, 310 (1987)
26. H. A. Mallot, E. Schulze, K. Storjohann: In *Proc. nEuro '88*, ed. by G. Dreyfus & L. Personnaz (Paris, 1988 in press)
27. J. J. Little, H. H. Bülthoff, T. Poggio: In *Proc. Image Understanding Workshop*, ed. by L. Baumann (Scientific Applications Interntl. Corp. Los Alamitos 1987)
28. H. A. Mallot, H. H. Bülthoff, J. J. Little: *Invest. Ophth. Vis. Sci. Suppl.* **29**, 398 (1988)

Pattern Formation on Analogue Parallel Networks

H.-G. Purwins and Ch. Radehaus

Institute for Applied Physics, University of Münster,
D-4400 Münster, Fed. Rep. of Germany

1. Introduction

In this contribution we give a review of some properties of real electrical analogue parallel networks consisting of simple electronic oscillators coupled linearely. The networks are the discretized versions of two-component reaction diffusion systems. Physical insight into the behaviour of the systems is obtained by interpreting one component as an activator and the other as an inhibitor. Interpreting the electrical circuit as an equivalent circuit for a continuous material we can apply the model also to gas discharge systems, semiconductor devices and other systems.

This article is the summary of a talk containing originally demonstration experiments. It is the aim of the authors to reflect this character as far as possible. In chapter 2 we describe the network and derive the network equations. In chapter 3 we give a qualitative discussion in terms of the activator inhibitor principle. In chapter 4 experimental results on pattern formation are described. In chapter 5 we discuss the results and draw some conclusions.

2. The Network and the Network Equation

The circuit of the investigated network is shown in Fig. 1a in the one-dimensional version. The single cell is a parallel resonance circuit with linear inductance L and capacity C and a nonlinear resistance R which is a three-terminal device. The resonant circuit is driven by a voltage source consisting of the load resistance R_0 and the ideal voltage source U_S. Adjacent cells are coupled linearly via the voltage and current coupling resistors R_U and R_I at the points A and B. The two-dimensional network is obtained by coupling adjacent chains at the points A and B by additional resistors R_U and R_I.

The nonlinear resistance R is realized by an equivalent circuit shown in Fig. 1b which has the characteristic e. g. S(I) of Fig. 1c. The characteristic is shifted by the current I'. S(I) can be approximated by a polynomial containing a linear and a cubic term. For an isolated cell we draw in Fig. 1c load lines which undergo a parallel shift when changing U_S. There are two types of load lines: those intersecting the nonlinear characteristic once and those intersecting three times. In Figs. 2a, b we give photographs of a single element and the two-dimensional version of the network R^2_{961} of

Springer Series in Synergetics Vol. 42: **Neural and Synergetic Computers**
Editor: H. Haken   ©Springer-Verlag Berlin Heidelberg 1988

Tab. 1: List of experiments showing various structures observed on one- and two-dimensional discrete periodic electrical networks R_n^m (n = number of cells, m = dimensionality). The networks are characterized in terms of the parameters of the corresponding continuum system described by (2-8). The characteristic of R for the experiments 1-4 is similar to S(I) of Fig. 1c and given in Ref.[18]. Experiments 1 and 3 are different in nonnormalized space coordinates due to the choice of R_U = 330 kΩ and 4 Ω, respectively. The numbers in parentheses indicate the reference where a more detailed description of the network can be found. The characteristic of R in experiments 5-7 is that of Fig. 1c.

Parameters and Refs.	bound. cond.	structure	Fig.	remark
Experiment 1 R_{961}^2 [18, 19] σ = 0, δ = 7.38, λ = 0.43 κ₁ = 0, κ₂ = 0	free →	homog.		all cells are oscillating with identical frequency large distance phase shift
	Dirichlet at selected points →	inhomog. stat.	5a,b,c, 6	quasi periodic spatial structures boundary is nucleation center
	free	inhomog. stat.	very simil. to 5a,b,c	multistability
Experiment 2 R_{961}^2 [18, 19] σ = 0. δ = 7.38, λ = 0.43 κ₁ ∈ [-3.1; 2.7], κ₂ = 0	free	inhomog. e. g. stat.	7a,b,c	statistically distributed inhomogeneities of the cells determine the structure
Experiment 3 R_{33}^1 [21] σ = 0' δ = 7.38, λ = 0.43 κ₁ = 0 κ₂ = 0	free (spec. grounding) →	inhomog. oscillatory		all cells oscillate with same frequency significant phase shifts between the cells
	free	inhomog. oscillatory	8	all clusters are separated by nonoscillatory regions; cells within one cluster are oscillating in phase; different clusters oscillate with different frequencies

Parameters and Refs.	bound. cond.	structure	Fig.	remark
Experiment 4 $\mathbf{R}^1_{33}$ [21] $\sigma = 0$, $\delta = 7.38$, $\lambda = 0.43$ $\kappa_1 = \hat{A} \sin 2\pi f t$, $\kappa_2 = 0$ $f = 175$ kHz	free	homog. oscillatory	9a, b, c	all cells oscillate in phase with f for large $\hat{A}$ ▼ (decreasing $\hat{A} \in$ [0.12; 0.17])
	free	inhomog. oscillatory	9d,e,f	no fixed phase relation intermittency spatial-temporal chaos
Experiment 5 $\mathbf{R}^1_{128}$ [1, 2] $\sigma \in$ [0.043, 0.064], $\delta = 1.85$, $\lambda = 0.423$, $\kappa_1 = 0$, $\kappa_2 = 0$	free	homog. stat.	10c,f	$\sigma > \sigma_c$ ▼ (decreasing σ)
	free	sine-like stat.	10a,b,d,e	for $\sigma < \sigma_c$ bifurcation into a new state
Experiment 6 $\mathbf{R}^1_{128}$ $\sigma \to \infty$, $\delta = 2727$, $\lambda = 0.423$ $\kappa_1 = -0.57$, $\kappa_2 = 0$	free	inhomog. travelling pulses	12a, b, c, d	Fitz-Hugh Nagumo nerve axon pulses
Experiment 7 $\mathbf{R}^1_{128}$ [1, 2] $\sigma = 0.14$, $\delta = 0.64$, $\lambda = 1.59$ $\kappa_1 \in$ [-1.93,0.97], $\kappa_2 = 7 \cdot 10^{-3}$	free	homog. stat.		small values of v, w ▼ (increasing κ_1)
	free	inhomog. stat.	13(1,2,3,4)	number of solitary filaments increases with κ_1 ▼ (increasing κ_1)
	free	homog. stat.	13(5)	▼ (decreasing κ_1)
	free	inhomog. stat.	13(6,7)	interacting filaments, number decreases with κ_1, ▼ (decreasing κ_1)
	free	homog. stat.		large values of v, w

Table 1. Applying Kirchhoff's laws to the two-dimensional version of Fig. 1a and indicating by i, j the cell in the i-th row and j-th column we obtain:

$$\frac{dI_{i,j}}{dt} = \frac{1}{L}\left[\frac{\gamma}{R_L}(I_{i-1,j} + I_{i+1,j} + I_{i,j-1} + I_{i,j+1} - 4I_{i,j}) - S(I_{i,j}) + U_{i,j}\right],$$

$$\frac{dU_{i,j}}{dt} = \frac{1}{C}\left[\frac{1}{R_U}(U_{i-1,j} + U_{i+1,j} + U_{i,j-1} + U_{i,j+1} - 4U_{i,j}) - I_{i,j} - \frac{1}{R_V}U_{i,j}\right.$$

$$\left. + \frac{U_S}{R_V + NMR_0} + \frac{R_0}{R_V(R_V + NMR_0)}\sum_{n,m=1}^{N,M}U_{nm}\right]. \tag{2-1}$$

S(I) =   nonlinear characteristic of R for I = $I_{i,j}$ and I´= 0 of Fig. 1c e. g.

γ = $11.5 \cdot 10^6 \Omega^2$ = current coupling constant [1, 2]

i, j; n, m = cell indices

N, M = number of cells in one row and one column

For the sake of an analytical representation we approximate the nonlinear characteristic by a cubic polynomial:

$$S(I_{i,j}) \approx U^* - \chi'(I - I^*) + \varphi'(I - I^*)^3. \tag{2-2}$$

For experiments 1 - 4 and 5 - 7 we have for χ' and φ' the values of 1.33 kΩ, 1.95 · 10^8V/A^3 and 1.27 kΩ, 1.65 · 10^8V/A^3. It turns out that the overall properties of the system do not change much if only the general form of a cubic polynomial of the kind (2-2) is retained.

Introducing the transformation

$$V_{i,j} = (I_{i,j} - I^*)(\varphi'/R_v)^{1/2}, \qquad W_{i,j} = (U^* - U_{i,j})(\varphi'/R_v^3)^{1/2},$$

$$\tau = (R_v/L)t, \quad \delta = R_v^2 C/L, \quad \lambda = \chi'/R_v, \quad d'_v = \gamma'/R_L R_v, \quad d'_w = R_v/R_U,$$

$$\kappa'_1 = \left(\frac{U_S - U^*}{R_V + NMR_0} - I^*\right)\left(\frac{\varphi'}{R_v}\right)^{1/2}, \qquad \kappa'_2 = \frac{R_0}{R_V + NMR_0} \tag{2-3}$$

we obtain the normalized equation

$$\frac{dV_{i,j}}{d\tau} = d'_v(V_{i-1,j} + V_{i+1,j} + V_{i,j-1} + V_{i,j+1} - 4V_{i,j}) + f(V_{i,j}) - W_{i,j},$$

$$\delta\frac{dW_{i,j}}{d\tau} = d'_w(W_{i-1,j} + W_{i+1,j} + W_{i,j-1} + W_{i,j+1} - 4W_{i,j}) + V_{i,j} - W_{i,j} - \kappa'_1$$

$$+ \kappa'_2\sum_{n,m}^{N,M}W_{n,m}, \tag{2-4}$$

$$f(V_{i,j}) \approx \lambda V_{i,j} - V_{i,j}^3. \tag{2-5}$$

140

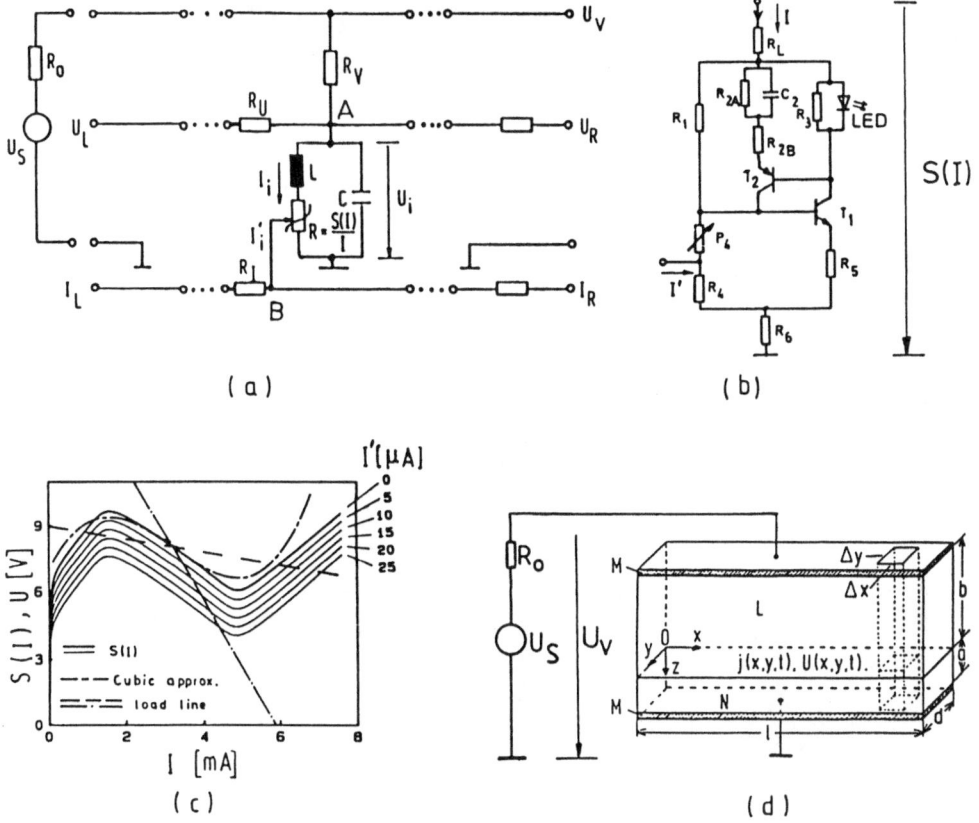

Fig. 1: The electrical circuit. a: A single cell of a one-dimensional periodic array biased by U_S via the load resistance R_0. U_L, I_L and U_R, I_R indicate the left and right hand boundary conditions. b: The equivalent circuit for R (See [1, 2] for details). c: Characteristic S(I) for R, cubic approximation for I' = 0 and typical load lines $U = U_S - (R_0 + R_V) I$ of the single cell. d: Continuum material consisting of a linear material L and a nonlinear material N with metallic electrodes M (see also [22]).

For a compact representation we interprete (2-1) as the discretized version of a continuum equation. For this reason we introduce discretization lengths corresponding to the distance Δx of adjacent elements in a row and Δy in a column and write (compare Fig. 1d)

$$j(i\Delta x, j\Delta y, \tau) = j(x, y, \tau) = I_{i,j} / (\Delta x \Delta y), \qquad j^* = I^* / (\Delta x \Delta y),$$

$$U(i\Delta x, j\Delta y, \tau) = U(x, y, \tau) = U_{i,j}.$$

$$(2\text{-}6)$$

Fig. 2: Photograph of the network R^2_{961}. a: A single cell with the LED on top. b: The two-dimensional network.

With the transformation

$$v = (j - j^*)(\varphi'/R_v)^{1/2}, \qquad w = (U^* - U)(\varphi'/R_v^3)^{1/2},$$

$$\tau = (R_v/L)t, \qquad \sigma = \gamma' R_U/R_I R_v^2, \qquad \delta = R_v^2 C/L, \qquad \lambda = \chi'/R_v,$$

$$\kappa_1 = \left[\frac{U_s - U^*}{R_v + NMR_0} - I^*\right]\left(\frac{\varphi'}{R_v}\right)^{1/2}, \qquad \kappa_2 = \frac{R_0 R_v}{R_u(R_v + NMR_0)} \qquad (2\text{-}7)$$

we obtain the continuum equation

$$v_\tau = \sigma \Delta v + f(v) - w,$$

$$\delta w_\tau = \Delta w + v - w - \kappa_1 + \kappa_2 \int_\Omega w d\Omega, \qquad (2\text{-}8)$$

$$f(v) \approx \lambda v - v^3, \qquad\qquad \sigma, \delta, \lambda, \kappa_2 > 0$$

which has (2-4) as a discretized version. Here Ω is the L-N interface in Fig. 1d.

(2-8) is a reaction diffusion equation containing two components and in addition an integral term. σ has the meaning of a diffusion constant of v measured in terms of the diffusion constant of w. δ is the relaxation time of w given in terms of the relaxation time of v. λ is determined by the slope of the nonlinear characteristic measured in units of R_V. κ_1 is zero if, for the single cell, U_S is chosen such that in the stationary state the nonlinear resistance R is operated in the inflection point. κ_2 and consequently the integral term vanishes for $R_0 = 0$.

It is useful to classify the behaviour of (2-8) according to various sets of parameters describing the uncoupled system. In Fig. 3a we see the uncoupled circuit which is described in terms of I and U. Without loss of generality we put $R_0 = 0$. According to (2-7) and (2-8) we get:

$$v_\tau = \lambda v - v^3 - w,$$

$$\delta w_\tau = v - w - \kappa_1.$$

(2-9)

The stationary solutions of the circuit Fig. 3a can be obtained from Fig. 3b by determining the intersection points of the nonlinear characteristic S(I) with the load

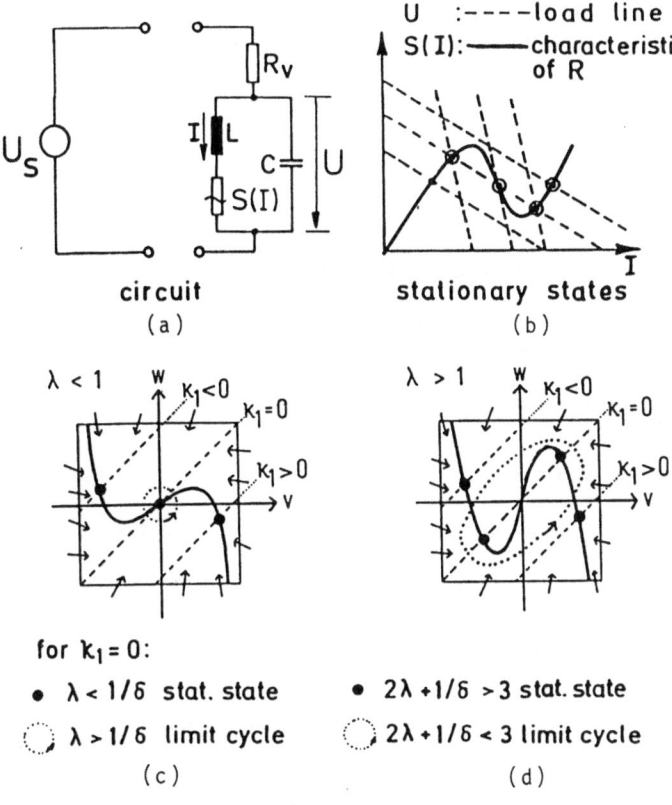

circuit
(a)

stationary states
(b)

for $k_1 = 0$:

● $\lambda < 1/\delta$ stat. state

◌ $\lambda > 1/\delta$ limit cycle

(c)

● $2\lambda + 1/\delta > 3$ stat. state

◌ $2\lambda + 1/\delta < 3$ limit cycle

(d)

Fig. 3: The uncoupled circuit. a: The circuit. b: Stationary states as intersection points of load line and nonlinear characteristic. c: Case of λ < 1 with one stationary state in normalized coordinates. d: Case of λ > 1 with the possibility of three stationary states presented again in normalized coordinates. The text in c and d corresponds to the possible stable motions.

line $U = U_S - R_V I$. Instead of a description in terms of physical components we switch to normalized coordinates in Figs. 3c,d. First we concentrate on load lines going through the inflection point, i.e. $\kappa_1 = 0$. In Fig. 3c we find for $\lambda < 1$ as stable motions a stationary state $(v,w) = (0,0)$ for $\lambda < 1/\delta$ and a limit cycle for $\lambda > 1/\delta$. For the case $\lambda > 1$ we obtain up to three stationary states. The stationary states $(v, w) = \pm(\sqrt{1-\lambda}, \sqrt{1-\lambda})$ are stable for $2\lambda + 1/\delta > 3$ and unstable for $2\lambda + 1/\delta < 3$. In the latter case we get a limit cycle. Due to a general choice of U_S we may obtain $\kappa_1 \neq 0$ in general. This leads to a parallel shift of the load lines in Figs. 3b, c, d.

For the case of $\kappa_2 = 0$ the system (2-8) is a nonlinear reaction diffusion equation. Such systems have been investigated in many branches of physics [3]. Related to the present article e.g. is the work of the schools of Prigogine [4, 5] and Haken [6,7] and the work of Field and Noyes [8]. In a closer context are the investigations of Fife [9], Murray [10], Gierer [11], Meinhardt [12], Maginu [13] and Rothe [14, 15]. These authors treat reaction diffusion systems in terms of activator and inhibitor. For the case $\kappa_1 = 0$, σ, $\delta \gg 1$ (2-8) is a special case of the Fitz-Hugh Nagumo equation for nerve axon pulse transmission [16,17]. This model has been realized physically as an electronic circuit by Nagumo [17]. Detailed investigations of (2-8), in connection with experimental work on electrical networks, have been performed by the Münster group; this article contains a partial summary of this work [1,2,18-21].Also semiconductor and gas discharge systems [19,20,22,23,24] described by (2-8) are investigated in Münster; however, the latter work, including investigations of pattern formation in homogeneous semiconductors [19,25] is not the subject matter of the present article.

3. The Activator Inhibitor Principle

We start with the general two-component reaction diffusion equation

$$v_\tau = \sigma \Delta v + f(v, w), \qquad v = v(\xi, \eta, \zeta, \tau),$$
$$\delta w_\tau = \Delta w + g(v, w), \qquad w = w(\xi, \eta, \zeta, \tau) \tag{3-1}$$

with stationary states (v_0, w_0) of the uncoupled system defined by

$$f(v_0, w_0) = 0, \qquad g(v_0, w_0) = 0. \tag{3-2}$$

In the neighbourhood of (v_0, w_0) v is an activator and w an inhibitor if the following local properties apply [9, 10]:

$$f_v(v_0, w_0) > 0, \quad g_v(v_0, w_0) > 0. \quad f_w(v_0, w_0) < 0, \quad g_w(v_0, w_0) < 0. \tag{3-3}$$

The solution (v_0, w_0) is stable if

$$f_v(v_0, w_0) + g_w(v_0, w_0) < 0. \quad f_v g_w(v_0, w_0) + f_w g_v(v_0, w_0) > 0 \tag{3-4}$$

is valid, otherwise it is unstable.

Returning to (2-8) with κ_1, $\kappa_2 = 0$ we see immediately that $(v_0, w_0) = (0,0)$ is a solution of the uncoupled system and in the neighbourhood of this solution v is activating and w is inhibiting both components. It is also obvious that this is still true for a certain range of values of κ_1 near zero. With these ideas in mind we obtain some physical insight into the behaviour of (2-8) with $\kappa_2 = 0$. Various cases can be discussed with the help of Fig. 4.

Case 1: $\delta \gg 1$, $\kappa_1 \approx 0$, free boundary: Starting from a stationary homogeneous distribution (v_0, w_0) we may assume a fluctuation of the activator v at ξ_0 in Fig. 4. For $\delta \gg 1$ the relaxation time of the inhibitor is much larger than that of the activator. Therefore, near ξ_0, with respect to w, the activator can grow fast by means of self-activation. For large values v suffers from self-inhibition and saturation occurs. However, after a time interval δ the inhibitor has become large enough to damp the activator. The phase lack between activator and inhibitor favours stable oscillatory behaviour.

Case 2: $\sigma \ll 1$, $\kappa_1 \approx 0$: As shown also in Fig. 4 we start again from a homogeneous state. A fluctuation at ξ_0 leads to an increase of the activator due to self-activation. The activator distribution will stay localized due to limited diffusion, while the inhibitor spreads out in space due to large diffusion. Very near to ξ_0 the inhibitor will not be strong enough to pull down the activator concentration. However, in a region further but not too far away from ξ_0 the inhibitor is able to do so. In summary we end up with

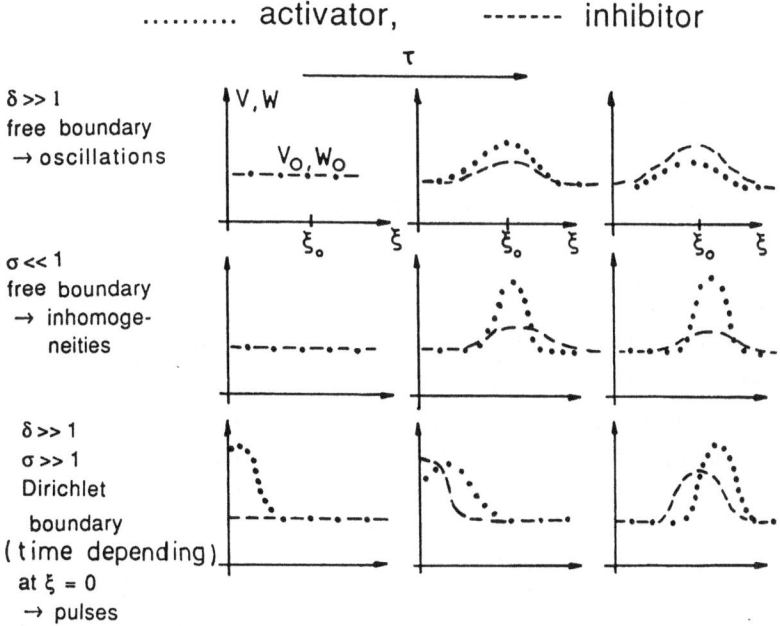

Fig. 4: Special cases for (2-8) with $\kappa_2 = 0$.

a local self-activation and a lateral inhibition favouring stable inhomogeneous spatial structures.

Case 3: $\delta \gg 1$, $\sigma \ll 1$, $\kappa_1 \approx 0$, Dirichlet conditions applied to oscillatory structure of case 1: In cases 1 and 2 the system acts as an activator inhibitor system which implies a choice of κ_1 such that in the uncoupled system of Fig. 3 the load line intersects the nonlinear characteristic near to the inflection point; consequently $|\kappa_1|$ is not too large. Applying Dirichlet conditions to the boundary, by using a fixed voltage, the region of the system near to the boundary may suffer a shift of the operation point to a region where no oscillations are possible. This is equivalent to a shift to large effective values of $|\kappa_1|$. Due to local activation and lateral inhibition this nucleation of a stationary inhomogeneous structure leads to the propagation of a front that runs over the lattice leaving behind a stable stationary inhomogeneous structure.

Case 4: $\delta \gg 1$, $\sigma \ll 1$, $\kappa_1 \neq 0$, free boundary and slow increase of κ_1: In cases 1 - 3 we assume that U_S, and therefore $\kappa_1 \approx 0$, has already been switched on or that the switching time has been short with respect to the internal relaxation times of the system. For small values U_S ($\kappa_1 \ll 0$) increasing adiabactically the system has stable operation points far from the inflection point, being continuously shifted to the latter by increasing U_S (κ_1) (Fig. 3). Increasing U_S, and therefore κ_1 further, we reach a value of κ_1 allowing for some region (favoured by inhomogeneity) to switch to the activator inhibitor type of behaviour. Due to local activation and lateral inhibition this region will have high activator values surrounded by an area with low activator concentration. Though the latter area might be favoured by inhomogeneities it will not switch to high activator values for moderately increased U_S because of lateral inhibition. Finally for large values of $\kappa_2 \gg 0$ the system has to switch to a practically homogeneous state.

Case 5: δ, $\sigma \gg 1$, Dirichlet conditions, $\kappa_1 \ll 0$: We consider first a system with free boundary conditions. In this case we expect qualitatively a linear behaviour and therefore a homogeneous stable stationary structure (v_0, w_0). Now applying Dirichlet conditions at one end of the system in Fig. 4, by applying a constant voltage to the boundary, we may create locally an effective value of $\kappa_1 \approx 0$, so that in this region v may act as an activator and w as an inhibitor. Due to self-activation v may increase near to the boundary. At the same time v spreads out in space due to large diffusion ($\sigma \gg 1$). After a time interval, δ and when the boundary voltage is removed the inhibitor is sufficiently large to damp out the activator and finally itself. However, further away from the boundary v can undergo self-activation and diffusion and after another time interval δ again v and w will be damped out successively. This process is continued and we end up with an activator pulse which seems to be pushed by an inhibitor pulse along the ξ-direction. This is precisely the model for the transmission of a nerve axon pulse in the Fitz-Hugh Nagumo model [16, 17].

146

4. Experimental results

In Table 1 we give a list of the patterns observed experimentally on one- and two-dimensional electrical networks of the kind described in chapter 2.

Experiment 1: This experiment corresponds to the case 3 with $\delta \gg 1$ and $\sigma \ll 1$ discussed in physical terms in chapter 3. $\kappa_1 = 0$ is realized by setting U_S to the appropriate value in a time interval much smaller than the typical relaxation time $\sqrt{LC} \approx 1\mu s$ of the network. For free boundaries the system exhibits nearly homogeneous spatial oscillations. Setting the voltage to zero at selected points of the boundary, these points act as nucleation centers for a stationary inhomogeneous structure and fronts travel over the oscillating network with constant velocity, leaving behind the stationary structure. Such structures can be seen in Figs. 5a, b, c and can be significantly different for different boundary conditions. Changing again to free boundary conditions these structures undergo only minor changes. Thus for the same set of parameters and free boundary conditions very many different structures can be stable, manifesting a very high degree of multistability. Fig. 6 shows a comparison of the voltage distribution between experiment and theory for three different voltage

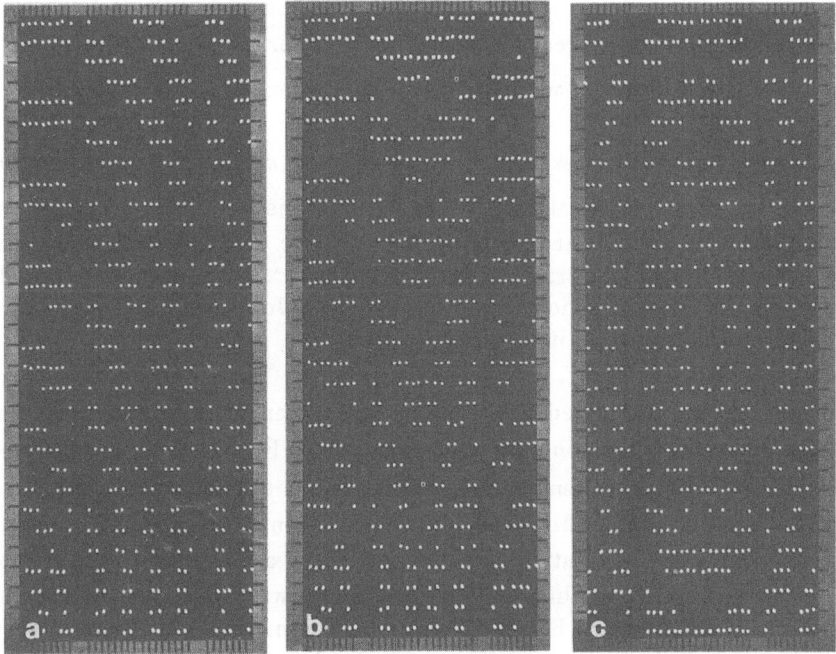

Fig. 5   Various stable stationary structures according to experiment 1 of Table 1. The light points are due to light emitting diodes indicating cells with high current. The boundary voltage is set to 0V at left hand bottom (a), left and right hand bottom (b) and center (c).

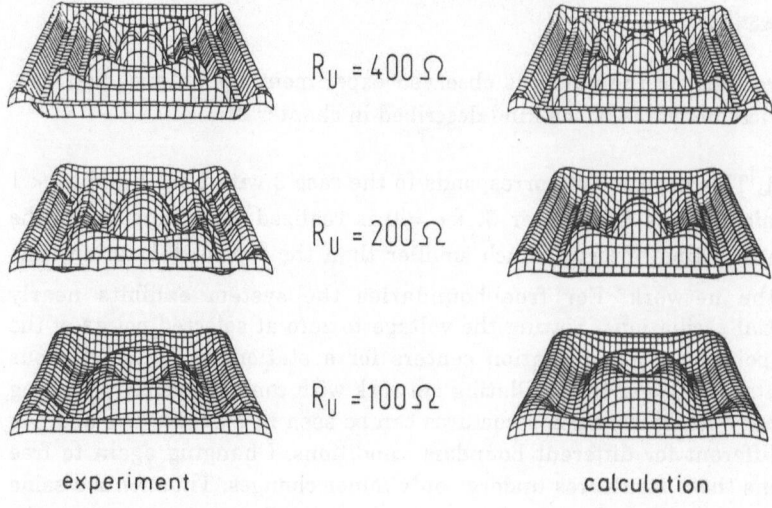

$R_U = 400\,\Omega$

$R_U = 200\,\Omega$

$R_U = 100\,\Omega$

experiment   calculation

Fig.6: Measured and calculated voltage patterns corresponding to experiment 1. The boundary points have been set to a voltage of 4.5 V. The calculations have been done by using (2-1) and measured values for S(I).

coupling resistances R_U. We note that the agreement between calculation and experiment is very good. Also we see that in nonnormalized coordinates the structures become smoother with decreasing R_U, i. e. increasing "voltage diffusion" constant. Interpreting the structures in terms of a characteristic wavelength we see that in Fig. 6 the wavelength scales with $R_U^{-1/2}$, which is also expected theoretically.

Experiment 2: In contrast to experiment 1 we increase κ_1 slowly with respect to $\sqrt{LC}$. This corresponds to an adiabatic change of κ_1, and is discussed physically as case 4 in chapter 3. The results are shown in Figs. 7a, b, c. The appearance of high current regions is somewhat arbitrary in space. Due to local activation and lateral inhibition every high current region is surrounded by a low current region. This structure is determined mainly by network inhomogeneities.

Experiment 3: In normalized coordinates this experiment is the same as experiment 1. In real space the coupling is smaller due to $R_U = 4\,k\Omega$, instead of $R_U = 330\,\Omega$ in experiment 1. The weakening of correlation due to the reduced coupling can be seen from the fact that, after setting $\kappa_1 = 0$ by switching on the appropriate value of U_S fast, there is poor phase correlation. Grounding the network at the point A of two different cells in Fig. 1a and switching off the ground afterwards we obtain a typical oscillatory structure, as shown in Fig. 8. The structure demonstrates a particularly nice example of self-organization where the correlation length is not long enough to correlate all cells. However, within certain regions we find cells with the same frequency with no phase shift. Different regions can have different frequencies and are separated by nonoscillating cells. This structure has not been obtained so far from theoretical calculations on the basis of (2-8) or (2-4).

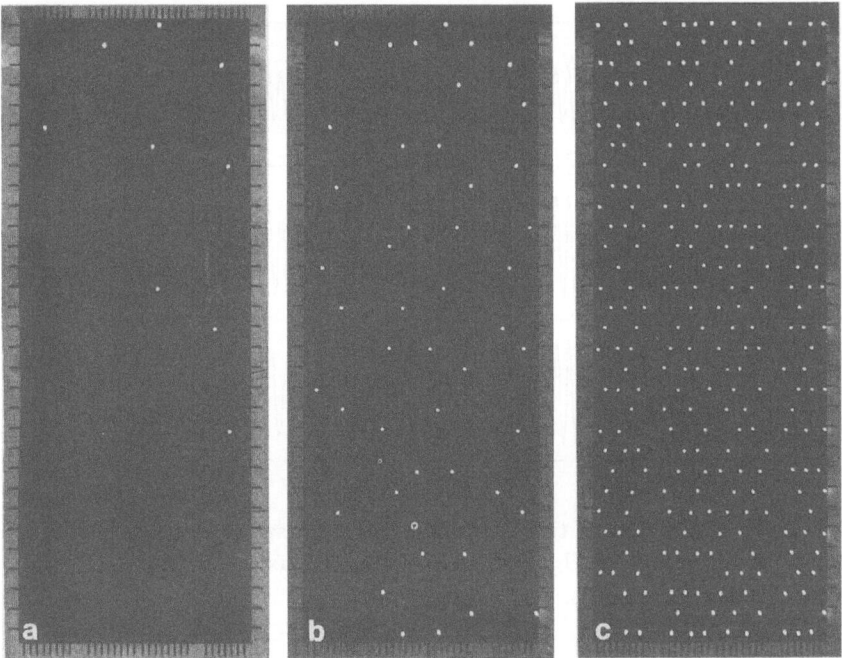

Fig. 7: Stable stationary current patterns corresponding to experiment 2 of Table 1. The light points are due to light emitting diodes indicating cells with high current flow (U_S = 12 V (a) ; 13 V (b); 25 V (c)).

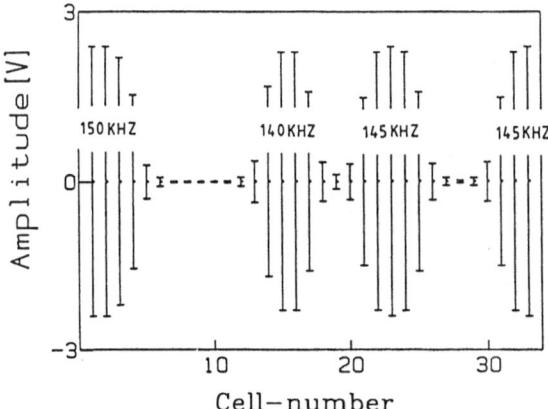

Fig. 8: Measured temporal spatial pattern corresponding to experiment 3 of Table 1.

Experiment 4: This is the same as experiment 3 except that $\kappa_1 = 0$ is replaced by $\kappa_1 = \hat{A} \sin (2\pi ft)$. For large amplitudes $\hat{A}$ the system exhibits forced oscillations as can be seen from Figs. 9a, b, c. Decreasing $\hat{A}$ there is competition between the external driving frequency f and the internal frequency ($\approx 1/\sqrt{LC}$) of the coupled system. As can be seen from Figs. 9d, e, f this leads to intermittancy type of chaotic behaviour.

149

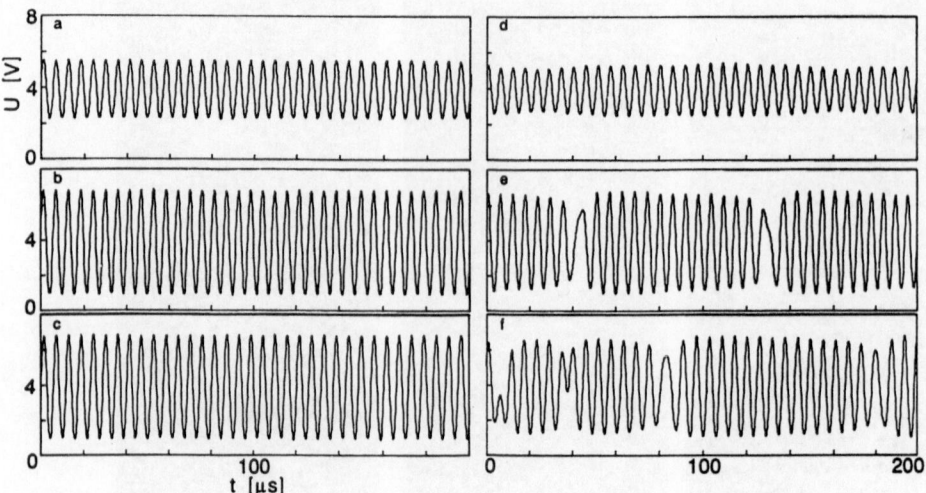

Fig. 9: Voltage as a function of time corresponding to experiment 4 of Table 1. (a, d): Driving voltage U_S. (b,e), (c,f): Voltage measured at point A of Fig. 1a of two different cells.

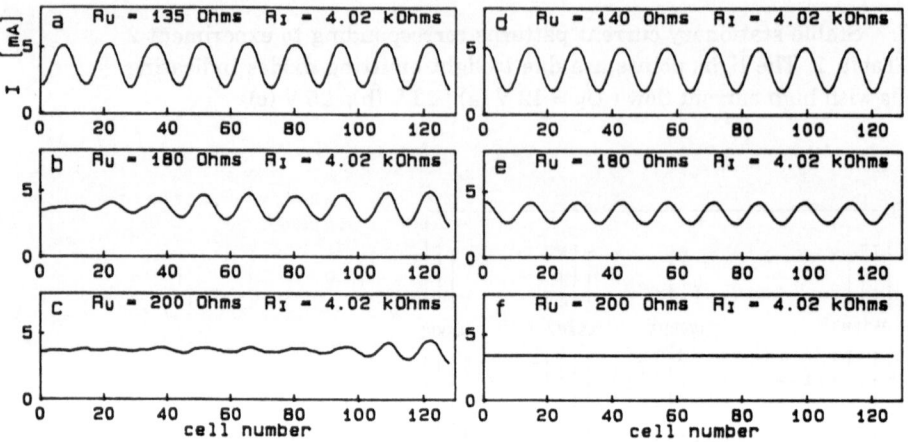

Fig. 10: Current as a function of cell number for variable voltage coupling R_U. (a, b, c): experiment. (d, e, f): Numerical calculation.

Due to the lack of phase correlation between different cells we may interpret this behaviour as a manifestation of spatial temporal chaos. Calculations of these structures are in progress.

Experiment 5: Here we demonstrate diffusion destabilization in the sense of Turing [26] as discussed in chapter 2 for the parameter set indicated in Table 1. The uncoupled system has (v, w) = (0,0) as stable solutions. By decreasing R_U e. a. by an increase of the inhibitor diffusion this state may be destabilized and eventually a sinusoidally modulated structure appears with a critical wave vector q_c and an amplitude $\hat{\mathfrak{i}}$. Precisely this is shown in Figs. 10c, b, a and 10f, e, d, experimentally and numerically. Bifurcation theory tells us that the spatial modulation in the inhomogeneous state has the form $\hat{\mathfrak{i}}\cos(q_c x)$. This result is confirmed by inspecting the experimental and numerical results of Fig. 10. Also we conclude from theory that $\hat{\mathfrak{i}} \sim \sqrt{\sigma}$. This is confirmed again in a rather large bifurcation parameter range by experiment and numerical calculation shown in Fig. 11.

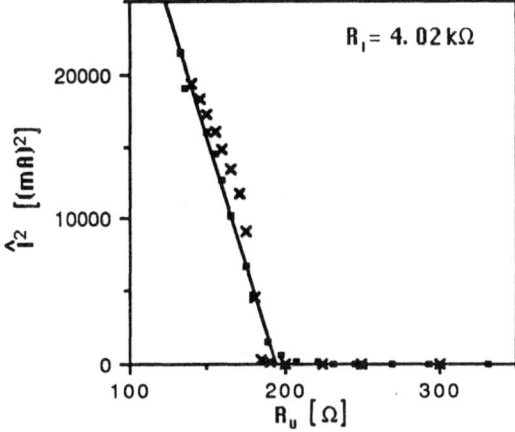

Fig. 11: Square of the amplitude of the sinusoidal modulation of the current of Fig. 10 (■ = experiment, x = calculation).

Experiment 6: In this case our system is operated in the limit of the Fitz-Hugh Nagumo model for nerve axon pulse transmission, also discussed in chapter 3 in connection with Case 5. In Fig. 12 we represent the pulse transmission along the one-dimensional chain for four approximately equidistant cells. We observe a well-defined solitary pulse initiated by the boundary voltage and travelling with constant velocity along the transmission line. Calculations of this observation are in progress.

Experiment 7: The last experiment is the only one where the integral term of (2-8) is non-zero [1,2]. Increasing the external voltage U_S gradually leads to a monotonic increase of κ_1. Initially we obtain a homogeneous state with increasing current. The first bifurcation leads to a filament as shown in Fig. 13(1). Increasing κ_1 further results in an increase in the number of filaments. The overall character of these filaments is of solitary kind and instead of increasing considerably in size the number of filaments in the system tends to increase. For large U_S the filaments start to interact as can be seen from e.g. Fig. 13(4). For even larger U_S the filaments become significantly wider. In this case solitary negative filaments are generated. The whole procedure can be reversed by decreasing κ_1, as can be seen from Figs. 13(6,7). Any generation of a filament is accompanied by a jump in the characteristic $I = I(U_V)$ of the total network, as can also be seen from Fig. 13. This characteristic has large

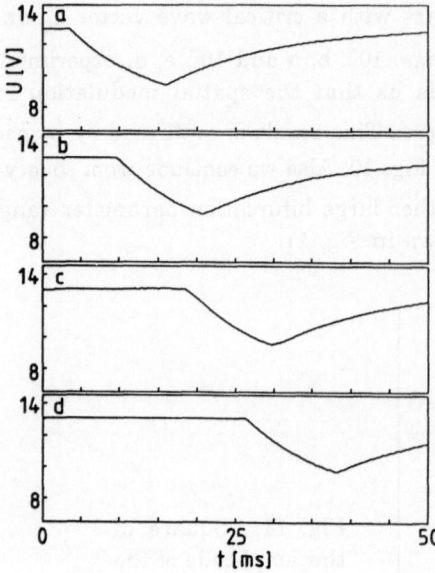

Fig.12: Voltage at point A of Fig. 1a measured as a function of time corresponding to experiment 6 of Table 1. Cell number: a: 1, b: 41, c: 85, d: 125.

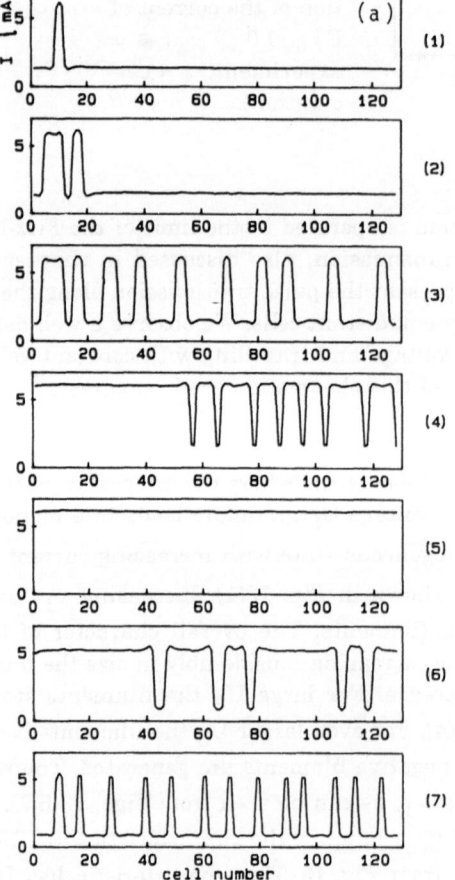

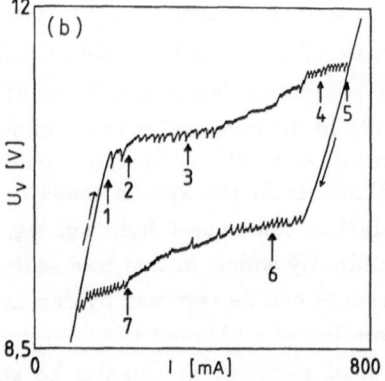

Fig. 13: Current distribution for various external voltages U_S and characteristic of the network corresponding to experiment 7. The numbers indentify the structures that correspond to selected points in the characteristic.

hysteresis. - At present there are no numerical results for solving (2-4) and (2-8) with the parameter of experiment 7. However, using (2-8) for a somewhat different set of parameters [19, 20, 22-24] and solving (2-8) numerically we observe all qualitative features of results described in the present experiment.

5. Discussion and Conclusions

In this paper we demonstrate that activator inhibitor reaction diffusion type electrical networks with two-component diffusion can be built. These networks show an impressive richness of spatial temporal and temporal-spatial structures. In many cases these structures can be described quantitatively in terms of (2-4) and (2-8). With respect to these cases the networks can be considered as rather fast analogue computers for solving (2-4) and to good approximation to (2-8). However, we also observe structures that are not understood theoretically so far. This is particularly true for the oscillatory regions of experiment 3 and the temporal spatial chaotic behaviour of experiment 4. With respect to such structures the investigated networks are very interesting physical objects.

Interpreting the electrical circuits as equivalent circuits for a continuous system we obtain a mathematical description of pattern formation in gas discharge [19, 20, 24] and semiconductor [20, 22, 23] systems. In this way it was, for the first time, possible to understand phenomenologically the principles of filament formation in semiconductor and planar gas discharge systems. Also we believe that memories and various other devices can be built on the basis of the principles outlined in the present article.

Networks of the kind described above are associative memories in the sense that the stable stationary structures are fixed points having a basin of attraction. If we set initial conditions which represent parts of a stable stationary pattern the dynamics of the system may pull the structure to the fixed points, thus completing the incomplete input pattern. If e. g., by modulating the diffusion coupling spatially, we implement a predetermined pattern to the network, we may be able to build up simple neural networks. Taking the integral term of (2-8) into account we also include long distance coupling, though in a very crude manner.

Acknowledgement

We are very grateful to J. Berkemeier, T. Dirksmeyer, M. Bode, R. Schmeling, N. Goerke, H. Janßen and H.-G. Kranz for help in preparing the demonstration experiments. We are also indepted to J. Kondakis and F.-P. Kaup for the careful preparation of the final figures.

6. Literature

1. R. Schmeling: Diplomarbeit, University of Münster (1988)
2. T. Dirksmeyer, R. Schmeling, J. Berkemeier and H.-G. Purwins: to be published
3. J. Smoller: Shock Waves and Reaction-Diffusion Equations (Springer, New York 1983)
4. G. Nicolis and I. Prigogine: Self-Organization in Non-equilibrium Systems (Wiley, New York 1977)
5. P. Glansdorff, I. Prigogine: Thermodynamic Theory of Structure, Stability and Fluctuations (Wiley, New York 1971)
6. H. Haken: Synergetik 2. Aufl. (Springer, Berlin 1983)
7. H. Haken: Advanced Synergetics (Springer, Berlin 1983)
8. R. Field, M. Burger (eds.): Oscillations and Traveling Waves in Chemical Systems (Wiley, New York 1985)
9. P. Fife: Mathematical Aspects of Reaction and Diffusion Systems, Springer Lecture Notes in Biomath., 28 (1979)
10. J. Murray: Lectures on Nonlinear Differential Equation Models in Biology (Clarendon Press, Oxford 1977)
11. A. Gierer: Prog. Biophys. Molec. Biol., 37, 1 (1981)
12. H. Meinhardt: Models of Biological Pattern Formation (Academic Press, London 1982)
13. K. Maginu: Math. Biology 7, 375 (1979)
14. F. Rothe: J. Math. Biology 7, 375 (1979)
15. F. Rothe: Nonlinear Analysis, Theory, Methods and Applications 5, 487 (1981)
16. R. Fitz-Hugh: Biophysical J. 1, 445 (1961)
17. J. Nagumo, S. Arimoto, S. Yoshizawa: Proc. IRI, 2061 (1962)
18. J. Berkemeier, T. Dirksmeyer, G. Klempt, H.-G. Purwins: Z. Phys. B. - Condensed Matter 65, 255 (1986)
19. H.-G. Purwins, G. Klempt, J. Berkemeier: Festkörperprobleme - Advances in Solid State Physics 27, 27 (1987)
20. H.-G. Purwins, C. Radehaus, J. Berkemeier: Z. Naturforsch. 43a, 17 (1988)
21. T. Dirksmeyer, M. Bode, H.-G. Purwins: to be published
22. C. Radehaus, K. Kardell, H. Baumann, D. Jäger, H.-G. Purwins: Z. Phys. B - Condensed Matter 65, 515 (1987)
23. H. Baumann, R. Symanczyk, C. Radehaus, H.-G. Purwins, D. Jäger: Phys. Lett. A 123, 421 (1987)
24. C. Radehaus, T. Dirksmeyer, H. Willebrand, H.-G. Purwins: Phys. Lett. A 125, 92 (1987)
25. K. Kardell, C. Radehaus, R. Dohmen, H.-G. Purwins, submitted for publication in Journal of Appl. Phys.
26. A. M. Turing: Phil. Trans. Roy. Soc. B 237, 37 (1952)

Optimization by Diploid Search Strategies

W. Banzhaf

Institut für Theoretische Physik und Synergetik der Universität Stuttgart,
Pfaffenwaldring 57/IV, D-7000 Stuttgart 80, Fed. Rep. of Germany

ABSTRACT

*We present some results on optimization of cost functions using
a population of parallel processors. Based on a simulated evolutionary
search strategy diploid recombination is introduced as a means for
maintaining variability in computational problems with large numbers
of local extrema. The new strategy is compared to some traditional
algorithms simulating evolution.*

1. Introduction

The problems we want to address here are those of optimization of certain complicated
cost functions. These cost functions are scalar functions defined over the M-dimensional
space of given variables which resemble a landscape with many hills and valleys. One is
interested in locating the one minimum (maximum) which is "optimal" or global in the
sense that no other local minimum is smaller than this. The number of possibilities and
thus the number of local minima determines the expenditure which has to be applied to
get a reasonable solution.

Often this number explodes in a combinatorial fashion and a nearly optimal solution is
the best that can be reached in finite time. The "travelling salesman problem" [1,2] is a
famous representative of this class of problems.

Evolution tells us a lot about how certain features may be optimized in a given environment. In fact, since the early sixties computer scientists, biologists and engineers are
working on different aspects of evolutionary optimization [3-10]. The present work is
intended to overcome some of the difficulties one has encountered in the last twenty-five
years, especially with respect to parameter tuning.

In evolutionary strategies, the mutation rate is subject to changes similar to in
simulated annealing techniques [11], where a cooling schedule for the temperature of the
process is necessary. At the early stages of an optimization process, it is advantageous
to have a large mutation rate. During the convergence, however, a large mutation rate
more and more turns into a disadvantage. Thus, a continuous rate adaption is commendable
which, in general, requires prior knowledge. Diploid search strategies get along without
any rate adaption due to their ability to "store" variance.

Before going into more detail of diploid search strategies which are part of the
evolutionary search strategies, let us consider some natural adaptive dynamical systems.
Table 1 shows three of them.

All of these systems are equipped with at least two internal time-scales. There are
fast relaxing variables describing the state of the system which evolve according to
dynamical laws from their initial states and - on the other hand - slowly changing
variables maintaining an overall control of the particular direction of evolution of the
fast variables [12,13]. Loosely speaking, adaptive systems are able to form internal
landscapes in their state spaces which may be interpreted as a reflection of the external
landscapes the systems are acting on.

Springer Series in Synergetics Vol. 42: **Neural and Synergetic Computers**
Editor: H. Haken   ©Springer-Verlag Berlin Heidelberg 1988

Table 1: Some adaptive dynamical systems in nature

Sort	Evolution	Immune System	Nervous System
Time scale	$\sim 10^7$ y	$\sim 10^2$ y	$\sim 10^0$ y
Internal separation $\Big\}$	$\Big\{$ genotype phenotype	$\Big\{$ antibody features antibody concentrations	$\Big\{$ synapses neurons

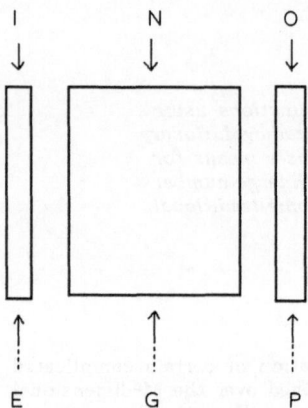

Fig. 1:
Analogy between supervised learning
and evolution: I=Input, N=Network,
O=Output, E=Environment, G=Genotype
P=Phenotype

To further illustrate the similarity between these systems let us give a more concrete analogy. Figure 1 shows a neural net performing some operations on input vectors which are transformed into output vectors according to the internal state of the net.

A supervised learning algorithm would consist of a comparison of the output vectors computed from N with the desired ones. Feeding back this information into the net will eventually change its state (the connections between neurons). Repeating this procedure many times with all the I/O-pairs that are given, N evolves into a certain state reflecting the given environment.

Re-interpreting the same events in evolutionary terms would read: A trial genotype G converts inflow of information/energy/matter into features of its phenotype. During its lifetime the phenotype is evalutated by the environment and this information is fed back via the gene pool so that a new genotype occurs in the next trial.

At this point we want to emphasize the similarity between a physicist's energy function, an error function in neural network learning, an inverse fitness function of the biologist and a cost function in optimization problems. All of these functions are bounded so that at least one minimum (maximum) must exist.

The philosophy we adopt in this paper is that we do not intend to establish a new model of evolution but we are interested in using the algorithmic structure of evolution known so far for optimization purposes.

2. Some remarks on evolution

Let us give a few basic ingredients of the strategy nature uses to optimize in her domain:

i) Nature works with *populations* of individuals. Adaption or optimization is performed on the total population, not on the individual. Consequently, the average feature is important, not the individual one. This means on the other hand that a parallel search is taking place in the space of possible feature combinations, cf. Fig. 2. If there is some way to communicate the state of individuals the probability to get stuck in a local minimum is very small. Another individual may provide the information necessary to escape from such a region.

As already mentioned in the introduction, adaption takes place by separating time scales of the individual phenotype τ_P and genotype τ_G :

$$\tau_P \ll \tau_G$$

ii) *Selection* between individuals assures continuous progress of a population with respect to its adaption to the environment due to changes in its gene pool.

Indeed, selection works similar to a gradient in which improvements are accepted and deteriorations rejected. It is not a steepest descent, but may be termed at least a descent method. As a consequence, one often encounters exponential decrease in progress due to big steps at the beginning and smaller steps towards the end of the process.

iii) *Recombination* is an important source of variability in a population. In fact, it is a message exchange or communication between individuals by means of genetic interaction. Figure 3 shows this interaction for the case of haploid and diploid recombination.

In the case of diploid recombination some criteria of how to determine the dominant and recessive features are required. These may be given by physical laws or (in our simulation) by random events.

iv) Another source of variation in a population is provided by *mutations*. Their effect is a diffusion-like motion allowing a drift of the individuals in feature space.

The mutational effects which are studied in more detail in the following are restricted to point mutations on a gene-string. This means that a certain feature is slightly changed at random. The other possible sorts of mutation such as those in the number or location of genes or even chromosome mutations are completely disregarded here.

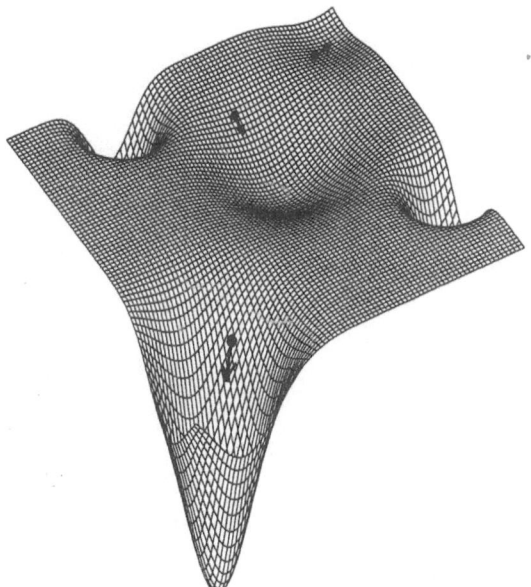

Fig. 2:
Parallel search of 3
different individuals
in a 2-dimensional
landscape

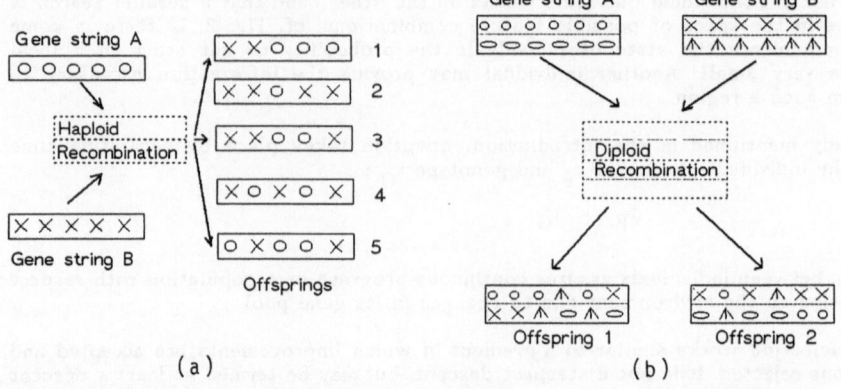

Fig. 3: Haploid (a) and diploid (b) recombination as information exchange

v) *Genes*, the "units of heredity", act together to generate the phenotype. Sometimes one gene controls more than one feature of the individual phenotype. In other cases different genes interact to generate a certain feature. Thus a complicated mapping from genotype to phenotype is performed by the molecular mechanisms at work.

3. A brief outline of the formalization

Let us briefly outline the possible formalization of the optimization task.
We introduce the genotype $\vec{G}$ as a string or N-dimensional vector

$$\vec{G} = \{G_1, \ldots, G_N\},$$

where the components G_i, $i = 1, \ldots, N$ are taken from a predefined symbol set

$$G_i \in \{\Gamma_j\}, \ j = 1, \ldots, \varkappa.$$

Accordingly, we introduce the phenotype $\vec{P}$:

$$\vec{P} = \{P_1, \ldots, P_M\},$$

where P_i is taken from the symbol set

$$P_i \in \{\Pi_j\}, \ j = 1, \ldots, \lambda$$

and constitutes another vector or string of dimension M.
The environment will be modeled similarly by the L–dimensional vector $\vec{E}$:

$$\vec{E} = \{E_1, \ldots, E_L\}.$$

To be consistent, we introduce the symbol set

$$E_i \in \{E_j\}, \ j = 1, \ldots, \mu.$$

We are free to fix the symbol sets mentioned above. We have chosen the real numbers in all cases, in contrast to other work [9,10] interested in modeling the evolutionary processes.

Now a mapping $\vec{f}$ has to be introduced to provide us with a relation between genotype and phenotype:

$$\vec{P} = \vec{f}(\vec{G}, \vec{E}). \tag{1}$$

Only the phenotype $\vec{P}$ is subject to an evaluation which is done by applying a scalar quality function Q on $\vec{P}$:

$$Q\,(\,\vec{P}\,) \;=\; Q\,(\,\vec{f}(\vec{G},\vec{E})\,)\;\leq\;a\;,\qquad\qquad(2)$$

which is bounded by a constant a.

The optimization problem may now be formulated in the following way: Find a vector $\vec{G}_{opt}\,\epsilon\,\Gamma^N$ such that after performing the mapping $\vec{f}$ the corresponding phenotype $\vec{P}_{opt}\,\epsilon\,\Pi^M$ has maximal quality:

$$Q\,(\,\vec{P}\,)\;=\;\text{Max\,!}\quad\text{for}\;\vec{P}=\vec{P}_{opt}\;.\qquad\qquad(3)$$

The optimization is done in an M-dimensional space!

In our particular application following in the next paragraph, we put the inverse question and look for the global minimum of a prescribed cost function.

For observational purposes we need some classification of the appearing phenotypes. Thus we classify the phenotypes into k different classes according to Table 2.

Table 2: Classification of diverse patterns into k different classes

Exemplars	Class
p, p', p" , ...	1
p̰, p̰̈, p̰̈ , ...	2
⋮	⋮
....	k

Then we are allowed to introduce "occupation numbers" $N_1, \dots, N_k$, $N_i\,\epsilon\,\mathbb{N}$ or "concentrations" $X_1, \dots, X_k$, $0\leq X_i\leq 1$ describing the state of the system. Sometimes constraints of the form

$$\sum_i N_i = N = \text{const.}\quad\text{or}\quad\sum_i X_i = X$$

are fulfilled.

The evolutionary processes cause a time dependence of the N_i's

$$N_1\,(t),\,\dots,\,N_k\,(t)$$

which consist of elementary processes of the kind

$$N_i \rightarrow N_i \pm 1\;.$$

Thus, a transition probability

$$W\,(\,N_1,\,\dots,\,N_j,\,\dots,\,N_i\pm 1,\,\dots,\,N_1,\,\dots,\,N_k\,) = A_i N_i + \sum A_{ij} N_j +$$

$$+\;\sum B_{ij}\,N_i\,N_j\;+\;\sum C_{ijl}\,N_i\,N_j\,N_l\;+\;\dots$$

159

may be specified and a master equation for the probability density $p(\vec{N},t)$ can be derived. This kind of dynamics has some similarities to the pattern recognition algorithm given in [14].

4 . The algorithms

We want to begin our description by introducing suitable notations for our variables. We use N processors in parallel, each searching in phenotype space for the global optimum. Their "programs" are specified by genes which are numbers parametrizing the algorithms to be performed by the processors.

The strategies we are calling diploid consist of string—like genotypes where Z is the

$$x_{i,a}^{\beta}(t) \quad , \quad \beta=1,...,N \quad , \quad i=1,...,Z \quad , \quad a=1,...,M$$

number of gene loci, N the number of individual processors and M the number of different strings present :
M = 1 haploid recombination , M = 2 diploid recombination and M ≥ 2 polyploid recombination.

To make use of the intrinsic parallel structure of evolution-like algorithms we had to introduce N individual processors equipped with different $\vec{x}_a^{\beta}$. Every genotype contains M strings each coming from one of its "parents". The numbers in $x_{i,a}^{\beta}$ then indirectly specify features of the individual β in component i. By applying the mapping function $\vec{f}$ which is in our case a simple projection operation to the dominant features, a phenotype $\vec{p}^{\beta}$ is generated:

$$\vec{p}^{\beta} = \vec{f}(\vec{x}_a^{\beta}) \quad . \tag{4}$$

Based on a quality function $Q(\vec{p})$, every processor gets its judgement

$$Q^{\beta} = Q(\vec{p}^{\beta}) \tag{5}$$

where an ordered set

$$\vec{Q}_p = \{ Q^{\beta}1, Q^{\beta}2, Q^{\beta}3 .... \} \tag{6}$$

with

$$Q^{\beta}1 \geq Q^{\beta}2 \geq .... Q^{\beta}N$$

of the qualities can be formed .

Now let us come to the dynamic part of the story. How are offsprings generated? Offsprings emerge from the M (from now on M equals 2 corresponding to diploid recombination) parent processors by recombination and mutation. For every component of the gene—string, a feature carried by one of the parent processors is chosen at random. No reference is made to the dominant or recessive character of this feature. Thus a double string of features for each offspring is generated. Then, another random choice is made to determine the dominant and recessive part of the offspring's gene—string. Finally, mutational effects $\Delta x_{i,a}^{\beta}$ of the following sort are added to the resulting string:

$$\Delta x_{i,a} = \sigma_{Ni} \exp\left[\sigma_N \sigma / \sqrt{10}\right] \qquad a=1,2 \tag{7}$$

where $\left.\begin{array}{c} \sigma_{Ni} \\ \tilde{\sigma}_N \end{array}\right\} = N(1,0)$ is normally distributed, σ is fixed.

As one can see, the mutation rate follows a log-normal distribution [8]. There is no updating of the mutation rate during the convergence process, in contrast to many other evolutionary search strategies. This technique is often applied due to needs for maintaining a good performance of the algorithm during the whole search process. Diploid strategies have a major advantage over haploid and even simpler ones: No parameter tuning is necessary during the process, an action which often would require considerable knowledge about the problem space.

Note that recombination of recessive genes gives enormous variability in the features of phenotypes and is therefore a really important source of variation in a population.

The descending genotypes which emerge in a fixed number D are then mapped onto the phenotype space resulting in D phenotypic offsprings

$$\vec{p}^{\gamma} \quad , \gamma = 1, \dots , D$$

each of which will have its quality $Q(p^{\gamma})$.
Now, the ordered set

$$Q_{pd} = \{ Q^{\gamma_1} , Q^{\beta_1} , Q^{\gamma_2} \dots \} \tag{8}$$

$$Q^{\gamma_1} \geq Q^{\beta_1} \geq Q^{\gamma_2} \dots$$

can be taken as a basis to form a new generation of genotypes. This is done by selecting the N best individuals, as given by the order in the set (8).

It remains to specify a last point: the starting conditions. These are given by gene strings chosen at random for all of the individuals, reflecting the fact that nothing at all is known about the landscape at the beginning of the optimization process.

The quality function we use constitutes the problem we want to solve with the algorithm. It will be described in the next paragraph.

5. Simulation results

We present simulation results concerning a particular cost function shown in Fig. 4. The aim of this choice is to show that diploid strategies are particularly well suited for problems with many local extrema. We use a 10-dimensional polynomial cost function which reads

$$Q(\vec{p}) = \sum_{i=1}^{10} a_i \, p_i^4 + \sum_{i=1}^{10} b_i \, p_i^2 + \sum_{i=1}^{10} c_i \, p_i \, . \tag{9}$$

A special choice of coefficients a_i , b_i , c_i is

$$a_i = 1 \qquad b_i = -4 \qquad c_i = 1.5 \quad \forall \, i \, , \tag{10}$$

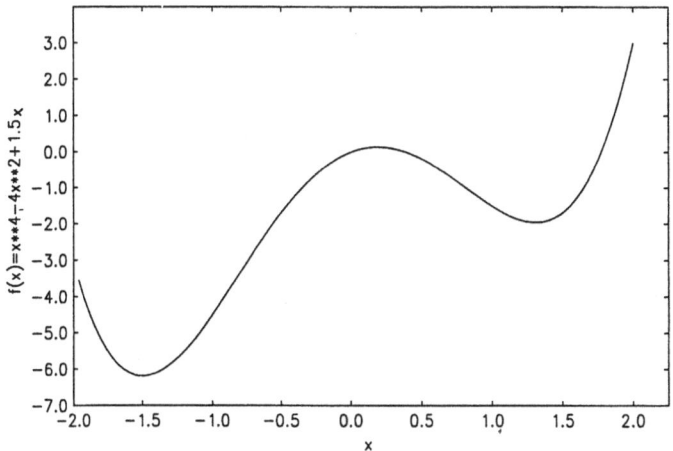

Fig. 4: Cost function (9) in one dimension. It possesses two local minima at $x_1 = - 3/2$, $x_2 = (3 + \sqrt{5})/4$ and one local maximum at $x_3 = (3 - \sqrt{5})/4$

which will guarantee the presence of 2 local minima in every dimension or 2^{10} minima in 10 dimensions. For the sake of simplicity, all coefficients are set equal. The 1024 local minima found here obey a binomial distribution if they are classified according to their quality. Of course, this cost function is rather arbitrary and one also should examine the behavior of the algorithm using other cost functions as in [15].

First of all, we present a comparison of tests of the behavior of different strategies. We tested strategies such as gradient search, haploid and diploid search and applied each strategy to 100 random starting states. In Fig. 5 we sketch the appearance frequency of the retrieved minima states after convergence over their quality. Although no parameter tuning was done in the diploid strategies they turned out to be very successful in locating the global minimum.

In the haploid cases, on the contrary, a tuning of the mutation rates was executed in every generation according to the following rule:
A characteristic individual mutation rate σ^β was introduced in (7)

$$\Delta x_i^\beta = \sigma_{Ni} \ \exp \left[\tilde{\sigma}_N \sigma / \sqrt{10} \right] \sigma^\beta(t) \qquad (11)$$

$$\sigma^\beta(t+1) = \sigma^\beta(t) \exp \left[\tilde{\sigma}_N \sigma / \sqrt{10} \right] \qquad (12)$$

starting with a fixed $\sigma^\beta(0)$.

The effect of different σ on the convergence velocity were studied elsewhere [16]. It turned out that diploid strategies are not sensible to variations in σ.
The result in gradient search just reflects the binomial distribution of minima since the method immeadiately gets stuck in a local minimum.

A few more words about the diploid strategy III:
Figure 6 is a record of the decrease in the cost function (9) of two exemplars. Abrupt transitions are visible. This is due to recombinations as the major source of improvements.

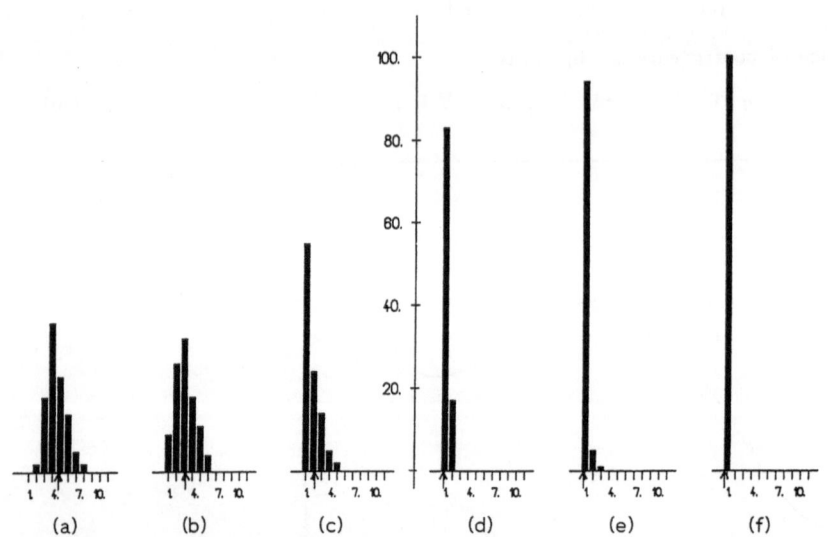

Fig. 5: Comparison of different search strategies. 100 minimum search trials are executed in any strategy. (a) Gradient search.(b) Haploid strategy: N = 1, M = 5. (c) Haploid strategy : N = 5, M = 10.(d) Diploid strategy: N = 20, M = 20. (e) Diploid strategy: N =20, M =40. (f) Diploid strategy: N =40, M = 20. Arrows indicate the average value.

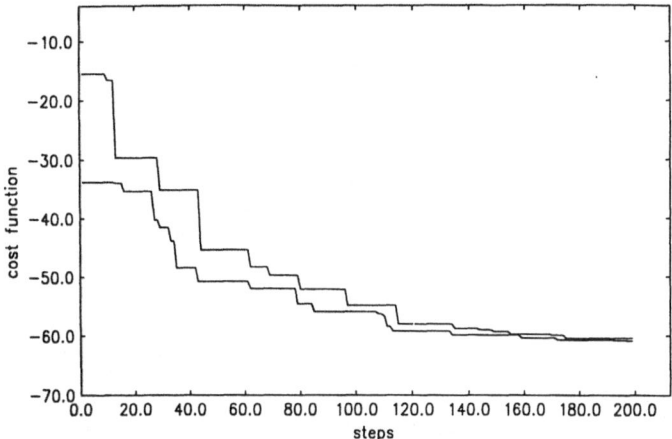

Fig. 6:
Decrease of the
cost function for
two processors
in strategy (f).
One processor
is a sample, the
other is the best
exemplar.

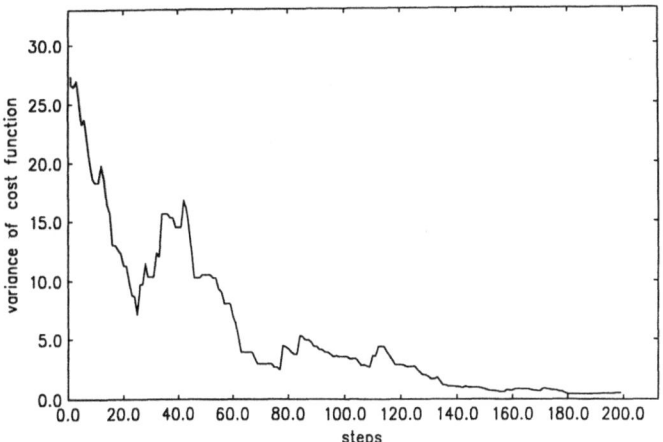

Fig. 7:
Variance of the
cost function
over all processors
in strategy (f)

Different individuals have distinct instants at which progress is made. On the other hand, they follow each other due to the fast radiation of improvements. Of course, during decrease of the cost function, not always the same processor will possess the most interesting state being the best exemplar.

Then the question arises, how can one get a reliable answer at all? Another signal is used to provide us with information about the overall state of the system: the state variance σ_s. As one can see even from Fig. 6, σ_s must decrease while the system is approaching the true minimum. Communication between individual processors via recombination allows progress of the population as a whole.
Sooner or later all of the processors will end up in the global minimum. This should be signalled very clearly by a dropping down of the processors' state variance of the processors.

Figure 7 shows the variance in cost function for the different processors (same run as in Fig. 6). Considerable fluctuations in the beginning are followed by a gradual decrease after about the 120th generation. From this point on, all processors are in the basin of the global minimum slowly approaching it from different directions. The "eruptions" of variance in the first 100 generations have their origin in successful recombinations . If one keeps track only of the best individual in every generation and observes the variance one can see a sharp increase in variance immediately after the best exemplar makes a leap followed by a falling off due to radiation after this event had occurred.

163

Summarizing, a sharp ascent of variance signals an enormous progress of single individuals, whereas a collapse in variance signals the end of development and an equally partitioned quality of solutions. The result may be measured at any processor since the final discrepancies between processors are very small.

Besides the variance of the cost function, other observables characterizing the population as a whole exist. "Allel" concentrations which may be identified with the $X_i(t)$ of section 4 can be measured without difficulty. The classification criterium used here is the basin of attraction in which each component is located. It is obvious that the deeper valley is termed "better". Figure 8 displays the concentration of the better allel at one of the 10 gene loci for the dominant and recessive part over the above run. The concentration beginning little below 0.5 (due to random initial conditions) is developing with smaller fluctuations toward the value 1.0 at ~130 generations. Compare the recessive locus where large fluctuations can be observed. Note the difference in generation scale! It is a random drift due to the fact that the recessive part does not underlie any selection pressure.

The only natural time scale we have at hand is the number of generations succeeding one after the other. A generation is defined as an ensemble of individuals, subject to an evaluation of its quality. This definition is also intended to include asynchronous updating. Asynchrony can be realized by judging and selecting after some descendants have been

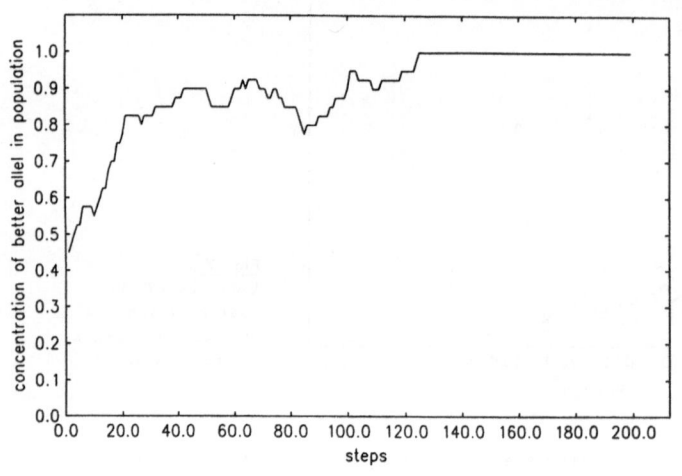

Fig. 8a:
Concentration of better allel in the population of strategy (f). Dominant part

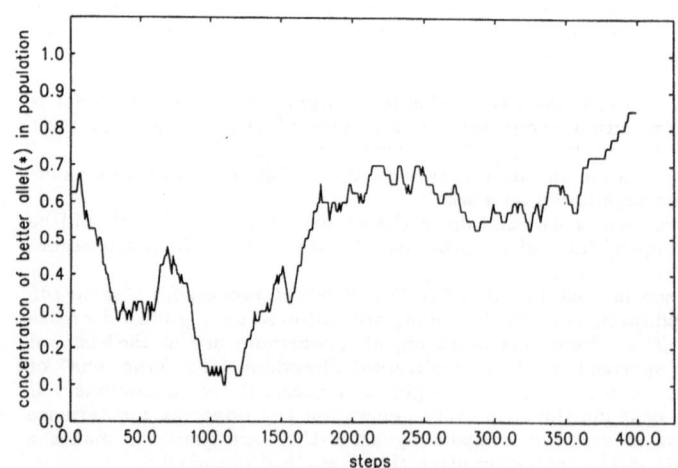

Fig. 8b:
Concentration of better allel in the population of strategy (f). Recessive part

generated. Then, selection should take place under all parents and descendants present at that time.

We emphasize that - at least in the present stage - time consumption considerations of the algorithms need not play an important role since we are concerned with the question of retrieval security of global minima, which in general is seen as a greater problem than acceleration of convergence.

6. Perspectives

Let us give an illustrationary example of what may be a useful phenotypic representation of a problem. The example we want to mention is taken from number theory [17].

Consider the representation of the irrational number π in terms of certain combinations of natural numbers. Number theory tells us that a representation called continued fraction [18] exists. Through an evolutionary algorithm one may wish to find continued fractions (CF) for π other than the well known regular one:

$$\pi = 3 + \cfrac{1}{7 + \cfrac{1}{15 + \cfrac{1}{1 + \cfrac{1}{292 + \cfrac{1}{\dots}}}}} \quad \cdots \quad \tag{13}$$

which is written shortly as

$$\pi = [3; 7, 15, 1, 292, \dots] . \tag{14a}$$

A quite natural choice for the (haploid) gene-string consists of the natural numbers $x_i \in \mathbb{N}$ where (14) reads

$$\vec{x} = (3, 7, 15, 1, 292, \dots) . \tag{14b}$$

Incidentally, a phenotype is introduced here also in the haploid case. The corresponding mapping from genotype to phenotype is the operation:
Interpret the gene - string (14b) as a continued fraction and form it:

$$p = CF (\vec{x}) \tag{15}$$

where now $p \in \mathbb{R}$.

The quality measure determining the selection criteria may be defined as

$$Q = (p - \pi)^2 . \tag{16}$$

Down to $Q \leq 10^{-4}$ many exemplars were found in 10-dimensional gene-strings. Three are given here:

$$\vec{x}_1 = (3, 7, 16, 232, 138, 15, 2, 187, 6, 6)$$

$$\vec{x}_2 = (3, 7, 16, 51, 0, 285, 296, 181, 98, 5)$$

$$\vec{x}_3 = (3, 7, 15, 1, 144, 281, 70, 237, 4, 222) .$$

During the process, p oscillates around π with smaller and smaller amplitudes. This contrasts with the conventional CF approach, where a kind of gradient strategy is followed by successively taking into account more and more terms in (13).

In the same way a non-regular continued fraction [17,18] for π may be found in every order of the quality measure. One of the exemplars found in our simulations is given below. It approximates π down to deviations of order $Q \leq 10^{-8}$

$$p = 3 + \left[\frac{22}{155 +} \ \frac{95}{253 +} \ \frac{1}{6 +} \ \frac{5}{1 +} \ \frac{8}{177 +} \ \cdots \right]$$

and was generated by a diploid strategy.

This example has roughly indicated the "holistic" character of a phenotype. It is a unity in our particular case generated by repeating the same operation with different operands. We suppose that the essential of a phenotype is just this emergence from a recursive application of certain "instruments" specified by the gene sequence.

In conclusion, a model for optimization was presented which had certain similarities to natural evolution. The basic entity was a population of processors which was adapted to solve the given optimization problem by means of diploid recombination and mutation. After generating descendants, selection took place under the guidance of the predefined function. It was shown that the security of discovering the global minimum of a relatively complicated cost function was greatly enhanced by introducing diploidy . We noticed that a distinction between genotype and phenotype of an individual was inevitable and could be seen as a separation of time scales in the evolutionary optimization process.

Besides exploring the behavior of diploid algorithms in different optimization tasks, a consistent formalization should be set up as a next step. In observing nature, however, other secrets of evolution may be uncovered in the future.

Acknowledgement

We thank the Volkswagenwerk Foundation, Hannover, for financial support.

References

1. M.R. Garey, D.S. Johnson: Computers and Intractability (W.H.Freeman, New York 1979)

2. E. Lawler: Combinatorial Optimization, Networks and Matroids (Rinehart and Winston, New York 1976)

3. H.J. Bremermann, M. Rogson, S. Salaff: In Biophysics and Cybernetic Systems, ed. by M. Maxfield, A. Callahan, L.J. Fogel (Spartan Books, Washington 1965)

4. D. Cohen: In Biogenesis, Evolution, Homeostasis, ed. by A. Locker (Springer Verlag, Berlin 1973)

5. I. Rechenberg: Evolutionsstrategie, (Holtzmann Froboog, Stuttgart 1973)

6. J.H. Holland: Adaption in natural and artificial Systems, (University of Michigan Press, Ann Arbor 1975)

7. V .Ebeling, R.Feistel: Ann. Phys. (Leipzig), 34, 81 (1977)

8. H.P. Schwefel: Numerical Optimization of Computer Models, (Wiley, New York 1981)

9. M. Eigen: Ber. Bunsenges. Phys. Chem. 89, 658 (1985)

10. P. Schuster: Chem. Scr. 26B, 27 (1986)

11. S. Kirkpatrick, C.D. Gelatt, M.P. Vecchi: Science 220 , 671 (1983)

12. M. Conrad: Adaptability (Plenum Press, New York 1983)

13. see: Evolution, Games and Learning- Models for Adaption in Machines and Nature, ed. by D. Farmer, A. Lapedes, N. Packard, B. Wendroff , Physica D 22, 1986

14. H. Haken: In Computational Systems – Natural and Artificial, ed. by H. Haken (Springer, Berlin, New York 1987)

15. D.H. Ackley: A Connectionist Machine for Genetic Hillclimbing (Kluwer Academic Publishers 1987)

16. W.Banzhaf: BioSystems, 1988, in press

17. M.R. Schroeder: <u>Number Theory in Science and Communication</u>, 2nd. ed. (Springer, Berlin, New York 1986)

18. O. Perron: <u>Die Lehre von den Kettenbrüchen</u>, 3rd. ed. (Teubner, Stuttgart 1977)

Part III

Perception and Motor Control

Movement Detectors of the Correlation Type Provide Sufficient Information for Local Computation of the 2-D Velocity Field

W. Reichardt[1], *R.W. Schlögl*[2], *and M. Egelhaaf*[1]

[1]Max-Planck-Institut für Biologische Kybernetik,
 Spemannstrasse 38, D-7400 Tübingen, Fed. Rep. of Germany
[2]Max-Planck-Institut für Biophysik,
 Kennedyallee 70, D-6000 Frankfurt/Main, Fed. Rep. of Germany

The projection of the velocity vectors of objects moving in three-dimensional space on the image plane of an eye or a camera can be described in terms of a vector field. This so-called *2-D velocity field* is time-dependent and assigns the direction and magnitude of a velocity vector to each point in the image plane. The 2-D velocity field, however, is a purely geometrical concept and does not directly represent the input site of a visual information processing system. The only information available to a visual system is given by the time-dependent brightness values as sensed in the image plane by photoreceptors or their technical equivalents. From spatio-temporal coherences in these changing brightness patterns motion information is computed. This poses the question about whether the spatio-temporal brightness distributions contain sufficient information to calculate the correct 2-D velocity field. Here we show that the 2-D velocity field generated by motion parallel to the image plane can be computed by purely local mechanisms.

In the literature on both biological motion processing and computer vision there is often stated that the 2-D velocity field cannot be computed by any local mechanism [1-6]. This conclusion is mainly based on approaches that implicitly regard a moving contour as nothing but a series of straight line segments which are each seen through a small aperture by some local motion analysing mechanism (see Fig.1). Information on the local curvature of the contour is, thus, not taken into account. From these mathematical approximations it is then concluded that all that a local mechanism can do is to determine the component of the local velocity vector perpendicular to the contour line, i.e. in the direction of the brightness gradient. If $F(x,y,t)$ =

Springer Series in Synergetics Vol. 42: **Neural and Synergetic Computers**
Editor: H. Haken   ©Springer-Verlag Berlin Heidelberg 1988

F[x+s(x,y,t); y+r(x,y,t)] represents the brightness of the moving pattern as a function of the spatial location x,y and time t, where s(x,y,t) and r(x,y,t) denote the time-dependent displacement of the pattern in the x- and y-direction, respectively, the mapping of the 2-D velocity vectors $v(x,y,t) = [ds(x,y,t)/dt; dr(x,y,t)/dt]$ on their components along the brightness gradient vectors $v^{\perp}(x,y,t)$ can be represented by the transformation

$$v^{\perp} = 1/(F_x{}^2 + F_y{}^2) \cdot \begin{bmatrix} F_x{}^2 & F_x F_y \\ F_y F_x & F_y{}^2 \end{bmatrix} \cdot v \ . \tag{1}$$

(The subscripts denote the partial derivatives of F with respect to x or y). Since this transformation is not one-to-one, an infinite number of velocity vectors is mapped onto the same $v^{\perp}$. This ambiguity is commonly refered to as the *aperture problem* [1-5]. Using this type of representation of motion information, the correct 2-D velocity field, therefore, cannot be measured locally. Instead, the correct 2-D velocity of moving objects or pattern segments can only be computed in a further stage of analysis by combining the motion measurements from different locations and taking some global constraints into account [3-7].

These conclusions, however, should not be generalized, as is often done in the literature [2-5], to motion detection schemes which yield different representations of motion information. In particular, the so-called aperture problem is just a by-product of a specific mathematical approximation of the spatio-temporal geometry of a moving contour. In contrast we will show in the following that the correct 2-D velocity field can, in principle, be calculated by purely local mechanisms without reference to additional global constraints.

Our approach to these problems differs from the aforementioned ones. It is based on a movement detection scheme, the so-called *correlation-type of movement detector*, which has originally been derived from experiments on motion vision in insects [8,9], but in the meantime has been shown to account for certain aspects in motion vision of other species including man [10-13]. The visual field is assumed to be covered by a two-dimensional array of local movement detectors which evaluate a kind of spatio-temporal cross-correlation between the light intensity fluctuations at two neighbouring points in space. More specifically, each movement detector has two spatially displaced input stages and consists of two mirror-symmetrical subunits (Fig.2). The input signal of one branch of each subunit is delayed in some way and multiplied with the instantaneous signal of the neighbouring input channel. The final output of the detector is then given by the difference between the outputs of the two subunits. Of course, a single movement detector senses only those components of motion which result in intensity changes along the orientation of its axis. Therefore, the outputs of a pair of *differently oriented* detectors at each retinal location are combined to a vector in order to obtain a two-dimensional representation of local motion (see Fig.3). The total of these local vectors, thus, represents a vector field which indicates the direction and magnitude of the local motion measurements.

Although computed by the movement detectors from the temporal modulations at their input stages, the field of local motion measurements can be related mathematically to the corresponding 2-D velocity field. This transformation can be described best on the basis of a formal approach which is characterized by a transition from an array of movement detectors with a discrete spatial sampling base to a continuous field of detectors with the distance between the neighbouring retinal inputs being infinitesimally small [14-17]. With the pattern function $F(x,y,t)$ and the 2-D velocity vectors $v(x,y,t)$ the local response vectors $v^*(x,y,t)$ of the movement detection system are given, in a first approximation, by the following transformation [16,17]

$$v^* = T \cdot v . \tag{2}$$

T represents a tensor which is proportional to the detector delay ϵ and has elements depending in a non-linear way on the pattern brightness function F and its first and second partial derivatives with respect to x and y

$$T = -\epsilon \begin{bmatrix} F_x{}^2 - F \cdot F_{xx} & F_x F_y - F \cdot F_{xy} \\ F_y F_x - F \cdot F_{yx} & F_y{}^2 - F \cdot F_{yy} \end{bmatrix} . \tag{3}$$

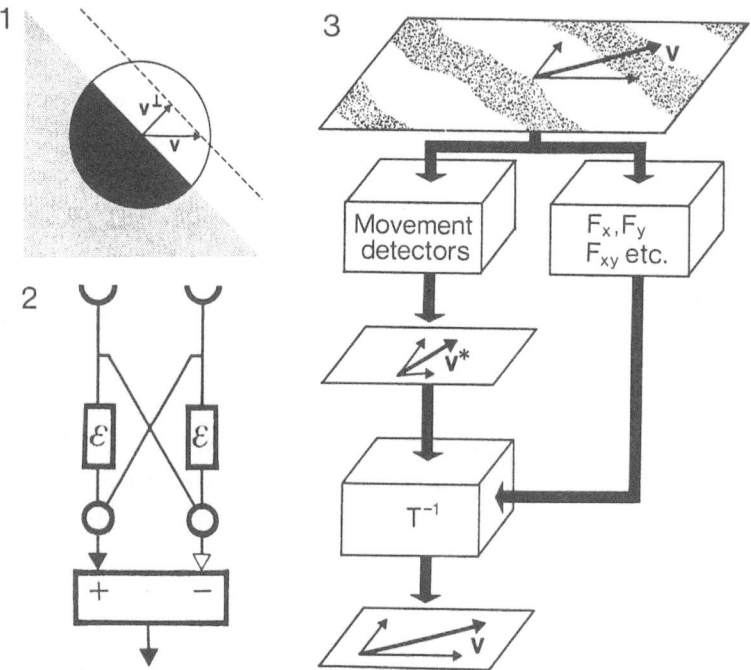

Fig. 1 An illustration of the so-called aperture problem in motion computation. A straight line segment moved with the velocity **v** is viewed through a local aperture. The only component of motion that can be computed is oriented perpendicular $v^\perp$ to the orientation of the segment.

Fig. 2 A local movement detector consisting of two spatially displaced input stages and two mirror-symmetrical subunits. The input signal of one branch of each subunit is delayed by a brief time interval ϵ and multiplied with the undelayed signal of the neighbouring input channel. The output of the detector is given by the difference of the outputs of the two subunits. Therefore, it forms a movement direction selective device.

Fig. 3 A possible procedure for computing the correct 2-D velocity locally. A contrast element of a brightness pattern F moves according to the vector **v** and a pair of motion detectors reponds. Their outputs form the two components of the response vector $v \cdot v^*$ and v^* are related by a two-dimensional tensor T, as described in the text. If the elements of the tensor T which contain spatial derivatives of the brightness pattern function are computed in parallel to the movement detector outputs and if the matrix of the tensor can be inverted, **v** may be computed.

By comparing the transformation of the 2-D velocity field described by equations (2) and (3) and equation (1), respectively, it is obvious that v^* usually deviates from the direction of the brightness gradient. This difference depends in a characteristic way on the curvature of the brightness function of the moving pattern. Moreover, the local response vectors, in general, also do not coincide with the correct 2-D velocity vectors. The occurrence of the second partial derivatives of the pattern brightness function with respect to the spatial coordinates might be surprising at first sight, since (in its discrete form) a movement detector samples the visual surround at only two spatial locations (see Fig.2). However, at least three points are necessary for an approximation of a second derivative. Due to the memory-like operation of the delay in one branch of each detector subunit (see Fig.2) three independent points of the pattern brightness function are simultaneously represented.

Because of the characteristic dependence of T on the curvature of the pattern brightness function, the map of v on v^* given by the transformation (2) is one-to-one for most stimulus patterns. This is the case, if the determinant of T does not vanish. T can then be inverted and equation (2) solved for v

$$v = T^{-1} \cdot v^*. \tag{4}$$

In this way the correct 2-D velocity field can be calculated by using only local information about the pattern (see Fig.3).

There is only one special class of brightness pattern functions for which T cannot be inverted at any spatial location and, consequently, leads to ambiguous local motion measurements. This is the case if the determinant of T vanishes. These pattern functions can be analyzed most conveniently by the substitution $F(x,y,t) = e^{q(x,y,t)}$, which is possible for all $F \geq 0$ [17]. In other words, $q(x,y,t)$ represents the logarithm of the brightness pattern function $F(x,y,t)$. Using this substitution the tensor in equation (3) reads

$$T = \epsilon \cdot e^{2q} \begin{bmatrix} q_{xx} & q_{xy} \\ q_{yx} & q_{yy} \end{bmatrix} \tag{5}$$

and its determinant vanishes if the following condition is satisfied:

$$\det T = q_{xx}\, q_{yy} - q_{xy}^2 = 0 \, . \tag{6}$$

Again, the subscripts denote partial derivatives with respect to x and y. The only solutions of this partial differential equation for x and y are spatial brightness distributions the logarithm of which represents so-called *developable surfaces* [18]. Intuitively, a developable surface is one that can be cut open and flattened out. Or more precisely, at any location on a developable surface one can find a tangent that lies in the surface and has the same surface normal for all of its points. Spatial brightness distributions which, on a logarithmic scale, can be described as *cylindrical*, *conical* or *tangent* surfaces are the only possible examples of developable surfaces. For a cylindrical surface the tangents are all parallel, for a conical surface the tangents intersect in a common point and for a tangent developable surface the surface makes contact with one point of an arbitrarily given space curve. In

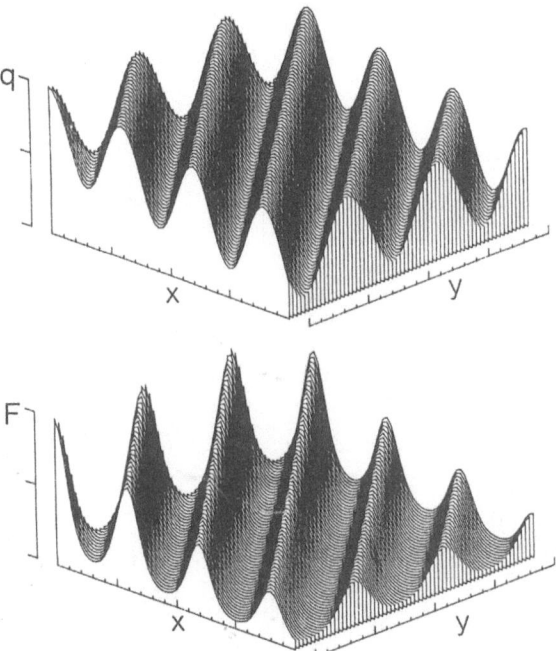

Fig. 4 An example of a developable surface of the *cylinder* class. The upper part of the figure shows the surface q = q(x,y), whereas the lower part the corresponding surface F = F(x,y).

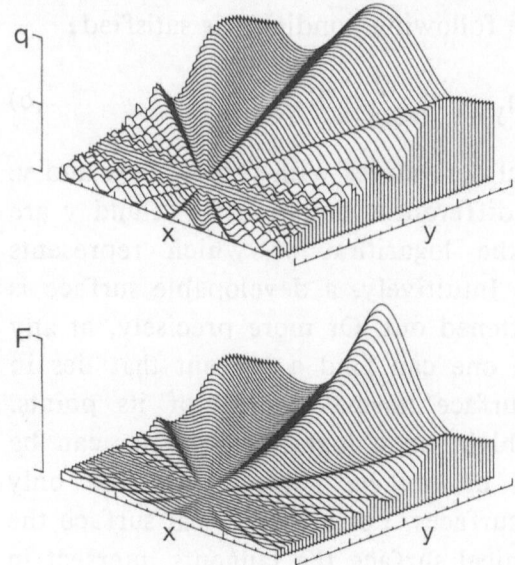

Fig. 5 An example of a developable surface of the *cone* class. The upper part of the figure shows the surface q = q(x,y) and the lower part the corresponding surface F = F(x,y).

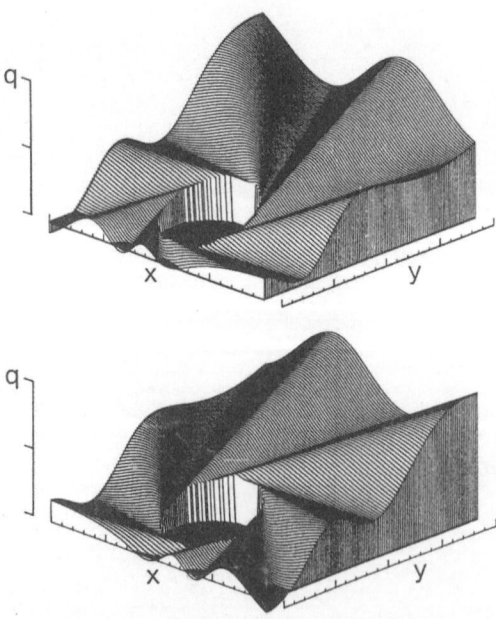

Fig. 6a An example of a developable surface of the *tangent* class.
a) The upper part of the figure shows the outer sheet and the lower part, the inner sheet of the surface q = q(x,y).

176

Figs. 4 to 6 examples of each of the three classes of developable surfaces are given.

Natural brightness patterns usually cannot be represented *globally* by developable surfaces and, therefore, do not solve equation (6) for all x and y. However, since we focus here on local mechanisms, the possibility that the determinant of T vanishes only *locally* has also to be taken into account. If this happens for particular values x,y and t, equation (2) cannot be solved at just this location and time. This suggests that apart from certain locations, the correct 2-D velocity field can usually be recovered. Only at these locations one is confronted with equivalent ambiguities in the local motion measurements as described above in connection with the so-called aperture problem. On the basis of the motion detection scheme used here these ambiguities are usually restricted to small segments of natural brightness patterns. Spatial integration over an array of local motion detectors is a simple means to overcome this remaining problem in most cases. It should be noted that these conclusions are solely based

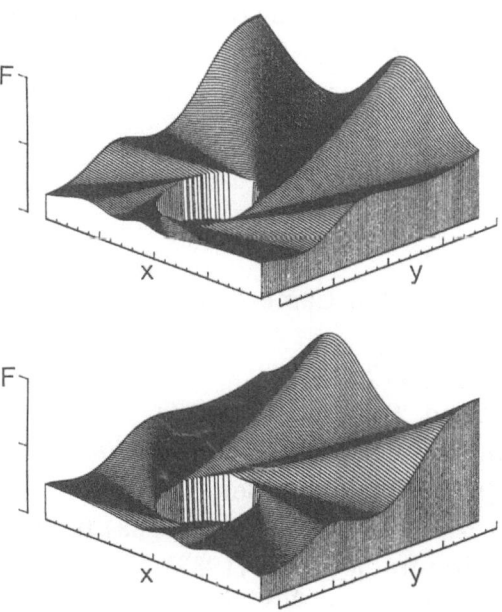

Fig. 6b An example of a developable surface of the *tangent* class. The upper part of the figure shows the corresponding outer sheet and the lower part the corresponding inner sheet of the surface F = F(x,y).

on a mathematical analysis. It was not intended here to address the problems which might arise when equation (4) is tried to be solved numerically.

Of course, the correct 2-D velocity field can only be computed from the local motion measurements by using equation (4), if the elements of T are explicitly known. These have to be derived from separate measurements in parallel to motion detection. It should be emphasized that while we have, in the correlation-type of biological movement detectors, a physiologically established and technically plausible implementation of the mechanism that yields v^*, it is beyond the scope of this article to propose algorithms which yield the elements of the tensor. A technical solution is comparatively simple, since algorithms approximating spatial derivatives are in common use in computer vision [1,19].

In conclusion, there is no principle reason why it should generally be impossible to compute the correct 2-D velocity field for moving stimulus patterns on the basis of local mechanisms alone. The correlation-type of movement detectors as derived from biological systems form an appropriate basis to accomplish this task. This is because it yields local motion measurements which usually contain sufficient information on the relevant properties of the stimulus pattern.

References

1. Horn, B.K.P.: Robot Vision. MIT Press, Cambridge, Mass., 1987)
2. Ullman, S.: Trends in Neurosci. 6, 177 (1983)
3. Hildreth, E.C.: Proc. R. Soc. Lond. B. 221, 189 (1984)
4. Hildreth, E.C., C. Koch.: Ann. Rev. Neurosci. 10, 477 (1987)
5. Koch, C.: In Neural Computers, ed. R. Eckmiller, Ch. von der Malsburg (Springer, Berlin, Heidelberg 1988) p. 101
6. Adelson, E.H., J.A. Movshon: Nature 300, 523 (1982)
7. Movshon, J.A., D.H. Adelson, M.S. Gizzi, W.S. Newesome: In Pattern Recognition Mechanisms, ed.C. Chagas, R. Gattass, C. Gross (Springer, Berlin, Heidelberg, New York, Tokyo 1985) p. 117

8.   Hassenstein, B., W. Reichardt: Naturforsch. 11b, 513 (1956)
9.   Reichardt, W.:In Principles of Sensory Communication, ed
      W.A. Rosenblith (John Wiley, New York 1961) p. 303
10.   van Doorn, A.J., J.J. Koenderink: Exp.Brain.Res. 45, 179 (1982)
11.   van Doorn, A.J., J.J. Koenderink: ibid. 45, 189 (1982)
12.   van Santen, J.P.H., G. Sperling: J. Opt. Soc. Am. A 1, 451
      (1984)
13.   Wilson, H.R.: Biol. Cybern. 51, 213 (1985)
14.   Reichardt, W., A. Guo: ibid. 53, 285 (1986)
15.   Egelhaaf, M., W. Reichardt: ibid. 56, 69 (1987)
16.   Reichardt, W.: J. Comp. Physiol. A 161, 533 (1987)
17.   Reichardt, W., R.W. Schlögl: in preparation (1988)
18.   Courant, R., D. Hilbert: Methods of Mathematical Physics,
      Vol.II (Interscience Publishers, New York, London 1962)
19.   Ballard, D.H., C.M. Brown: Computer Vision. (Prentice-Hall,
      Inc., Englewood Cliffs, New Jersey 1982)

Order in Time: How the Cooperation Between the Hands Informs the Design of the Brain

J.A.S. Kelso and G.C. deGuzman

Center for Complex Systems, Florida Atlantic University,
Boca Raton, FL 33431, USA

1. Introduction

Our interest is in identifying the laws or principles that govern how the nervous system produces coordinated activity. To find these laws, we do not think it so useful to treat the brain as a general purpose machine, capable of producing arbitrary outputs to arbitrary inputs. Rather, our strategy is to place the organism in particular contexts, viewing the brain more as a special purpose device, that is temporarily self-organized for specific behavioral functions. Just as the boundary conditions must be carefully chosen for studies of other physical systems ('the system must be prepared') so too it is important to select a class of tasks that affords insights into the relationship between brain and behavior. When this is done we believe it is possible to see how surface simplicity–in the form of laws–may arise from deep complexity. Along with new data presented here, the evidence draws us to the following inescapable conclusion: *In tasks–that involve ordering in time, the high-dimensional nervous system* (approximately 50 neurotransmitters, 10^3 cell types, 10^{14} neurons and neuronal connections) *restricts its effective degrees of freedom to the phase- and frequency-locking phenomena of coupled nonlinear oscillators.* It lives, in other words, on a low dimensional manifold and is governed by low-dimensional, dynamical laws. This design may be seen at multiple levels of description [1,2] and appear in a variety of functional behavioral patterns including coordination, learning, memory, perception-action patterns and even intentional behavior (see e.g., [3,4]). In the next section we briefly summarize previous and recent results on temporal patterns and our theoretical understanding of them. Then we extend the treatment to an important case, namely forms of temporal organization in which the intrinsic frequencies in the system are not the same. Overall, our results–in addition to supporting the main hypothesis mentioned above–serve to unify a number of physical, biological and psychological phenomena. By definition, the laws of temporal order are *abstract*, though realizable (and hence measurable), on many different scales of observation.

Springer Series in Synergetics Vol. 42: **Neural and Synergetic Computers**
Editor: H. Haken   ©Springer-Verlag Berlin Heidelberg 1988

2. Order in time: Dynamic patterns of behavior

For over a decade, we have studied how human beings perform tasks that fundamentally involve *temporal order*. Since it is part of the behavioral repertoire of many species, rhythm may be viewed as a general case. However, ordering in time is a common, if not universal property of living things and plays a number of important roles also in technological devices, including computers (e.g.,[5]). Experiments by ourselves and others have revealed quite remarkable temporal constraints between the two hands when they are functioning together. Only a restricted set of phase relations appears to exist, even in skilled musicians; people who have had the two halves of the brain surgically separated in order to control epileptic seizures are strongly attracted to just two phase relations, in-phase (homologous muscle groups contracting together) and anti-phase (homologous muscle groups alternately contracting); the anti-phase relation is less stable than the in-phase relation. Spontaneous shifts from anti- to in-phase pattern occur when a control parameter, frequency, is increased. The opposite is not observed over the same frequency range; accompanying such changes in temporal pattern are enhanced fluctuations of the relative phase and a strong increase in the relaxation time of the anti-phase pattern; the switching time–a theoretically founded measure–is also determined by the relative stability of the patterns (for review see [3]).

Using the theoretical concepts and tools of nonlinear dynamics and a detailed mapping between theory and experiment, these results may be understood as follows [6,7]: observed temporal patterns are characterized by a collective variable, relative phase, which serves as an *order parameter* in the sense of synergetics [8]; stable patterns are mapped onto point attractors (i.e., zero dimensional, relaxational dynamics) for the collective variable; the stability and stationarity of the patterns are measured and interpreted in terms of the stochastic dynamics of relative phase with certain time scale relations; the switching among patterns is due to the loss of stability of the attractor for anti-phase motion, a (nonequilibrium) phase transition; the dynamics of the switching process are well-characterized by the stochastic dynamics of relative phase.

All observed features of these dynamic patterns, i.e., multistability, transitions, hysteresis and so forth, have been derived using nonlinearly coupled limit cycle oscillators perturbed by stochastic forces [6, see also 9,10]. There is strong evidence suggesting that similar features in *neuronal patterns* may be understood and synthesized in an identical fashion (cf. [1,2]), thus permitting a linkage among levels of observation.

Because we have a good grasp of the dynamics of certain intrinsically stable temporal patterns (i.e., those that are produced without the direct influence of the environment), it has proved possible to extend the dynamic pattern strategy to other situations, including learning, memory and intentional behavior patterns (see e.g., [4]). As one example–which we use to illustrate a general principle–consider the case in which a (temporally) structured environment modifies patterns of hand coordination. The core idea is to treat the environment as forces acting on the dynamics of the collective variable, relative phase, simultaneously with the intrinsic dynamics in which no structured environment is present [11]. These forces can either *cooperate* or *compete* to produce the resulting behavioral pattern. For example, when the required relative phase (experimentally, the relative phase that the subject must produce is varied between two pacing lights operating at a common frequency in 10 steps between 0 deg. and 360 deg.) corresponds to one of the basic intrinsic patterns (in-phase and anti-phase), the systematic deviation between required and actually produced relative phase is small and variability is low. However, when the required relative phase is not compatible with the intrinsic patterns, much less accurate and more variable performance results [12,13 and Figs. 1 and 3 of ref. 11].

Theoretically, environmental information is captured as a contribution to the collective variable dynamics attracting the system toward the required behavioral pattern. The concrete model of the *intrinsic dynamics*, i.e., a dynamical description in terms of relative phase and its stochastic behavior *without* environmental information present, is [7]

$$\dot{\phi} = -\frac{dV}{d\phi} + \sqrt{Q}\xi_t \, , \tag{1}$$

where $V(\phi) = -a\cos(\phi) - b\cos(2\phi)$ and ξ_t is Gaussian white noise of zero mean and unit variance. The model, when environmental information specifying a required relative phase, ψ, between the pacing stimuli (see [11] for details) is included, reads

$$\dot{\phi} = -\frac{dV_\psi}{d\phi} + \sqrt{Q}\xi_t \, , \tag{2}$$

where now $V_\psi = V(\phi) - c\cos((\phi - \psi)/2)$. The new model parameter, c, represents the strength of environmental influence on the intrinsic dynamics (Eq. 1).

An intuitive way to see cooperative and competitive effects in this model of environmentally specified behavioral patterns is to plot the potential of Eq. 2 for three required relative phases, 0, $\pi/2$, and π (see figure 1). Using model parameters corresponding to the intrinsic dynamics in the bistable régime, and a value of the parameter c that is consistent with TULLER and KELSO'S experimental data, the following features emerge: 1) when the required relative phase coincides with one of the basic intrinsic patterns, $\phi = 0$ or π, the minimum of the potential is exactly at the required relative phase and its shape is well-articulated (less so for $\phi = \pi$ than $\phi = 0$). Clearly, this case reflects the *cooperation* of extrinsic and intrinsic dynamics; 2) on the other hand, when the environmentally required relative phase does not correspond

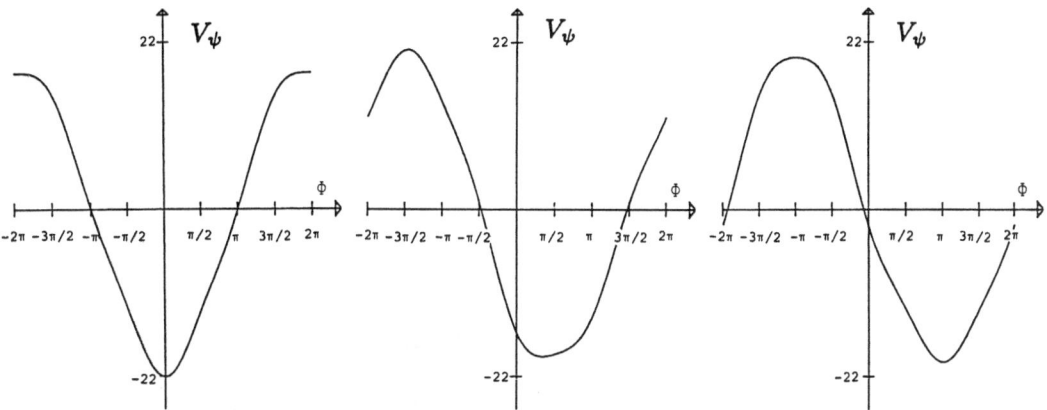

Fig. 1. The potential of Eq. 2 for $a = b = 1Hz$ and $c = 20Hz$; the required relative phase is 0 (left), $\pi/2$ (middle) and π (right). See text for details (from [11]).

to one of the intrinsic patterns, a *competition* between the two forces pulls the minimum away from the required relative phase (Fig. 1, middle). The potential is deformed, and a wider, less articulated minimum results (see also e.g., von der MALSBERG'S [14] and GROSSBERG'S [15] use of cooperation and competition in neural networks).

There are many more aspects of these theoretical results that have led to further empirical predictions regarding environmentally specified and learned patterns of temporal coordination. For present purposes, we note that the influence of environmental information may also be understood at the level of the component oscillators. That is, the dynamics of relative phase can once again be derived from the underlying oscillatory dynamics and their coupling [16]. It is important to emphasize again the fundamentally *abstract* nature of the order parameters and the purely *functional* coupling among the oscillators that gives rise to these macroscopic behavioral patterns. Our theoretical framework thus affords an understanding of how similar *patterns* may arise from a variety of mechanisms.

Finally, since temporal patterns between the hands occur in parallel, and parallel neural computation is the topic of this book, it is intriguing to inquire what happens when the two cerebral hemispheres are divided through surgical section of the corpus callosum. Such "split brain" subjects are of special interest because previous research suggests that if pacing stimuli specifying each hand's timing are projected to separate hemispheres then (because control of finger movements is strongly lateralized) the temporal patterns produced should conform to these external phasing requirements. In fact, the intrinsic patterns (in-phase and anti-phase) turn out to be even more dominant than in normal subjects. Fig. 2 shows sets of ten histograms for two such split-brain subjects (experimental details are provided in [12,13]), in which the

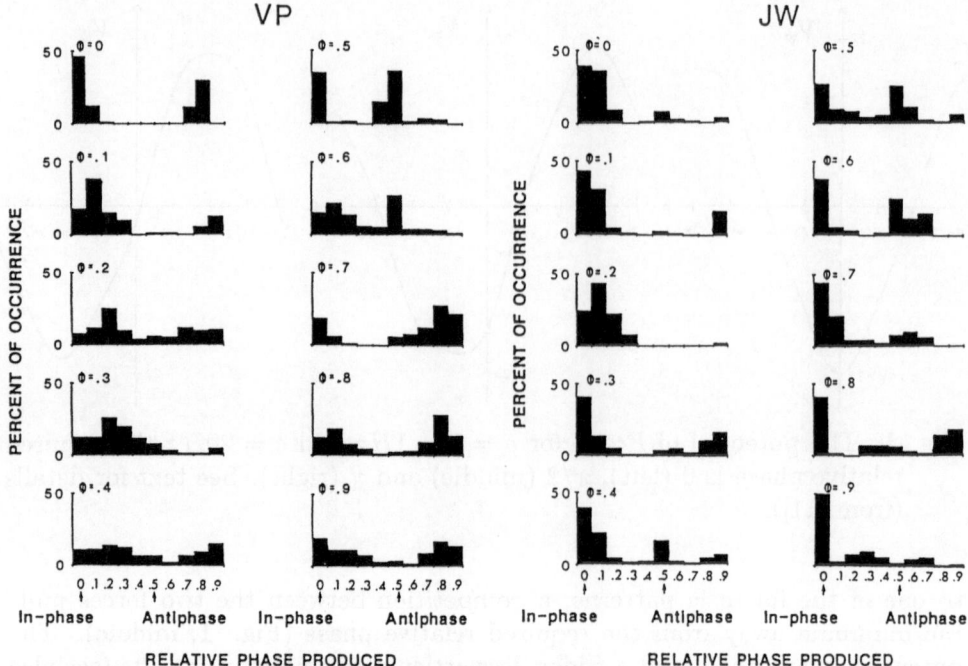

Fig. 2. Frequency distributions of two 'split-brain' patients, VP and JW. The relative phase produced is plotted as a function of the required relative phase (top left of each histogram). See text for details (from [12,13].

relative phase produced is expressed as a function of the required relative phase between the hands. Both subjects, but especially JW, show a strong dominance of the intrinsic dynamics. Magnetic Resonance Imaging (MRI) reveals that the corpus callosum of JW is completely split whereas VP has some residual callosal fibers in the area of the splenium and rostrum [17]. JW shows an especially strong bias toward synchrony at all required phases. However, he is not simply moving his fingers at the same frequency and ignoring the pacing lights, because anti-phase movements appear at required phases of 0.4, 0.5 and 0.6. Anti-phase patterns are all but absent at required phases closer to the in-phase mode.

Performance of split-brain subjects reflects the persistence of the intrinsic dynamics even when environmental demands specify other patterns. This strong reliance on the intrinsically stable patterns produces less flexible, not more flexible intermanual coordination. Environmental information arriving in visual cortex thus cannot overcome the intrinsic patterns, which likely reflect the organization of subcortical neural networks. Viewed from Eq. 2, the extrinsic dynamics only weakly, if at all, perturb the order parameter dynamics of the intrinsic patterns in split-brain subjects.

Our purpose in discussing this research on perception-action patterns is that it has led us to a quite explicit formulation of information in biological systems [4]: *Information is only meaningful and specific to a biological system to the extent that it contributes to the order parameter dynamics, attracting the system to the required (e.g., perceived, learned, memorized, intended) behavioral pattern* (see Fig. 2). We may incorporate this axiom to the earlier set of theoretical propositions (themselves generalizations based on an understanding of the intrinsic patterns, their stability and change) as a new part of dynamic pattern theory [2,3,18].

3. "New" kinds of ordering in time

Concepts of collective variables (order parameters) characterizing entrained states, of shifts in behavioral patterns interpreted as phase transitions, of pattern stability, loss of stability, fluctuations, time scales and so forth have only recently been shown to be useful for understanding pattern generation and coordination in biological systems (see also [19]). They have not yet permeated psychological thinking. Psychologists nevertheless have long been interested in problems of temporal order both in perception and production (e.g., [20-22]), and temporal coherence in neural networks has been recognized as a crucial property since HEBB [23]; see also [14,15]. Static, usually binary, hierarchical schemes have dominated explanations of temporal phenomena in psychology. A base interval is often posited which may be broken up in various ways to account for why some durational relationships among stimuli are perceived (or produced) better than others (e.g., [21]). POVEL [24], for example, shows that temporal stimulus sequences in which the durations vary in a 2:1 ratio are imitated or reproduced better than other, more complicated combinations. He concludes that presented temporal sequences must fit a preferred 'mental schema' if subjects are to reproduce or imitate correctly. It is speculated, following FRAISSE [25] that the 2:1 schema predominates because a large proportion of Western music tone durations relate as this interval ratio.

DEUTSCH [26] aimed to determine if the accuracy of producing parallel patterns with the two hands (finger-tapping) depended on the temporal relation between the stimuli (tone blips) presented to each ear of the subject. If independent timing mechanisms existed, parallel production, according to DEUTSCH, should not depend on the temporal relations between elements. DEUTSCH [26] found less interference (evaluated in terms of timing variability) when the stimuli were harmonically related (e.g., 2:1, 3:1) than when they were more complicated (e.g., 4:3, 5:2). Given the experimental paradigm it is difficult to imagine how these"parallel" effects could not have occurred. The general result however, that performance accuracy is inversely related to the 'pattern complexity' is of considerable interest for what we shall present next.

Here we shall provide experimental evidence for 'parallel effects' in an experiment which is strongly biassed *against* finding them! Consistent with the work reviewed in Section 2, our findings suggest a universal design for temporal ordering in the nervous system. Moreover, mysterious aspects of previous studies, including a) why certain 'internal schema' such as 2:1 are favored; b) the existence of performance biases toward these, i.e., errors in production are systematically related to the proximity (in terms of durational relations) to favored internal structures; and c) the underpinnings of 'hierarchical complexity' descriptions, will become clear, hopefully, as a result of this work.

4. Mode-locking

When nonlinear oscillators are combined to synthesize phase-locked patterns (e.g., [6,9,10]), the coupling that is responsible for the phase-locking can, in most cases, also generate synchronization even if the eigenfrequencies of the oscillators are not identical. Such synchronization or entrainment is crucial to temporal self-organization: What looks like a single macroscopic temporal pattern often turns out to be the collective consequence of mutual entrainment among many constituent oscillatory processes (see, e.g., [27] for many examples).

Forms of entrainment at frequency-ratios other than 1:1 (e.g., 2:1; 3:1; 2:3...) are of course possible. The parameter regions corresponding to these mode-locked states are called *Arnold Tongues*, the width of which reflects the stability of the mode-lockings (e.g., [28]). Figure 3 shows an example. The natural space on which to conceive the combined motion of two coupled oscillators is the product of two circles, i.e., a 2-torus (e.g., [29]). Unlocked behavior corresponds to ergodic trajectories covering the whole torus, never closing.

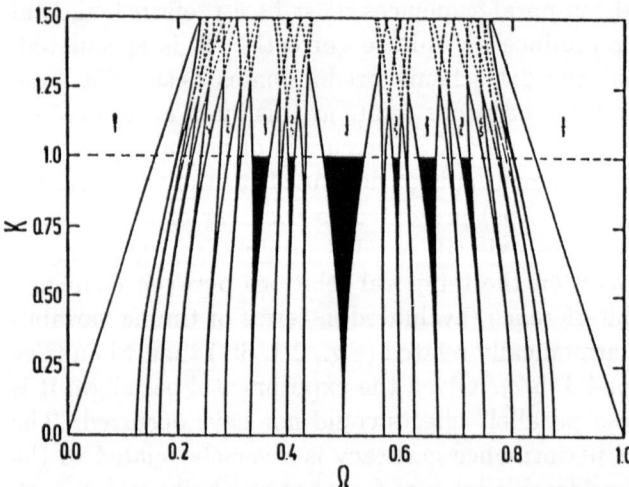

Fig. 3. Phase diagram for the sine map plotted as a function of the nonlinearity parameter, K and the winding number Ω. Arnold Tongues correspond to locked rational values for the winding number. Black regions are mode-lockings manipulated in our experiment (adapted from [28]).

Mode-lockings, however, correspond to a limit cycle attractor embedded in the torus whose dynamics may be further reduced to the circle map. Though it is difficult to obtain explicitly from differential equations, the circle map (which maps the circumference of a circle, $0 < \theta < 1$ onto itself) represents an infinite family of coupled oscillators. Much recent numerical work has been performed on this map: at the transition from quasiperiodicity to chaos its properties are believed to be universal (e.g., [30]). Interesting scaling behavior has also been noted in the sub-critical mode-locked region below the transition to chaos [31].

Even though different frequency-ratios have been observed phenomenologically in a variety of biological systems, e.g., in locomotor-respiratory rhythms [32], cardiac rhythms [33], electrically stimulated giant squid axons [34] and even speech-hand movement rhythms [35], direct experimental manipulation of mode-lockings or transitions among mode-lockings has not always been feasible or possible. On the other hand, there are many numerical simulations in the biological oscillator literature.

In the following experiment we introduce techniques to directly manipulate mode-lockings, using a custom made apparatus especially configured for driving and monitoring rhythmical movements of the two hands simultane-

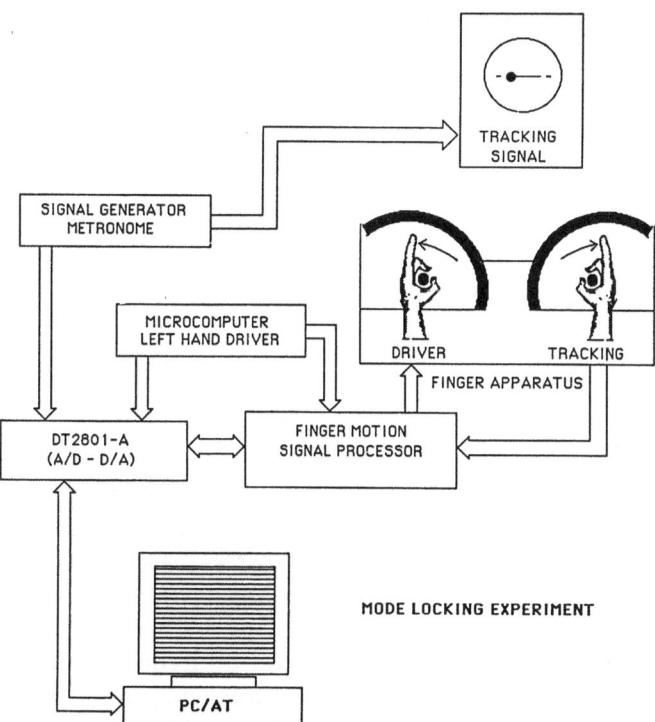

Fig. 4. Diagram of experimental configuration for studying mode-locking in human movement.

ously (cf. [36] for description). The task for the subject was simply to track a visual signal moving back and forth on an oscilloscope screen at 1.5 Hz or, in another condition, 2 Hz. with the right index finger. After about 15 sec., the display was turned off and the subject instructed to maintain the frequency and amplitude of the hand movement set by the visual tracking signal for another 70 sec. Meanwhile, the left index finger was driven passively by a torque motor, itself driven by a sinusoidal waveform, which was output by a Macintosh computer. The subject was instructed to ignore the passively driven finger and to concentrate on the tracking task. Six different driving frequencies were used, corresponding to frequency ratios of 4:3, 3:2, 5:3, 2:1, 5:2, 3:1. The order of driver frequencies was randomized for each tracking frequency condition; two runs per driver-tracking ratio were carried out. All data were acquired at 200 Hz using a PC/AT with 12-bit resolution. An interactive program called CODAS allowed us to do A-D conversion directly on to a hard disk while monitoring all channels in real time. It also enabled us to acquire data over long runs, a necessary feature in order to observe component trajectories beyond the transient stage. Fig. 4 gives a diagram of the experimental set-up.

Figure 5 shows the average (over 5 subjects) of the mean frequency and variability (standard deviation) plotted against the required ratio during the 'free run' part of the experiment, i.e., with the visual signal turned off. Were there no influence of the passive driver on the tracking hand, its mean frequency and variability should be unchanged across the manipulated ratios.

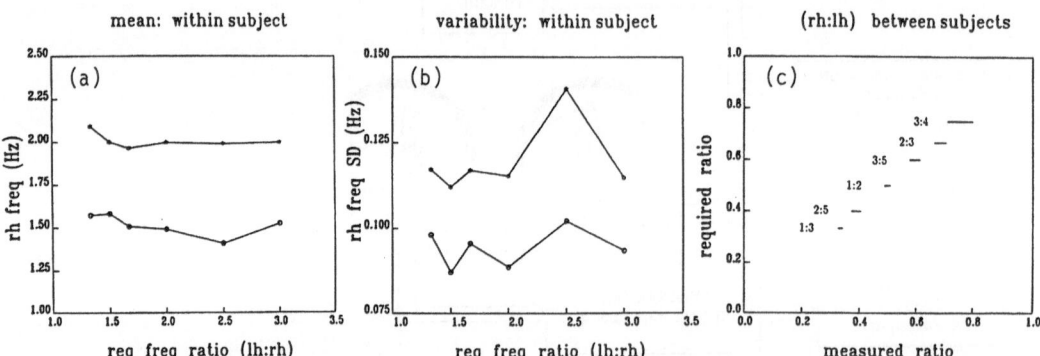

Fig. 5 (a). Average of mean frequencies of 5 subjects as a function of the required ratios (4:3, 3:2, 5:3, 2:1, 5:2, 3:1) in the two tracking conditions: 2.0 Hz (*) and 1.5 Hz (o). (b) Average of the standard deviations of the means in (a). (c) Modified Devil's Staircase based on between-subject statistics. Measured ratios are based on average over all frequency data from two runs and from 5 subjects under the same frequency ratio conditions. The widths (standard deviation) are computed on the basis of frequency rather than frequency ratio. We take the *inverse* of the width as a measure of stability.

However, *both* dependent variables are significantly affected by the frequency ratio. When the ratios between the driver and tracking frequencies are 4:3 and 3:2 (in the 1.5 Hz tracking condition), the actually produced frequency shifts in the direction of 1:1–a nearby attractor. Mean frequency is also underestimated (in the direction of 3:1) when the frequency ratio is 5:2. Characteristic effects on variability are also present. The simple frequency ratios (2:1, 3:1, 3:2) are significantly less variable than the other conditions. The 5:2 condition is less stable than either the 4:3 and 5:3 ratio which are not statistically different from each other. In the typical devil's staircase, the width of the 'steps' reflects the stability of the mode-lockings. The wider they are, the more stable they are in the face of parameter changes. In our modified version of the devil's staircase shown in Figure 5c, the width also reflects the stability of different mode-lockings, i.e., as a variability measure. The *inverse* of the step width in our Figure 5c, however, may be conceived as reflecting the standard devil's staircase picture (see e.g., [28,31].

In summary, either subjects maintain the frequency ratio but are highly variable, or they are less variable, but are attracted toward a nearby, more stable state. This latter feature is easily seen by examining the distribution of frequencies produced, for example, when the ratio is 4:3. Figure 6 shows a bimodal distribution of the frequency ratios in the first run (4:3 and 1:1) suggesting the coexistence of two competing states. With practice, however, the influence of the 1:1 state is attenuated, and the distribution is consistent with the 'required' 4:3 ratio. Based on our discussion in Section 2, this result may be interpreted roughly as follows: the 1:1 ratio corresponds to the *intrinsic*

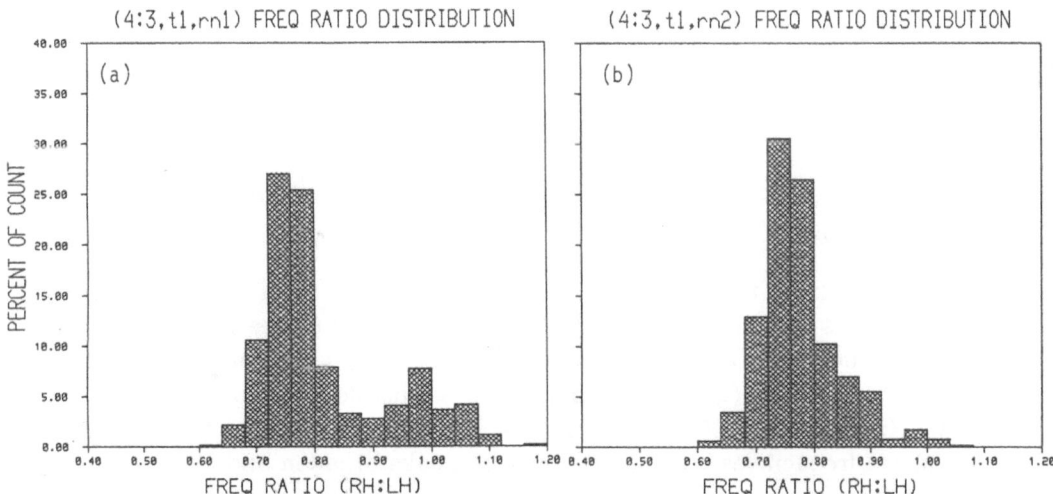

Fig. 6. Frequency distribution of the 4:3 mode in the 2.0 Hz tracking condition. The bimodal distribution in the first run (a) shows two competing modes 4:3 and 1:1. In (b), the smaller peak is attenuated and the required task 4:3 dominates.

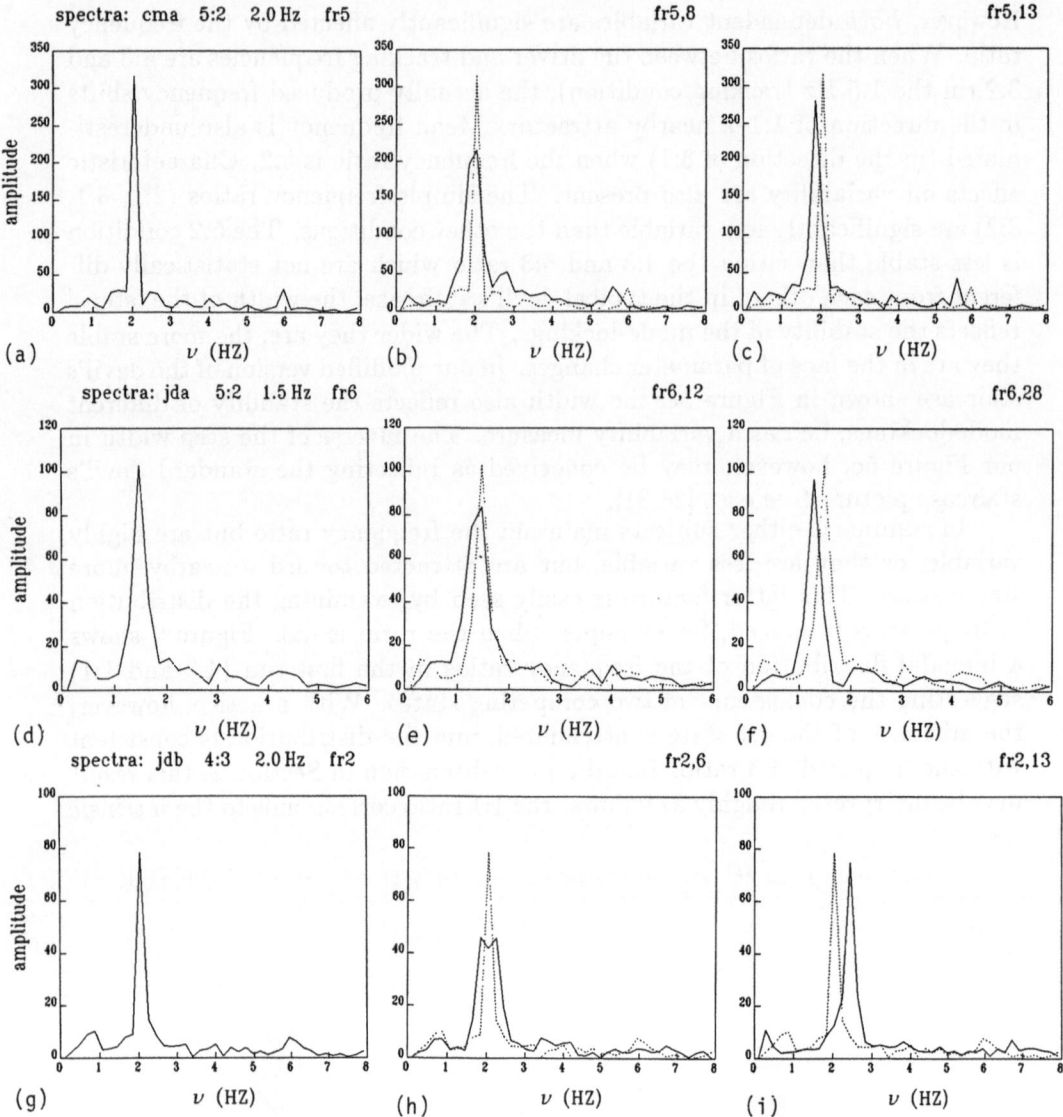

Fig. 7. Transitions to different modes within a run. Each spectral frame corresponds to 5 seconds of right-hand data and is representative of the system's behavior from that frame to the next chosen one. In the "training period", (a), (d), (g) the tracking signal is visible. The peak values at 2.05 Hz, 1.47 Hz, 2.05 Hz for a,d,g respectively, accurately reproduce the required frequencies. In (b) and (e), the peaks broaden and shift toward lower frequencies (solid lines) 1.85 Hz and 1.26 Hz showing attraction (c) and transition (f) toward 3:1. The characteristic broadening is again evident in (h). Here the attraction to 2.44 Hz is from 4:3 toward 1:1 (i) For ease of visualization, spectra displayed in left column are reproduced as dotted lines on all other spectral displays.

dynamics which persist in the presence of the new task, a 4:3 ratio. However, with learning or adaptation, the influence of the intrinsic dynamics disappears (at least temporarily) as the required, new state emerges. This kind of short-term "learning"–which is not mediated by feedback or knowledge of results–appears to involve the creation of an attractor corresponding to the required task, and an accompanying reduction of the influence of the order parameter dynamics for the intrinsic patterns (see also [11]).

The results of Figures 5 and 6 are based on complete data runs. Short term spectral analysis (5 sec. of data, with overlap of 2.5 sec.), however, reveals transitions in mode-locking *within* a run (see Figure 7). Characteristic spectral broadening, reduction in amplitude of the previously dominant frequency and amplification of higher harmonics precedes these transitions which are always to nearby, more stable attractors (e.g., 5:2 goes to 3:1; 4:3 to 1:1). It is worth pointing out that the phase of the driven hand was set arbitrarily with respect to the tracking hand. However, phase-locking between the two signals was frequently present, particularly for the simple ratios. We checked this aspect quantitatively using relative phase histograms, and by inspection of the relative phase in time. Figure 8 shows a 3-D plot of a 2:1 condition in which the position and velocity of the driver hand occupy the x and y axes, and position of the tracking hand is plotted continuously in the z-coordinate. Two bands of phase concentration on the driven cycle show up clearly, corresponding to 0 and π. Thus, the tracking hand visits these phase states for prolonged periods of time. However, phase wandering also occurs. Monitoring relative phase in time, we observe epochs of locking, wandering and then locking again–often to a different phase-locked mode (see Figures 9 and 10). The dynamics are rich, and

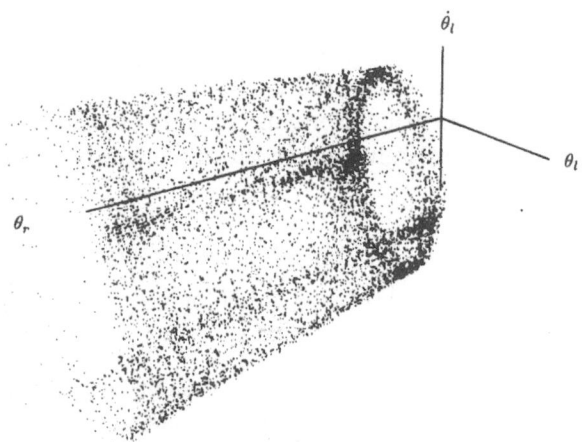

Fig. 8. Phase space portrait of a 2:1 case. $(x, y, z) = (\theta_l, \dot\theta_l, \theta_r)$ where $\theta_{l,r}$ is the angular position of L,R fingers. The dark bands show phase concentration around 0 and π. The length of these bands is due to flattening of the right hand trajectories at the crests and troughs.

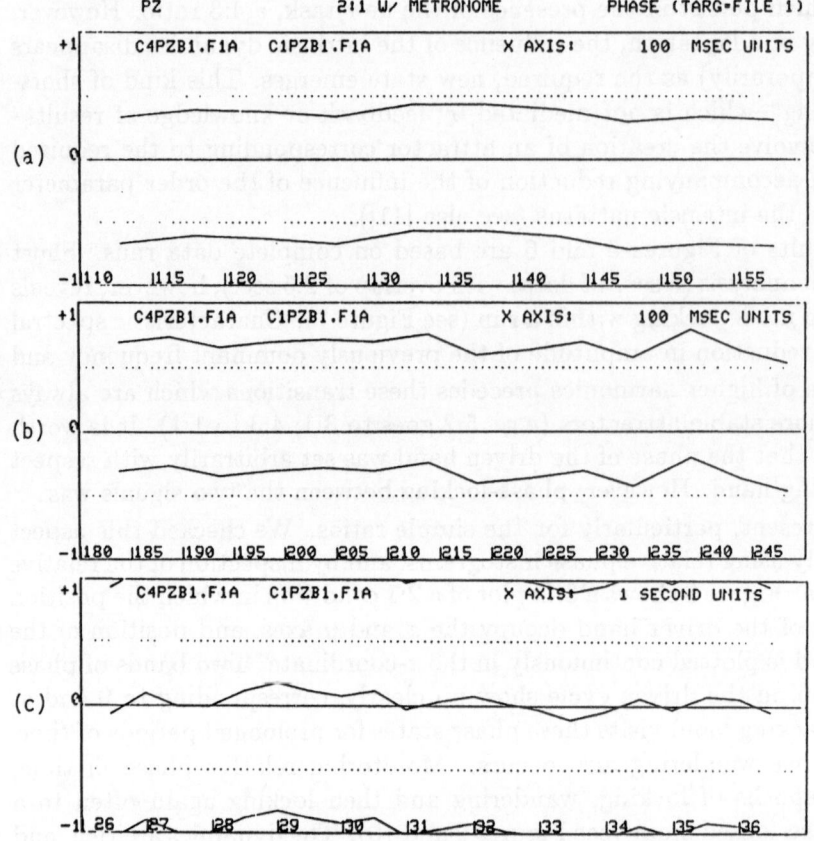

Fig. 9. Relative phase of right versus left hand for a 2:1 case. The units of the ordinate are $\pm 2\pi$ radians. (a) Final 5 seconds before visual metronome was turned off. (b) and (c) Phase locked modes around $3\pi/4$ and 0 respectively. This shows transition to in-phase mode. (d) Phase wandering. (e) and (f) Phase-locking and transition from in- to anti-phase mode. The preference for 0 and π indicates influence of the intrinsic dynamics as discussed in [4].

not always stationary. MANDELL [37] has captured this ephemeral property of neurobiological dynamics elegantly:

> "Neurobiological dynamical systems lack stationarity: their orbital trajectories visit among and jump between attractor basins from which (after relatively short times) they are spontaneously 'kicked out' by their divergent flows".

Many mathematical models of periodically driven oscillators, or weakly coupled nonlinear oscillators exist in the literature which are capable of producing the mode-locking phenomena we have observed here–as well as many

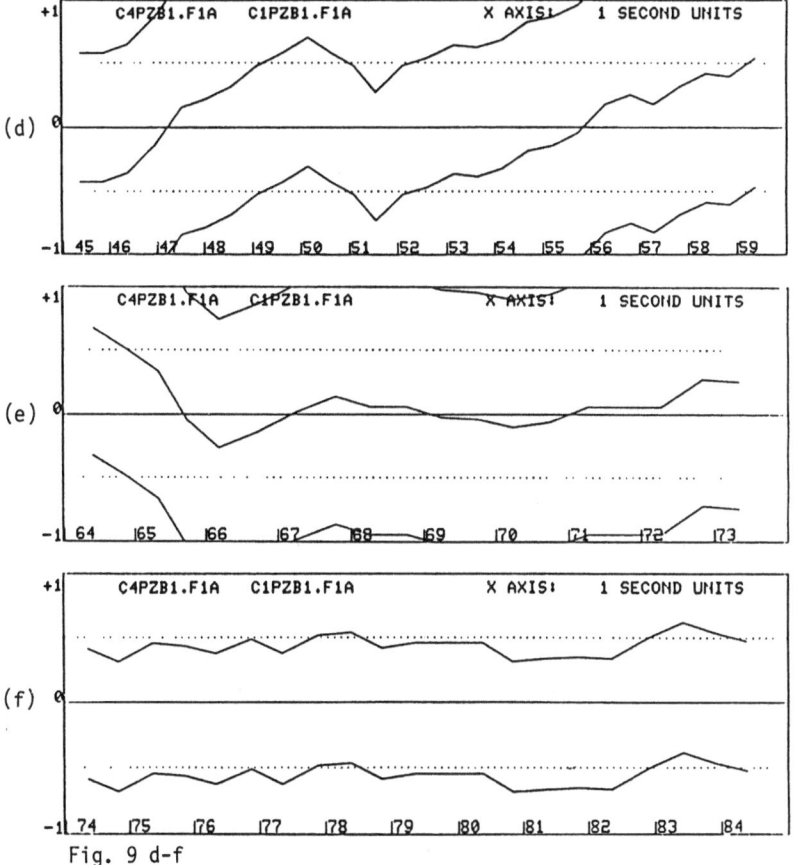

Fig. 9 d-f

other effects, including quasiperiodicity, period doubling bifurcations, and chaos (e.g., [33]; see also [38] for review). There is probably not much insight to be gained by writing a specific model for the present data. The circle map which uses the key notion of rotation number (corresponding here to the phase-locking of the tracking hand onto a rational multiple of the driver frequency), provides a generally valid description. On the experimental side, much more is possible, of course. We have not perturbed the frequency ratios to determine the width of the mode-lockings. Nor have we varied the amount of coupling between driver and tracking hand systematically, or studied transitions among mode-locked states in detail. Many other methods to determine the relative stability of the mode-lockings are available that have proved useful in previous work e.g., calculation of relaxation time, switching time and so forth. The present experiment, however, has directly observed mode-locking and transitions between mode-locked states in a biological system, and therefore constitutes an important generalization of the theory of dynamic patterns we have developed thus far. We have shown that the mode-lockings are differentially stable and that the more stable (less variable) mode-lockings attract less stable (more variable) ones.

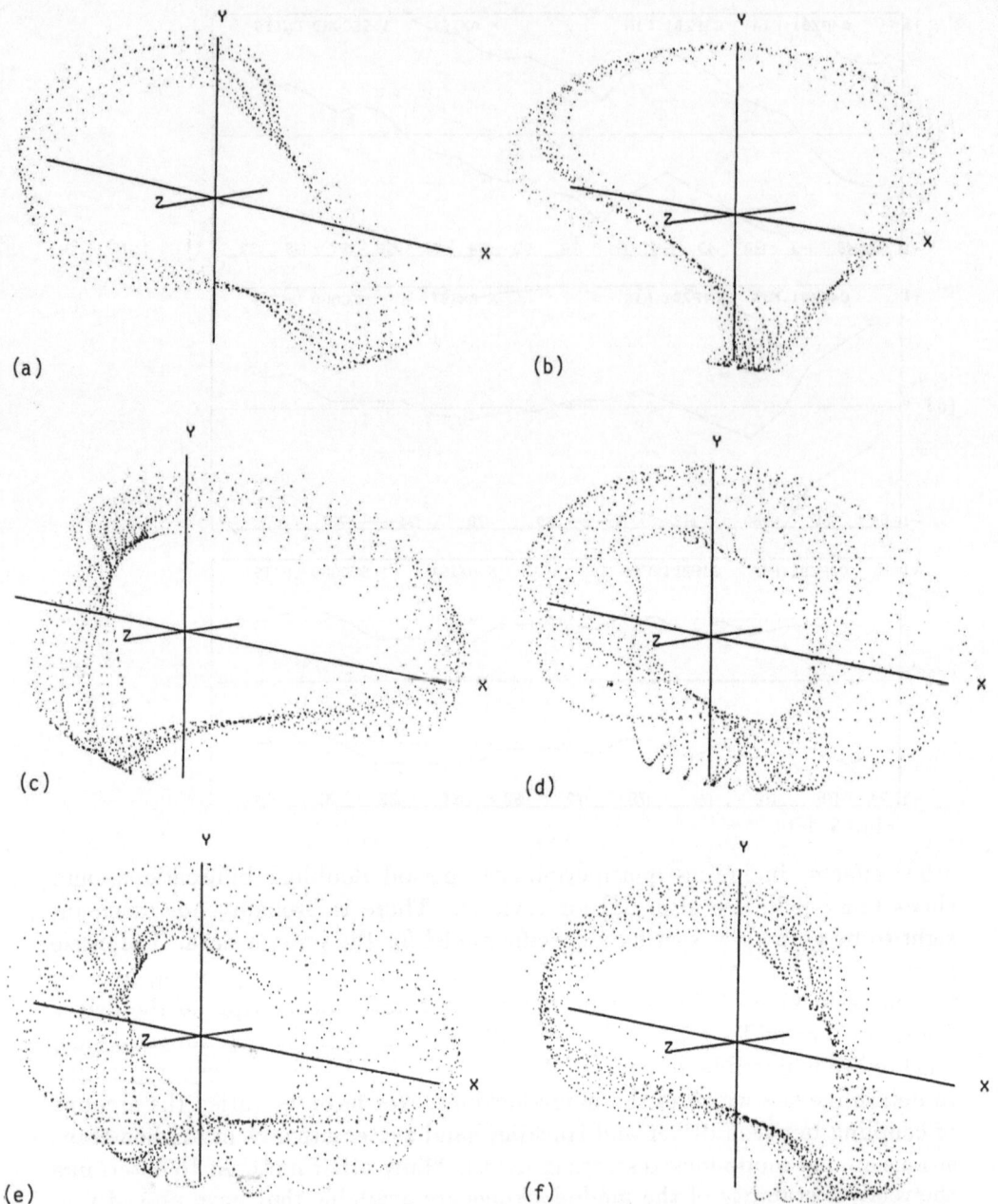

Fig. 10. Trajectory on a torus (surface not shown) for the same 2:1 case. The torus is centered at (0,0,0) and has rotational symmetry about the z-axis. The time frames are identical to that of Fig. 9. Longitudinal and azimuthal angles correspond to right and left hand oscillations, respectively. Asymetries in the trajectory are due to half-cycle variations in the right hand movement.

5. Conclusion

With respect to temporal order, patterns and their change are well- characterized by the concepts of collective variables, control parameters, stability, loss of stability and so forth. Previously, we speculated that the fundamental basis of temporal organization in the nervous system lies, not at any priviledged scale of analysis, but in the coupling of dynamics at all scales [1,2]. The evidence and analysis presented here—*using hand movements as a window*— suggests that the nervous system is a high-dimensional, dissipative dynamical system that compresses its relevant degrees of freedom to the phase- and frequency-locking phenomena of coupled nonlinear oscillators. Such phenomena are crucial to self-organization and pattern formation in a number of fields [8] and are being used as architectural principles for synergetic computers (e.g., [39]). Though *measurable* in different biological systems, at different levels of observation and in different functional behaviors, the fundamentally *abstract* nature of the dynamics that govern ordering in time, cannot be understated.

6. Acknowledgements

J.A.S.K. is grateful to I. Procaccia and C.D. Jeffries for introductory discussions to the dynamics of circle maps at an earlier conference organized by H. Haken (see ref. [9]). This research was supported by a NIMH (Neurosciences Research Branch) Grant MH42900-01 and a grant from the U.S. Office of Naval Research.

References

1. J.A.S. Kelso, G. Schöner, J.P. Scholz, H. Haken: *Phys. Scripta. 35*, 79 (1987)
2. G. Schöner, J.A.S. Kelso: *Science. 239*, 1513 (1988)
3. J.A.S. Kelso, G. Schöner: *Human Movement Sci.* (in press, 1988)
4. G. Schöner, J.A.S. Kelso: In *Dynamic Patterns in Complex Systems*, ed. by J.A.S. Kelso, A.J. Mandell, M.F. Shlesinger (World Scientific, Singapore, 1988) p. 77
5. P.W. Anderson, *Science, 177*, 193 (1973)
6. H. Haken, J.A.S. Kelso, H. Bunz: *Biol. Cyber. 51*, 347 (1985)
7. G. Schöner, H. Haken, J.A.S. Kelso: *Biol. Cyber. 53*, 247 (1986)
8. H. Haken: *Synergetics: An Introduction* (Springer, Berlin, Heidelberg, 1983) 3rd edit.
9. J.A.S. Kelso, J.P. Scholz: In *Complex Systems: Operàtional Approaches in Neurobiology, Physics and Computers* ed. by H. Haken (Springer, Berlin, Heidelberg, 1985) p. 124
10. B.A. Kay, J.A.S. Kelso, E.L. Saltzman, G. Schöner: *J. Exp. Psychol: Hum. Perc. & Perf. 13*, 178 (1987)

11. G. Schöner, J.A.S. Kelso: *Biol. Cyber,* *58,* 71 (1988)
12. B. Tuller, J.A.S. Kelso: Paper presented at Psychonomic Society, Boston, Mass (1985)
13. B. Tuller, J.A.S. Kelso: *Experimental Brain Res.* (submitted)
14. C. von der Malsberg: In *Self-Organizing Systems* ed. by F. E. Yates (Plenum, New York, London 1987) p. 265
15. S. Grossberg: In *Competition and Cooperation in Neural Networks* ed. by S. I. Amari, M.A. Arbib (Springer, New York, 1982)
16. G. Schöner, J.A.S. Kelso: *Biol. Cyber,* *58,* 81 (1988)
17. M.S. Gazzaniga, J.D. Holtzman, M.D.F. Deck, B.C.P. Lee: *Neurol. 35,* 1763 (1985)
18. J.A.S. Kelso, G. Schöner: In *Springer Proc. Phys. 19,* 224 (Springer, Berlin, Heidelberg, 1987)
19. H. Frauenfelder: *Ann. N.Y. Acad. Sci. 504,* 151 (1987)
20. K.S. Lashley: In *Cerebral Mechanisms in Behavior* ed. by L.A. Jeffress (Wiley, New York, 1951)
21. J.G. Martin: *Psychol. Rev. 79,* 487 (1972)
22. M. Reiss-Jones: *Psychol. Rev. 83,* 323 (1976)
23. D.O. Hebb: *The Organization of Behavior* (Wiley, New York, 1949)
24. D.-J. Povel: *J. Exp. Psychol. Hum. Perc. & Perf. 7,* 3 (1981)
25. P. Fraisse: *Les Structures Rhythmiques* (Publ. Univ. de Louvain, Louvain, 1956)
26. D. Deutsch: *Perc. & Psychophys. 35,* 331 (1983)
27. Y. Kuramoto: *Chemical Oscillations, Waves and Turbulence* (Springer, Berlin, Heidelberg, 1984)
28. P. Bak. T. Bohr, M.H Jensen: *Phys. Scripta, T9,* 50 (1984)
29. R. H. Abraham & C.D. Shaw: *Dynamics: The Geometry of Behavior* (Ariel Press, Santa Cruz, 1982)
30. S.J. Shenker: *Physica 5D,* 405 (1982)
31. R.E. Ecke, J.D. Farmer, D.K. Umberger: Preprint, 1988
32. D.M. Bramble: *Science, 219,* 251 (1983)
33. M.R. Guevara, L. Glass: *J. Math. Biol. 14,* 1 (1982)
34. R. Guttman, L. Feldman, E. Jakobsson *J. Membrane Biol. 56,* 9 (1980)
35. J.A.S. Kelso, B. Tuller, K.S. Harris: In *The Production of Speech* ed. by P.F. MacNeilage (Springer, New York, 1983)
36. J.A.S. Kelso, K.G. Holt: *J. Neurophysiol. 43,* 1183 (1980)
37. A.J. Mandell: Personal communication (1988)
38. F.C. Hoppensteadt: *An Introduction to the Mathematics of Neurons* (Cambridge Univ. Press, London, 1986)
39. H. Shimizu. Y. Yamaguchi, K. Satoh: In *Dynamic Patterns in Complex Systems* ed. by J.A.S. Kelso, A.J. Mandell, M.F. Shlesinger (World Scientific, Singapore, 1988) p. 42

Self-Organizing Neural Architectures for Eye Movements, Arm Movements, and Eye-Arm Coordination

D. Bullock and S. Grossberg

Center for Adaptive Systems, Mathematics Department, Boston University,
111 Cummington Street, Boston, MA 02215, USA

1. Introduction: Self-Organization of Eye-Arm Coordination

Many important sensory-motor skills arise as a result of a self-organizing process whereby a biological organism actively moves within a fluctuating environment. The present chapter discusses recent results concerning several key steps in this process: the adaptive transformation of a target position light on the retina into a target position computed in head-centered coordinates; the transformation of a target position computed in head-centered coordinates into a target position command to move an arm to the corresponding location in space; and the conversion of the target position command into a synchronous trajectory of the arm's many components that executes this command in real-time. These transformations involve several different brain regions, such as posterior parietal cortex, precentral motor cortex, and the basal ganglia. Our concern in this chapter will be primarily with the underlying functional and computational issues. Extended discussions of empirical data and predictions are provided elsewhere [1–4], but some recent supportive data are mentioned, such as data concerning the GROSSBERG AND KUPERSTEIN [4] model of how head-centered target position maps are learned in posterior parietal cortex.

These transformations are part of a larger self-organizing sensory-motor system. For example, we will note the multiple uses of motor outflow (brain to motor plant) signals and inflow (motor plant to brain) signals. In particular, outflow motor commands are used to generate internally controlled active movements, whereas inflow motor signals are used to update motor commands in response to externally controlled passive movements. We will also suggest that the internal teaching signals that are used to learn a target position map in head coordinates are themselves derived from a more primitive visually reactive eye movement system which bases its learning of accurate movements upon externally derived visual error signals.

2. The Problem of Infinite Regress

In all, such a multi-component self-organizing sensory-motor system faces a serious problem of infinite regress: What prevents learning in one subsystem from undoing the learning in a different subsystem? If parameters in several subsystems can all change due to learning, what prevents a global inconsistency from developing due to the very fact that individual subsystems, by needing to be specialized to deal with part of an adaptive problem, do not have complete information about the problem as a whole?

To avoid the problem of infinite regress, it is essential to correctly characterize the several computational problems that an adaptive sensory-motor system is designed to solve. Only by correctly characterizing the parts can one hope to successfully piece together these parts into a competent whole. Such a system is thus like a jigsaw puzzle in which each part needs to be formed correctly in order to complete the puzzle.

We have found that infinite regress can be avoided through a mechanistically characterized developmental sequence. In it, obligatory eye movements to flashing or moving lights on the retina are supplemented by attentionally mediated movements towards motivationally interesting sensory cues. These movements are supplemented once again by predictive eye movements that form part of planned sequences of complex movement synergies capable of ignoring the sensory substrate on which they are built. Each of these categories of eye movement requires one or more types of learning in order to achieve high accuracy. The movement systems wherein attention and intention play an increasingly important role base their adaptive success upon the prior learning of the more primitive, visually reactive types of movement.

As the eye movement system is learning to achieve increasing accuracy and autonomy, learned correlations between eye and arm movement commands are also forming, and enable a visually attended object to be grasped. The simple planned arm movements that are generated in this fashion provide the foundation for learning complex sequences of skilled arm movements, and for dexterous eye-arm coordination.

3. Learning an Invariant Self-Correcting Target Position Map in Head Coordinates: Cooperative Synthesis of Globally Consistent Rules

Among the most important types of problems in neural network theory are those which concern the adaptive emergence of recognition invariants. Many different type of invariants can be identified; for example, the emergent invariants encoded by an ART architecture (see CARPENTER AND GROSSBERG [5]) enable the architecture to group all exemplars that share certain similarity properties into a single recognition category. As in ART, a number of other types of invariants are learned through a match-regulated process which gates on and off the learning mechanisms that enable individual events to be grouped together adaptively. Such invariants are called *match invariants* to distinguish them from invariants that arise (say) due to a passive filtering process.

Another model of a match invariant process was developed by GROSSBERG AND KUPERSTEIN [4]. This model describes the learning of an invariant self-regulating target position map in egocentric, or head-centered, coordinates. This problem arises when one considers how a visual signal to a moveable eye, or camera system, can be efficiently converted into an eye-tracking movement command.

To solve this problem, the positions of target lights registered on the retina of an eye need to be converted into a head-coordinate frame. In order to convert the position of a light on the retina into a target position in head coordinates, one needs to join together information about the light's retinal position with information about present position of the eye in the head (Figure 1). GROSSBERG AND KUPERSTEIN [4] suggested that this type of transformation is learned. Otherwise, the retinal system and the eye position system—which are widely separated in the brain and designed according to different internal constraints—would have to be pre-wired with perfectly chosen parameters for their mutual interaction. This model shows how such a transformation can be learned even if parameters are coarsely chosen initially and if significant portions of either system are damaged or even destroyed. This

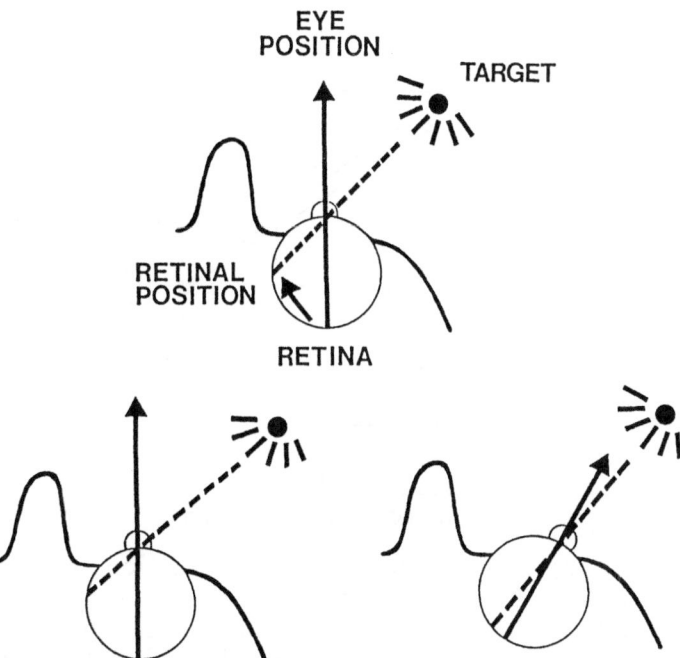

Figure 1. Many combinations of retinal position and eye position can encode the same target position.

type of learning exhibits properties which are of general interest in other biological movement systems, in cognitive psychology, artificial intelligence, and in the design of freely moving robots. In particular, the model can be applied more generally to show how multiple sources of information can cooperate to generate globally consistent rules.

The most important properties of this transformation are that it is many-to-one, invariant, and self-regulating. As Figure 1 illustrates, many combinations of retinal position and eye position correspond to a single target position with respect to the head. When a single target position representation is activated by all of these possible combinations, the transformation is said to be invariant (Figure 2a). The key difficulty in understanding how such an invariant transformation is learned arises from its many-to-one property. The many-to-one property implies that each retinal position and each eye position can activate many target positions in head coordinates (Figure 2b). Even after learning takes place, each pair of retinal and eye positions can activate many target positions in head coordinates, but only the correct target position should receive the maximal total activation.

What prevents learning due to one pair of retinal and eye positions from contradicting learning due to a different pair of positions? In

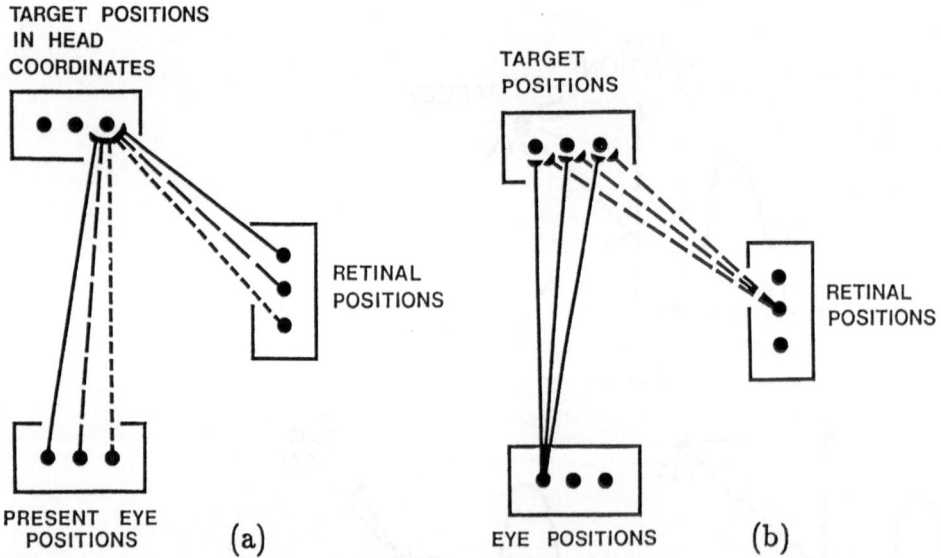

Figure 2. (a) When the many combinations of retinal position and eye position that correspond to each fixed target position all activate the same internal representation of that target position in head coordinates, the ensemble of such head coordinate representations is said to form an invariant map. (b) Every eye position and retinal position can send signals to many target position representations.

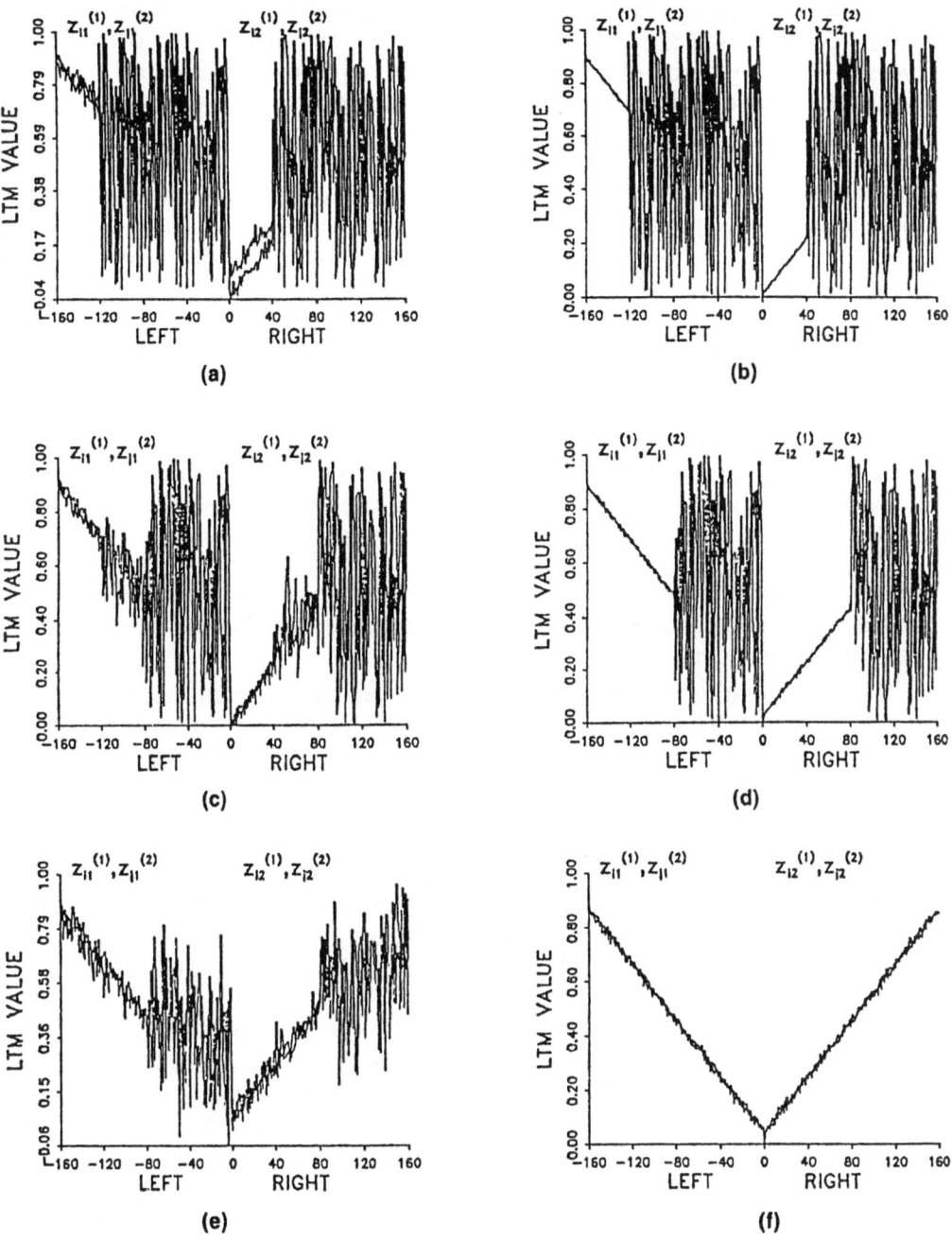

Figure 3. Expansion of LTM spatial maps due to increase of the number of light positions and eye positions being correlated. The learned LTM values in corresponding positions agree, thereby illustrating map self-regulation. Initial values of LTM traces were randomly chosen. (Reprinted with permission from [4].)

particular, if pairing retinal position R_1 with eye position E_1 strengthens the pathways from these positions to target position T_1, then why does not future pairing of R_1 with a different eye position E_2 continue to maximally excite T_1 instead of the correct target position corresponding to R_1 and E_2? How is a globally consistent rule learned by a network, despite the fact that all computations in the network are local? How can a target position map be *implicitly* defined, such that each eye position and retinal position, taken separately, activates a large number of target positions, yet in combination always maximally activate the correct target position?

The property of self-regulation means that the map can correct itself even if a large fraction of the retinal positions and/or eye positions are destroyed, or if their parameters are otherwise altered through time. Destruction of a single retinal position eliminates all the combinations which that position made with all eye positions to activate target positions. In a similar fashion, destroying a single eye position can disrupt all target positions with which it was linked. A self-regulating map must thus be able to reorganize all of its learned changes to maintain its global self-consistency after removal of any of its components. More generally, this property shows how cooperating events can continue to generate self-consistent rules after some of these agents undergo modification or even destruction.

The self-regulation property is illustrated by the computer simulation from GROSSBERG AND KUPERSTEIN [4] that is summarized in Figure 3. Each row in Figure 3 depicts learning of target positions corresponding to a different number of retinal and eye positions. More combinations of positions are represented in each successive row. The first column in each row depicts an intermediate learning stage, and the second column depicts a late learning stage. The abscissa plots topographic positions across 1-dimensional retinal and eye position maps, whereas the ordinate plots the sizes of the adaptive path strengths, or learned long term memory (LTM) traces, in the pathways from these maps to the target position map.

The LTM traces in Figure 3 were randomly chosen before learning began. A comparison of panels (b), (d), and (f) shows that the LTM traces can reorganize themselves when more combinations of positions are associated in such a way as to preserve that part of the map which was learned when fewer combinations of positions were associated. This self-regulation property also holds when more combinations are replaced by fewer combinations, or if the initial LTM traces are not randomly chosen (Figure 4).

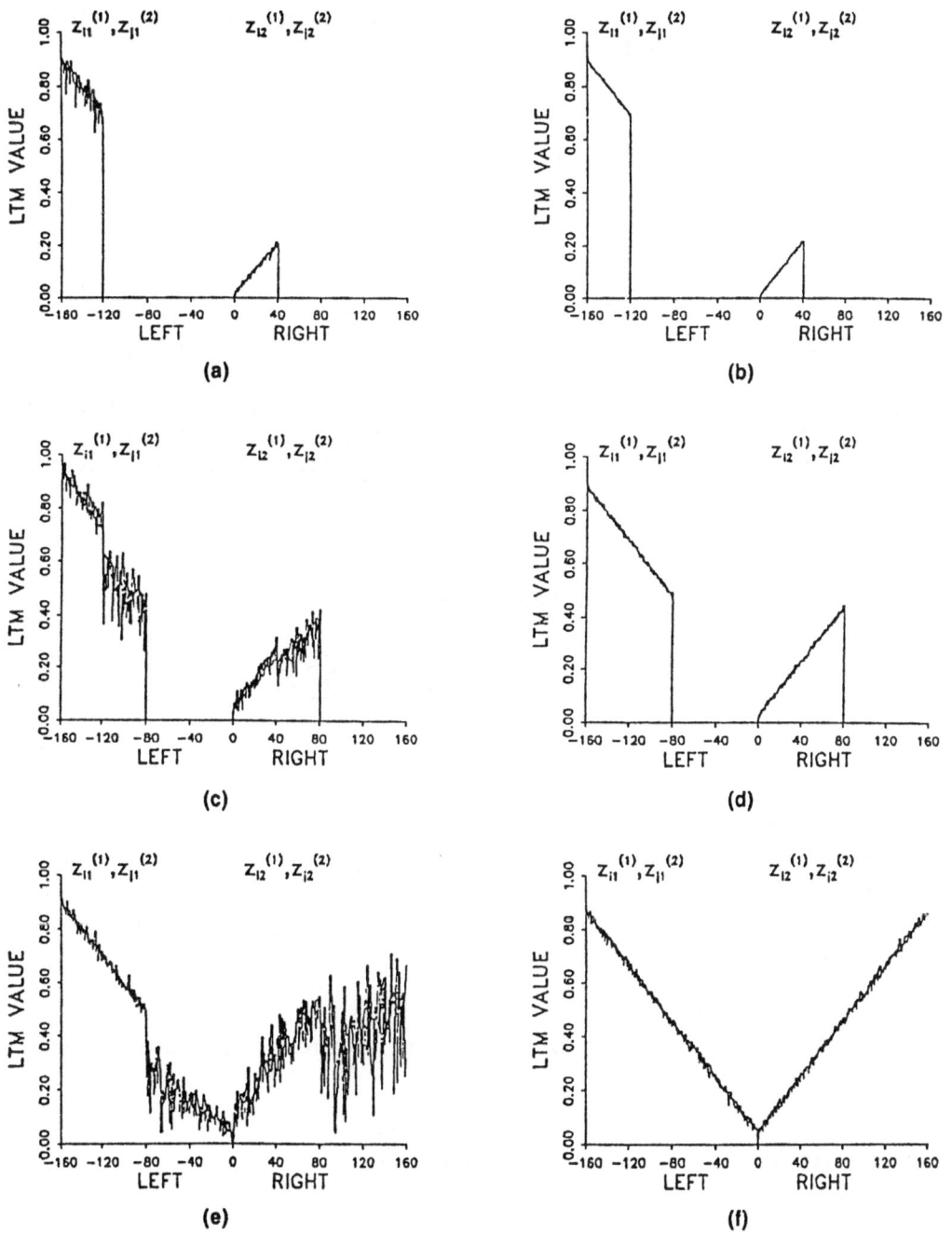

Figure 4. Same as in Figure 3, except initial values of LTM traces were chosen equal to zero. Note that learned spatial maps in Figures 3 and 4 agree, thereby illustrating the ability of map learning to overcome noise. (Reprinted with permission from [4].)

4. Presynaptic Competition for Long Term Memory: Self-Regulating Competitive Learning

The invariance and self-regulation properties of the target position map, or TPM, are due to the fact that all the LTM traces whose pathways project to a single TPM cell readjust themselves in a compensatory fashion when any one of these LTM traces changes due to learning, as required to deal with the many-to-one constraint depicted in Figure 2. GROSSBERG AND KUPERSTEIN [4] suggested that the synaptic endings in which these LTM traces are computed contain autoreceptors [6–12]. In a network whose cells contain autoreceptive synapses, when transmitter is released by one synaptic ending, a portion of it can undergo reuptake via the autoreceptors of other active and nearby synaptic endings. Reuptake has an inhibitory effect on the LTM trace of each active synaptic ending. Thus autoreceptors realize a type of presynaptic competition among all the LTM traces whose pathways converge upon the same cell within the TPM. Autoreceptors hereby mediate a novel type of self-regulating competitive learning [13].

Such an LTM trace obeys an equation of the form:

Autoreceptive Associator

$$\frac{d}{dt} z_{jk} = \epsilon S_j \left[-F z_{jk} + G x_k - H \sum_{r=1}^{n} S_r z_{rk} \right].$$ (1)

In equation (1), z_{jk} is the LTM trace in the pathway from the jth cell in the retinotopic map or eye position map to the kth cell in the TPM; S_j is the signal emitted by the jth cell into this pathway; and x_k is the activity of the kth TPM cell. The terms ϵ, F, G, and H are constants. Equation (1) says that reuptake via autoreceptors of a fraction of released transmitter, as in term $-H \sum_{r=1}^{n} S_r z_{rk}$, inhibits the growth of the corresponding LTM trace.

5. Model Equations for Short Term Memory and Long Term Memory

The model equations are defined as follows.

Short Term Memory (STM) Traces

Let

$$\frac{d}{dt} x_k = -x_k + I_k + z_k,$$ (2)

$k = 1, 2, \ldots, m$, where

$$z_k = \sum_{i=1}^{n} \sum_{j=1}^{j_i} S_j^{(i)} z_{jk}^{(i)}.$$ (3)

Long Term Memory (LTM) Traces
Let

$$\frac{d}{dt}z_{jk}^{(i)} = \epsilon S_j^{(i)}\left[-Fz_{jk}^{(i)} + Gx_k - Hz_k\right].\tag{4}$$

Equation (2) describes the activation of the STM traces x_k at the target position map by an internal teaching input I_k and the total read-out of LTM via term z_k. The teaching vector $(I_1, I_2, \ldots, I_m)$ is activated at the end of an eye movement in a manner that will be summarized below. The system's task is to learn how to reproduce this vector in the STM spatial pattern $(x_1, x_2, \ldots, x_m)$.

The term $S_j^{(i)}$ in equation (3) is the output signal from the jth node of the ith sampling map. In the present application, there are two sampling maps: the eye position map (EPM) that stores the position of the eye in the head before the eye movement begins; and the retino-topic map (RM) that stores the position of the selected light on the retina to which the eye movement is directed. Each of the stored locations in these sampling maps reads-out divergent signals to the TPM, as in Figure 2b. An LTM trace $z_{jk}^{(i)}$ exists at the end of each sampling pathway from the jth node of the ith sampling map to the kth node of the TPM. Term z_k in equations (2) and (3) summarizes the problem faced by the system in parsing the teaching pattern $(I_1, I_2, \ldots, I_m)$ among multiple sampling maps via the LTM traces $z_{jk}^{(i)}$, such that any combination of read-outs from these maps synthesizes the correct target pattern $(x_1, x_2, \ldots, x_m)$ via the LTM vector $(z_1, z_2, \ldots, z_m)$. This self-regulating property is achieved via the normalizing effect of the au-toreceptive term z_k upon the learning by each LTM trace in equation (4).

A full description of the model requires that a lawful relationship be specified between the sampling signals $S_j^{(i)}$ and the teaching signals I_k. Two types of relationships were simulated on the computer, but an additional mathematical analysis of the system [4] also indicated that these results will significantly generalize. In order to understand the logic leading to these simulated relationships, we briefly digress to indicate how the teaching signals are generated.

6. Invariant Target Position Derived from Non-Invariant Visual Error Signals

The initial position of the light on the retina and the initial position of the eye in the head are stored in the sampling map before an eye movement begins. Early in the learning process, the eye movement is triggered by a visually reactive eye movement system. This system is sensitive to rapid changes in lights registered on the retina. After a movement is completed, the teaching vector $(I_1, I_2, \ldots, I_m)$ is derived from an outflow signal that characterizes the position of the eye. However, the task of this TPM-learning system is to learn a correct *target* position, not just any final eye position of a movement.

In the complete theory [4,3,14], such accuracy is due to the fact that the visually reactive movement system is sensitive to visual error signals that compute whether or not the eye movement enabled the eye to foveate the target light. These error signals are used to trigger a learning process that enables visually reactive movements to become accurate. When these movements become accurate, the final eye position generates an internal representation of target position that is used as a teaching vector to learn an invariant TPM.

Our theory assumes that learning of an invariant TPM takes place while the system is making visually reactive movements. Thus an invariant TPM can be learned because the visually reactive movement system can correct its movement errors.

Once an invariant TPM is learned, it stimulates further learning within the attentionally modulated and predictive movement systems of which it forms a part. As a result of this learning, the attentive and intentional movement systems can benefit from the learned parameters of the reactive movement system, yet compete effectively with that system to enable adult movements to come increasingly under the control of an individual's plans, rather than the vicissitudes of environmental fluctuations.

7. Parallel Maps of Eye Position: An Application of Competitive Learning

In order to carry out these computations, two distinct representations of eye position information need to be computed in parallel and stored at different times during the eye movement cycle. The first eye position representation, which is denoted by EPM_I for *initial eye position map*, stores the position of the eye before the movement begins. The second representation, which is denoted by EPM_T for *target eye position map*, reads-out the eye position after the movement terminates. The EPM_I is the source of eye position sampling signals which combines with sampling signals from the retinotopic map (RM). The EPM_T is the recipient of the teaching vector $(I_1, I_2, \ldots, I_m)$ in equation (2). Thus the model describes how an EPM_T can automatically be transformed into an invariant TPM, by building upon the learning of the (non-invariant) visually reactive movement system.

The theory assumes that both EPMs are derived from outflow signals that define the present position of the eye-in-the-head [4]. These outflow signals are computed in motor coordinates that calibrate how much each of the six muscles that hold each eye in the head is contracted. Such a six-dimensional motor vector is transformed into an EPM via the mechanism of competitive learning [4,13]. The most extreme form of competitive learning compresses a multidimensional vector into the choice of a single node in the EPM. Different nodes represent different motor vectors in such a map. A less extreme form of competitive learning converts a motor vector into a unimodal activity profile within the TPM. Such a unimodal teaching vector for the EPM_T is used in one of the simulated models.

At the other extreme, the competitive learning mechanism does not distort the read-out of the motor vector. In effect, the motor vector is itself the teaching vector for the EPM_T. Due to the organization of motor commands into agonist-antagonist muscle pairs, the agonist activation increases (decreases) as the antagonist activation decreases (increases). This suggests the relevance of simulating an EPM_T in which the teaching vector components I_k are a linear function of position k. This situation defines the second model that was simulated. Some of the results from the linear map simulations are depicted in Figures 3 and 4.

8. An Actively Gated Learning Process

A further point needs to be made about the timing of input storage and reset within the maps EPM_I and EPM_T. Map EPM_I stores its sampling representation during the postural state and is insensitive to changes in eye position during movement. Map EPM_I updates its representation when the eye returns to the postural state after movement terminates. Map EPM_T also registers its teaching signal when a movement terminates. On the other hand, map EPM_I still stores the eye position that was stored before the movement began during a time interval subsequent to the movement while map EPM_T is registering the eye position attained by the movement. In this way, the EPM_I and RM maps sample and learn the target position at EPM_T. Thus the storage and reset of these maps needs to be actively gated on and off during the movement cycle. GROSSBERG AND KUPERSTEIN [4] discussed the types of burster cells, pauser cells, and internal matching events that could automatically regulate such cycle-specific processing.

9. Learning of Linear and Unimodal Target Position Maps

The Linear TPM is specified as follows in the model.
Linear TPM

On every trial, one population $v_i^{(1)}$ in RM and one population $v_j^{(2)}$ in EPM_I is randomly chosen. Let $i = i_n$ be the random index chosen from the set $\{1, 2, \ldots, p\}$ of RM cell indices, and $j = j_n$ be the random index chosen from the set $\{1, 2, \ldots, q\}$ of EPM_I set indices on each trial n. Thus

$$S_i^{(1)} = \begin{cases} 1 & \text{if } i = i_n \\ 0 & \text{if } i \neq i_n \end{cases} \tag{5}$$

and

$$S_j^{(2)} = \begin{cases} 1 & \text{if } j = j_n \\ 0 & \text{if } j \neq j_n \end{cases} \tag{6}$$

where $n = 1, 2, \ldots, N$. Let EPM_T consist of two cell populations V_1 and V_2 (for example, to represent the target position of an agonist-antagonist muscle position pair). Define I_{1n} and I_{2n} to be the teaching inputs to V_1 and V_2, respectively, on movement trial n. Suppose that

$$I_{1n} = L(i_n + j_n) \tag{7}$$

and

$$I_{2n} = K - I_{1n}. \tag{8}$$

Thus both I_{1n} and I_{2n} are linear functions of the total sampling index $i_n + j_n$ and I_{2n} decreases as I_{1n} increases. The simulation results are depicted in Figures 3 and 4 at an intermediate stage (column 1) and a late stage (column 2) of learning in response to increasing numbers of RM and EPM_I combinations (40 cells in each map in row 1, 80 in row 2, and 160 in row 3). A statistical analysis of learning accuracy is provided in [4].

Unimodal Gaussian TPM

In this simulated version of the model, each map RM, EPM_I, and EPM_T contained 40 cells. The EPM_T was represented by a unimodal spatial map such that

$$I_{kn} = \exp\left[\frac{-(k - k_n)^2}{\lambda}\right] \tag{9}$$

where k_n equaled the largest integer less than or equal to $\frac{1}{2}(i_n + j_n)$. In this situation, map learning was again excellent. Figure 5 depicts the unimodal spatial distributions of the LTM values, after learning, corresponding to each RM and EPM_I position.

Subsequent to the publication of GROSSBERG AND KUPERSTEIN [4], ANDERSON and his colleagues [15] described receptive fields in posterior parietal cortex which displayed linear and unimodal properties, and ZIPSER AND ANDERSON [16] hypothesized linear and unimodal Gaussian teaching signals in a back propagation simulation of map formation. It has elsewhere been argued that, in principle, a back propagation model cannot be used as a brain model because of its use of a non-local transport of LTM traces to control its learning process [13]. In addition, the back propagation model is forced to assume the existence of a hidden layer of interneurons between the RM and EPM_I sampling levels and the EPM_T target level. It is nonetheless of interest that the identical types of teaching signals were used in [16] as in [4], especially since their use in [4] predicted key properties of the data subsequently reported in [15]. A finer probe of the present

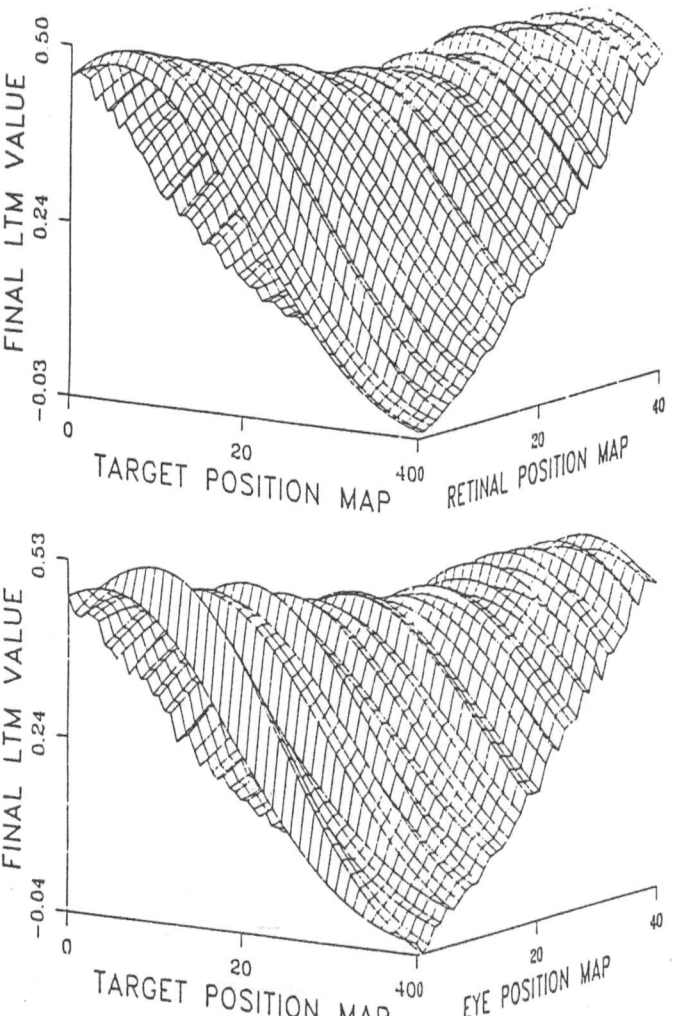

Figure 5. Final LTM values when randomly chosen RM and EPM_I positions sample teaching vectors distributed as Gaussian gradients across a spatial topography in EPM_T. The graphs depict the spatial distributions of LTM values across EPM_T corresponding to each RM and EPM_I position. (Reprinted with permission from [4].)

model would be to test its prediction that catecholaminergic autoreceptors govern the transmitter system (1) which controls self-regulating TPM learning in the posterior parietal cortex.

10. Learning an Invariant Intermodal Transformation between the Eye-Head and the Hand-Arm Target Position Maps

The previous sections indicated how a flashing light on the retina can be adaptively transformed into an invariant TPM computed with respect to the head. An extension of this analysis suggests how the target light can be transformed into an invariant TPM computed with respect to the body, by also incorporating information about the position of the head with respect to the body. In the previous case of learning an invariant TPM of a light's target position with respect to the head, additional information was needed in the form of (1) a sampling map EPM_I to compute the position of the eye in the head before the movement began, and (2) a teaching vector delivered to EPM_T to compute the target position of the eye with respect to the head after the movement was completed.

Analogous information is needed to compute an invariant TPM with respect to the body. In the present instance, this TPM computes target position commands of the hand-arm system. The two sampling maps are the EPM_T which computes the position of target lights with respect to the head and an EPM_I which computes the head position with respect to the body before the hand-arm movement begins. The teaching vector to the TPM uses outflow signals of the hand-arm system to compute the hand-arm position after its movement is complete.

The net effect of this learning process is to compute an invariant transformation from target positions of the eye-head system to target positions of the hand-arm system. This process is of critical importance in sensory-motor control because many arm movements are activated in response to visually seen objects that the individual wishes to grasp. This transformation must be learned because, as the arm grows, the motor commands which move it to a fixed position in space with respect to the body must also change in an adaptive fashion. Our central problem is thus formulated as follows: How is a transformation learned and adaptively modified between the parameters of the eye-head system and the hand-arm system so that an observer can touch a visually fixated object? The following discussion adds further constraints to our analysis.

Following PIAGET'S [17] analysis of *circular reactions*, let us imagine that an infant's hand makes a series of unconditional movements, which the infant's eyes unconditionally follow. As the hand occupies a variety of positions that the eye fixates, a transformation is learned from the parameters of the hand-arm system to the parameters of the eye-head system. A reverse transformation is also learned from parameters of the eye-head system to parameters of the hand-arm system. This reverse transformation enables an observer to intentionally move its hand to a visually fixated position.

How do these two sensory-motor systems know what parameters are the correct ones to map upon each other? This question raises the fundamental problem that many neural signals, although large, are unsuitable for being incorporated into behavioral maps and commands. The learning process needs to be actively modulated, or gated, against learning during inappropriate circumstances.

In the present instance, not all positions that the eye-head system or the hand-arm system assume are the correct positions to associate through learning. For example, suppose that the hand briefly remains at a given position and that the eye moves to foveate the hand. An infinite number of positions are assumed by the eye as it moves to foveate the hand. Only the final, intended, or expected position of the eye-head system is a correct position to associate with the position of the hand-arm system.

Learning of an intermodal motor map must thus be prevented except when the eye-head system and the hand-arm system are near their intended positions. Otherwise, all possible positions of the two systems

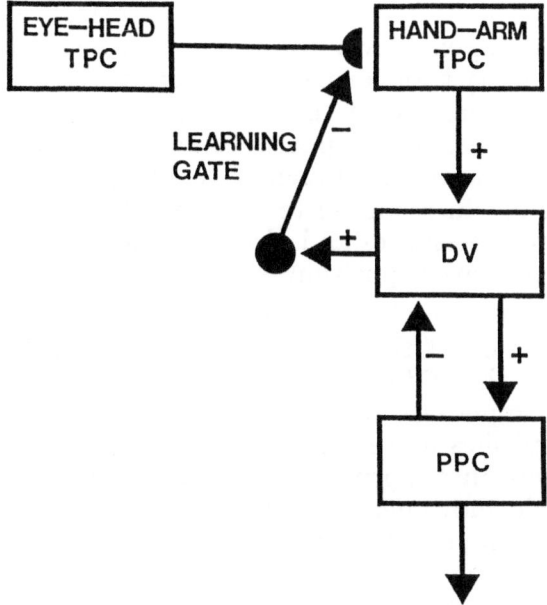

Figure 6. Learning in sensory-motor pathways is gated by a DV process which matches TPC with PPC to prevent incorrect associations from forming between eye-head TPC's and hand-arm TPC's.

could be associated with each other, which would lead to behaviorally chaotic consequences. Several important conclusions follow from this observation [1,4,18]:

(1) All such adaptive sensory-motor systems compute a representation of target position (also called expected position, or intended position). This representation is the TPC (target position command) of the TPM, or target position map.

(2) All such adaptive sensory-motor systems also compute a representation of present position. This representation is the PPC (present position command).

(3) During movement, target position is matched against present position. Intermodal map learning is prevented except when target position approximately matches present position (Figure 6). A *gating*, or modulator, signal is thus controlled by the network at which target position is matched with present position. This gating signal enables learning to occur when a good match occurs and prevents learning from occurring when a bad match occurs.

11. Trajectory Formation using Difference Vectors: Automatic Compensation for Present Position

The above discussion of how *intermodality* sensory-motor transformations are learned also sheds light upon how *intramodality* movement trajectories are formed. Intermodality transformations associate TPC's because only such transformations can avoid the multiple confusions that could arise through associating arbitrary positions along a movement trajectory. TPC's are not, however, sufficient to generate intramodality movement trajectories. In response to the same TPC, an eye, arm, or leg must move different distances and directions depending upon its present position when the target position is registered.

PPC's can be used to convert a single TPC into many different movement trajectories. Computation of the difference between target position and present position at the difference vector, or DV, stage can be used to automatically compensate for present position. Such automatic compensation accomplishes a tremendous reduction in the memory load that is placed upon an adaptive sensory-motor system. Instead of having to learn whole movement trajectories, the system only has to learn intermodality maps between TPC's. As shall be shown below, the DV can be used to automatically and continuously update the PPC movement commands from which the trajectory is formed. In summary, the types of information that are used to learn *inter*modality commands during motor development are also used to generate *intra*modality movement trajectories.

12. Vector Integration to Endpoint (VITE) during Trajectory Formation

We now specify in greater detail a model of how TPC's, PPC's, and DV's interact with each other through time to synthesize a movement trajectory. Each PPC generates a pattern of outflow movement signals to arm system muscles (Figure 7). Each such outflow pattern acts to move the arm system towards the present position which it encodes. Thus, were only a single PPC to be activated, the arm system would come to rest at a single physical position. A complete movement trajectory can be generated in the form of a temporal succession of PPC's. Such a movement trajectory can be generated in response to a single TPC that remains active throughout the movement or to a succession of TPCs that are instated intermittently through time.

This process of continuous updating proceeds as follows. At every moment, a DV is computed from the fixed TPC and the PPC (Figure

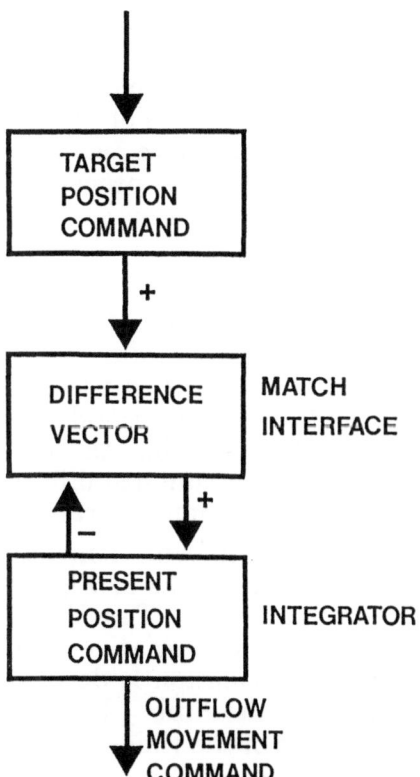

Figure 7. A match interface within the motor command channel continuously computes the difference between the target position and present position, and adds the difference to the present position command.

7). This DV encodes the difference between the TPC and the PPC. In particular, the DV is computed by subtracting the PPC from the TPC. Because a DV computes the difference between the TPC and the PPC, the PPC equals the TPC only when all components of the DV equal zero. Thus, if the arm system's commands are calibrated so that the arm attains the physical position in space that is coded by its PPC, then the arm system will approach the desired target position in space as the DV's computed during its trajectory approach zero. This is accomplished as follows.

At each time, the DV computes the direction and amplitude which must still be moved to match the PPC with the TPC. Thus the DV computes an error signal that is used to continuously update the PPC in such a way that the changing PPC approaches the fixed TPC by progressively reducing the vector error to zero. In particular, the match interface at which DV's are computed sends excitatory signals to the stage where PPC's are computed. This stage integrates these vector signals through time. The PPC is thus a cumulative record of all past DV's, and each DV brings the PPC closer to the target position command.

In so doing, the DV is itself updated due to negative feedback from the new PPC to the match interface (Figure 7). This process of updating present positions through vector integration and negative feedback continues continuously until the PPC equals the TPC. This process of Vector Integration to Endpoint (VITE) gives the model its name [1,2]. Several important conclusions follow from this analysis of the trajectory formation process.

Two processes within the arm control system do double duty: A PPC generates feedforward, or outflow, movement signals *and* negative feedback signals which are used to compute a DV. A DV is used to update intramodality trajectory information *and* to gate intermodality learning of associative transformations between TPC's. Thus the DV continuously updates the PPC when the arm is moving *and* disinhibits the intermodality map learning process when the arm comes to rest.

In Figure 7, the PPC is computed using outflow information, but not inflow information. The PPC negative feedback shown in Figure 7 is an "efference copy" of a "premotor" command [19]. The VITE model's use of efferent feedback distinguishes it from an alternative class of models, which propose that present position information is derived from afferent feedback from sensory receptors in the limb. Further differences are introduced by the VITE model's use of the time-varying multiplicative GO signal to control read-out of the PPC at variable speeds.

13. Intentionality and the GO Signal: Motor Priming without Movement

The circuit depicted in Figure 7 embodies the concept of intention, or expectation, through its computation of a TPC. In models of cognitive processing, such as the ART models [5], the analog of a TPC is a learned top-down expectation. The complete movement circuit embodies intentionality in yet another sense, which leads to a circuit capable of variable speed control. The need for such an additional process can also be motivated through a consideration of eye-hand coordination [1,18].

When a human looks at a nearby object, several movement options for touching the object are available. For example, the object could be grasped with the left hand or the right hand. We assume that the eye-head system can simultaneously activate TPC's in several motor systems via the intermodality associative transformations that are learned to these systems. An additional "act of will," or GO signal, is required to convert one or more of these TPC's into overt movement trajectories within only the selected motor systems.

There is only one way to implement such a GO signal within the circuit depicted in Figure 7. This implementation is described in Figure 8. The GO signal must act at a stage intermediate between the stages which compute DV's and PPC's: The GO signal must act after the match interface so that it does not disrupt the process whereby DV's become zero as PPC's approach the TPC. The GO signal must act before the stage which computes PPC's so that changes in the GO signal cannot cause further movement after the PPC matches the TPC. Thus, although the GO signal changes the outputs from the match interface before they reach the present position stage, the very existence of such processing stages for continuous formation of a trajectory enables the GO signal to act without destroying the accuracy of the trajectory.

The detailed computational properties of the GO signal are derived from two further constraints. First, the absence of a GO signal must prevent the movement from occurring. This constraint suggests that the GO signal multiplies, or *shunts*, each output pathway from the match interface. A zero GO signal multiplies every output to zero, and hence prevents the PPC from being updated. Second, the GO signal must not change the direction of movement that is encoded by a DV. The direction of movement is encoded by the *relative* sizes of all the output signals generated by the vector. This constraint reaffirms that the GO signal *multiplies* vector outputs. It also implies that the GO signal is *nonspecific*: The *same* GO signal multiplies each output signal from the matching interface so as not to change the direction encoded by the vector.

When the GO signal equals zero, the present position signal is not updated. Hence no overt movement is generated. On the other hand, a zero GO signal does not prevent a TPC from being activated, or a

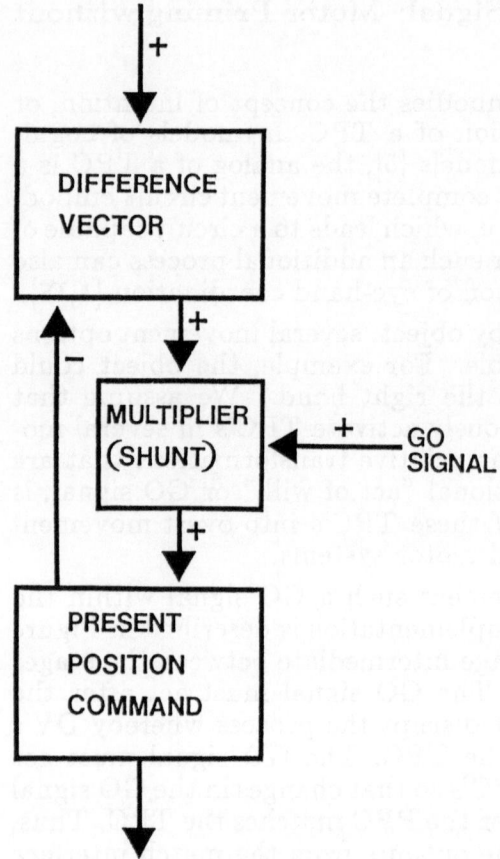

Figure 8. A GO signal gates execution of a primed movement vector and regulates the rate at which the movement vector updates the present position command.

DV from being computed. Thus a motor system can become ready, or primed, for movement before its GO signal turns on. When the GO signal does turn on, the movement can be rapidly initiated. The size of the GO signal regulates overall movement speed. Larger GO signals cause faster movements, other things being equal, by speeding up the process whereby directional information from the match interface is integrated into new PPC's. In models of cognitive processing, such as the ART models [5], the functional analog of the GO signal is an attentional gain control signal.

GEORGOPOULOS, SCHWARTZ, AND KETTNER [20] have reported data consistent with this scheme. In their experiment, a monkey is trained to withhold movement for 0.5 to 3 seconds until a lighted target dims. They reported that cells with properties akin to DV cells computed a direction congruent with that of the upcoming movement during the waiting period. These data support the prediction that the neural stage where the GO signal is registered lies between the DV stage and the PPC stage. HORAK AND ANDERSON [21] have provided data suggesting that the GO signal is computed in the globus pallidus. See [2] for further discussion.

14. Synchrony, Variable Speed Control, and Fast Freeze

The circuit in Figure 8 possesses qualitative properties of synchronous synergetic movement, variable speed control, and fast freeze-and-abort. We apply the circuit properties that each muscle synergist's motor command is updated at a rate that is proportional both to the synergist's distance from its target position and to a variable-magnitude GO signal, which is broadcast to all members of the synergy to initiate and sustain the parallel updating process.

To fix ideas, consider a simple numerical example. Suppose that, prior to movement initiation, muscle synergist A is 4 distance units from its target position and muscle synergist B is 2 distance units from its target position. In that case, if the mean rates at which PPC's are updated for the two synergists are in the same proportion as the distance (i.e., 2:1), then the updating of synergist A will take 4/2 time units while the updating of synergist B will take 2/1 time units. Thus both processes will consume approximately 2 time units. Although the PPC updating process occurs at different rates for different synergists, it consumes equal times for all synergists. The result is a synchronous movement despite large rate variations among the component motions.

Changing the magnitude of the GO signal governs variable speed control. Because both of the updating rates in the example (2 and 1) are multiplied by the same GO signal, the component motions will remain synchronous, though of shorter or longer duration, depending on whether the GO signal multiplier is made larger or smaller, respectively. In general, the GO signal's magnitude varies inversely with duration and directly with speed. Finally, if the value of the GO signal remains at zero, no updating and no motion will occur. Thus rapid freezing can be achieved by completely inhibiting the GO signal at any point in the trajectory.

15. VITE Model Equations

In its simplest form, the VITE circuit obeys the
Difference Vector

$$\frac{d}{dt}V_i = \alpha(-V_i + T_i - P_i) \tag{10}$$

and

Present Position Command

$$\frac{d}{dt}P_i = G[V_i]^+, \tag{11}$$

where $[V_i]^+ = \max(V_i, 0)$. Equations (10) and (11) describe a generic component of a target position command $(T_1, T_2, \ldots, T_n)$, a difference vector $(V_1, V_2, \ldots, V_n)$, and a present position command $(P_1, P_2, \ldots, P_n)$ in response to a time-varying velocity command, or GO signal $G(t)$. The difference vector computes a mismatch between target position and present position, and is used to update present position at a variable rate determined by $G(t)$ until the present position matches the target position.

Such a scheme permits multiple muscles, or other motor effectors, to contract synchronously even though the total amount of contraction, scaled by $T_i(0) - P_i(0)$, may be different for each effector. This property of movement synchrony and duration invariants is now illustrated using a computer simulation.

16. Computer Simulation of Movement Synchrony and Duration Invariance

In simulations of synchronous contraction, the same GO signal $G(t)$ is switched on at time $t = 0$ across all VITE circuit channels. We consider only agonist channels whose muscles contract to perform the synergy. Antagonist channels are controlled by opponent signals [1]. We assume that all agonist channels start out at equilibrium before their TPC's are switched to new, sustained target values at time $t = 0$. In all agonist muscles, $T_i(0) > P_i(0)$. Consequently, $V_i(t)$ in (10) increases, thereby increasing $P_i(t)$ in (11) and causing the target muscle to contract. Different muscles may be commanded to contract by different amounts. Then the size of $T_i(0) - P_i(0)$ will differ across the VITE channels inputting to different muscles.

Figure 9 depicts a typical response to a faster-than-linear $G(t)$ when $T_i(0) > P_i(0)$. Although $T_i(t)$ is switched on suddenly to a new value T_i, $V_i(t)$ gradually increases-then-decreases, while $P_i(t)$ gradually approaches its new equilibrium value, which equals T_i. The rate of change $\frac{dP_i}{dt}$ of P_i provides a measure of the velocity with which the muscle group that quickly tracks $P_i(t)$ will contract. Note that $\frac{dP_i}{dt}$ also gradually increases-then-decreases with a bell-shaped curve whose decelerative portion ($\frac{d^2P_i}{dt^2} < 0$) is slightly longer than its accelerative portion ($\frac{d^2P_i}{dt^2} > 0$), as has also been found in a number of experiments [22,23].

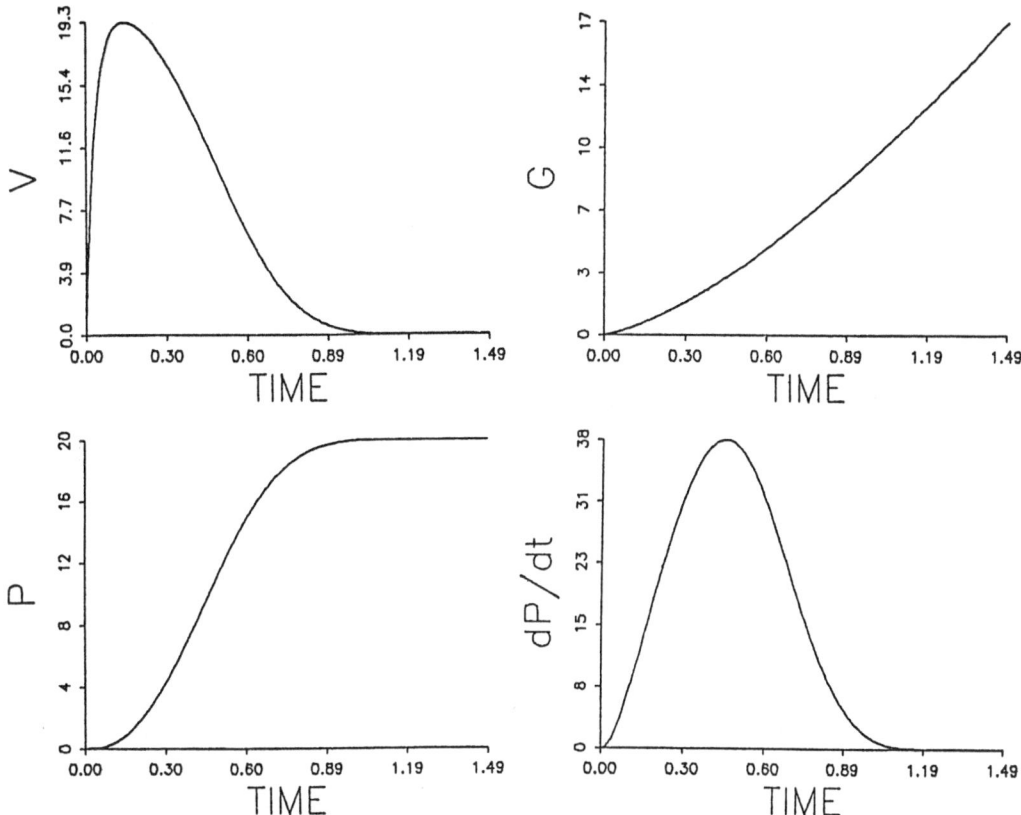

Figure 9. The simulated time course of the neural network activities V, G, and P during an 1100 msec movement. The variable T (not plotted) had value 0 at $t < 0$, and value 20 thereafter. The derivative of P is also plotted to allow comparison with experimental velocity profiles. Parameters for equations (10) and (11) are $\alpha = 30$, $G(t) = t^{1.4}$. (Reprinted with permission from [1].)

Figure 10 demonstrates movement synchrony and duration invariance. This figure shows that the V_i curves and the $\frac{dP_i}{dt}$ curves generated by widely different $T_i(0) - P_i(0)$ values and the same GO signal $G(t)$ are perfectly synchronous through time. This property has been proved mathematically given any nonnegative continuous GO signal [1]. The simulated curves mirror the data collected from vector cells in precentral motor cortex [24,25]. These results demonstrate that the PPC output vector $(P_1(t), P_2(t), \ldots, P_n(t))$ from a VITE circuit dynamically defines a synergy which controls a synchronous trajectory in response to any fixed choice $(T_1, T_2, \ldots, T_n)$ of TPC, any initial positions $(P, (0), P_2(0), \ldots, P_n(0))$, and any GO signal $G(t)$.

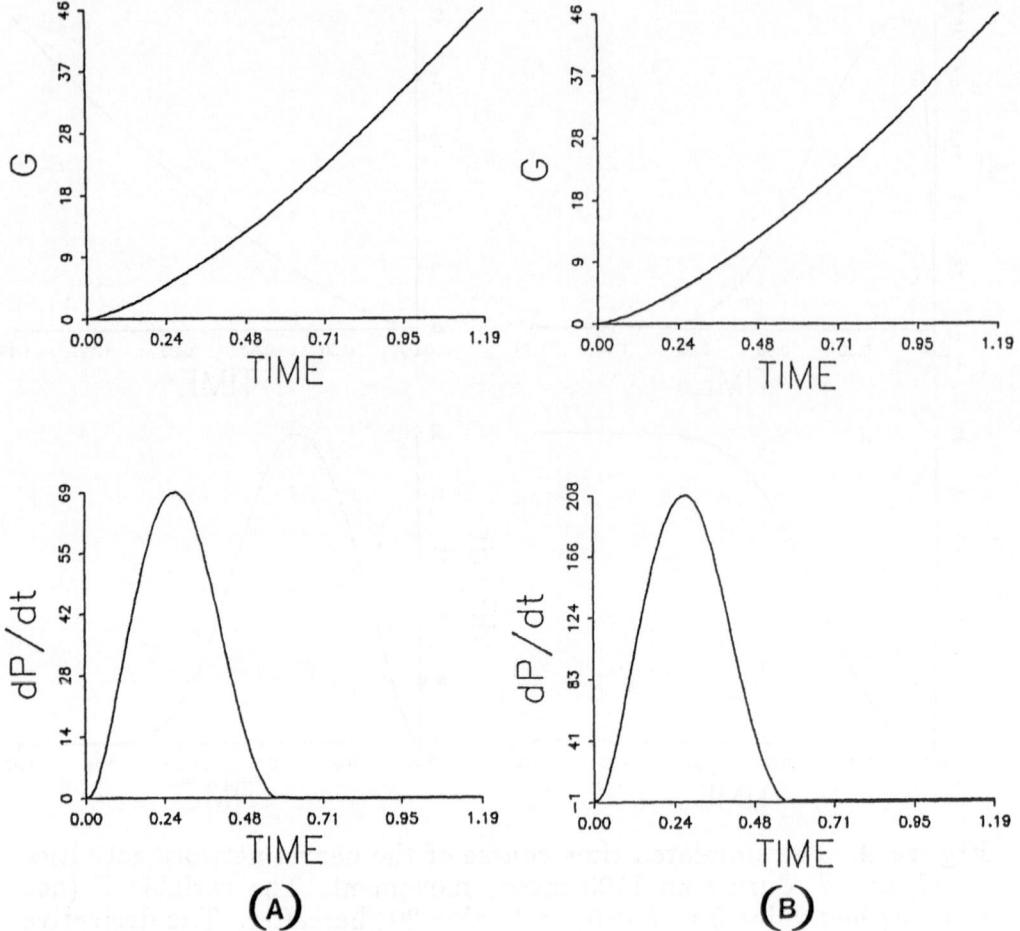

Figure 10. With equal GO signals, movements of different size have equal durations and perfectly superimposable velocity profiles after velocity axis rescaling. (A, B): GO signals and velocity profiles for 20 and 60 unit movements lasting 560 msec. (See Figure 9 caption for parameters.) (Reprinted with permission from [1].)

A wide variety of other important behavioral and neurophysiological data about arm movements have also been quantitatively explained using the VITE model [1,2]. Despite these successes, the VITE model as described above is far from complete. It is merely one module in a larger system for the control of planned and passive arm movements. In order to emphasize the need to define correct system parts in order to piece together the total jigsaw design of this system, we end the chapter by extending our conclusions concerning the manner in which the PPC is computed.

17. Updating the PPC using an Inflow Signal during Passive Movements

When you are sitting in an armchair, let your hands drop passively towards your sides. Depending upon a multitude of accidental factors, your hands and arms can end up in any of infinitely many final positions. If you are then called upon to make a precise movement with your arm-hand system, this can be done with great accuracy. Thus the fact that your hands and arms start out this movement from an initial position which was not reached under active control by an outflow signal does not impair the accuracy of the movement.

Much evidence suggests, however, that comparison between target position and present position information is used to actively move the arms and that, as in Figures 7 and 8, this present position information is computed from outflow signals. In contrast, during the passive fall of an arm under the influence of gravity or other external forces, changes in outflow signal commands are not responsible for the changes in position of the limb. Since the final position of a passively moving limb cannot be predicted in advance, it is clear that inflow signals must be used to update present position when an arm is moved passively by an external force, even though outflow signals are used to update present position when the arm moves actively under neural control.

This conclusion calls attention to a closely related issue that must be dealt with to understand the neural bases of skilled movement: How does the motor system know that the arm is being moved passively due to an external force, and not actively due to a changing outflow command? Such a distinction is needed to prevent inflow information from contaminating outflow commands when the arm is being actively moved. The motor system uses internally generated signals to make the distinction between active movement and passive movement, or postural, conditions. Computational gates are opened and shut based upon whether these internally generated signals are on or off.

Two problems motivate our model of PPC updating by inflow signals, which is summarized in Figure 11. First, the process of updating the PPC during passive movements must continue until the PPC registers the position coded by the inflow signals. Thus a difference vector of inflow signals minus PPC outflow signals updates the PPC during passive movements. We denote this difference vector by DV_p to distinguish it from the DV which compares TPC's with PPC's. At times when $DV_p = 0$, the PPC is fully updated. Although the DV_p is not the same as the DV which compares a TPC with a PPC, the PPC is a source of inhibitory signals, as will be seen below, in computing both difference vectors.

Second, PPC outflow signals and inflow signals may, in principle, be calibrated quite differently. We suggest how PPC outflow signals are

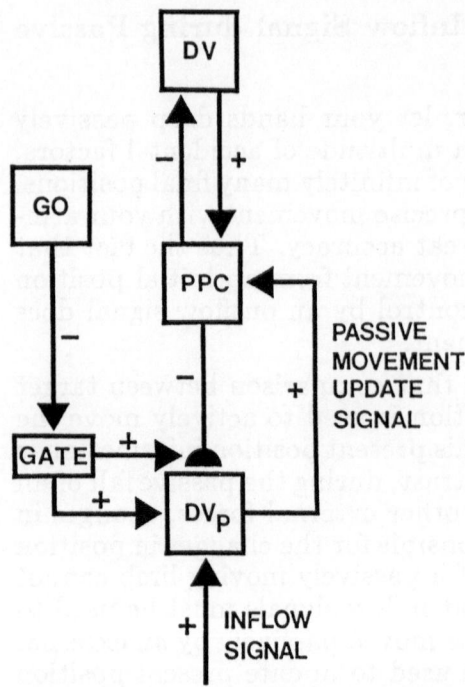

Figure 11. A passive update of position (PUP) circuit. An adaptive pathway $PPC \rightarrow DV_P$ calibrates PPC-outflow signals in the same scale as inflow signals during intervals of posture. During passive movements, output from DV equals zero. Hence the passive difference vector DV_P updates the PPC until it equals the new position caused by any passive movements that may occur due to the application of external forces.

adaptively recalibrated until they are computed in the same numerical scale as the inflow signals to which they are compared. We also show that this adaptive coordinate change automatically computes a DV_p which updates the PPC by the correct amount.

The model circuit in Figure 11 for adaptively computing this DV_p is called the *passive update of position (PUP) model.* In this model, the PPC sends inhibitory outflow signals towards the outflow-inflow match stage where the inflow signals are registered. It is assumed that this DV_p stage is inhibited except when the movement command circuit is inactive. A simple way to achieve this property is to assume that the GO signal in the VITE circuit inhibits the outflow-inflow match stage. Thus the mismatches of outflow and inflow signals that occur during every active movement do not erroneously update the outflow-inflow match stage. In addition, the GO signal is assumed to inhibit learning at the LTM traces which multiply the PPC signals on their way to the outflow-inflow match stage.

After a movement is over, both the outflow-inflow match stage and the LTM traces are released from inhibition. Typically, the PPC represents the same position as the inflow signals, but perhaps in a different numerical scale. The learning laws of the PUP circuit define LTM traces which change until the PPC *times* the LTM trace equals the inflow signal. After a number of such learning trials during stable posture, $DV_p = 0$ and the PPC signals are rescaled by the LTM traces to correctly match the inflow signals.

During a passive movement, the PPC does not change, but the inflow signal may change. If the DV_p becomes positive, it causes an increase in the PPC until the DV_p decreases to 0 and the PPC is correctly updated by the inflow signals. If the DV_p becomes negative, then the DV_p of the opponent muscle can decrease the PPC until a match again occurs.

When a VITE circuit is augmented by a PUP circuit, we find the following system of equations:

Difference Vector

$$\frac{d}{dt}V_i = \alpha(-V_i + T_i - P_i),\qquad(12)$$

Present Position Command

$$\frac{d}{dt}P_i = G[V_i]^+ + G_p[M_i]^+,\qquad(13)$$

Active and Passive Gates

$$GG_p = 0,\qquad(14)$$

Outflow-Inflow Match (Passive Difference Vector)

$$\frac{d}{dt}M_i = -\beta M_i + \gamma I_i - z_i P_i,\qquad(15)$$

Adaptive Coordinate Change

$$\frac{d}{dt}z_i = \delta G_p(-\epsilon z_i + [M_i]^+).\qquad(16)$$

18. Learning an Adaptive Coordinate Change

The match functions M_i in (15) rapidly compute a time-average of the difference vector between inflow (γI_i) and gated outflow $(z_i P_i)$ position codes. Thus

$$M_i \simeq \frac{1}{\beta}(\gamma I_i - z_i P_i). \tag{17}$$

If the inflow signal γI_i exceeds the gated outflow signal $z_i P_i$, then $[M_i]^+ > 0$ in (17). Otherwise $[M_i]^+ = 0$. The *passive gating function* G_p in (13) is positive only when the muscle is in a passive, or postural, state. In particular, $G_p > 0$ only when the GO signal $G(t) \simeq 0$ in the VITE circuit. In Figure 11, it is assumed that a signal $f(G(t))$ inhibits a tonically active source of the gating signal G_p. Thus G_p is the output from a "pauser" cell, which is a tonically active cell whose output is attenuated during an active movement. If both G_p and $[M_i]^+$ are positive in (16), then $\frac{d}{dt}P_i > 0$. Consequently, P_i increases until $M_i = 0$; that is, until the gated outflow signal $z_i P_i$ equals the inflow signal γI_i. At such a time, the PPC is updated to match the position attained by the muscle during a passive movement. To see why this is true, we need to consider the role of function z_i in (15) and (16).

Function z_i is a long term memory (LTM) trace, or associative weight, which adaptively recalibrates the coordinates of inflow signals until they are in the same scale as outflow signals. Using this mechanism, a match between inflow and outflow signals accurately encodes a correctly updated PPC. Adaptive coordinate recalibration proceeds as follows.

In equation (16), the learning rate parameter δ is chosen to be a small constant to assure that z_i changes much more slowly than M_i or P_i. The passive gating function G_p also modulates learning, since z_i can change only at times when $G_p > 0$. At such times, term $-\epsilon z_i$ describes a slow forgetting process which prevents z_i from getting stuck in mistakes. The forgetting process is much slower than the process whereby z_i grows when $[M_i]^+ > 0$. Since function M_i reacts quickly to its inputs γI_i and $-z_i P_i$, as in (17), term $[M_i]^+ > 0$ only if

$$\gamma I_i > z_i P_i. \tag{18}$$

The outflow signal P_i is multiplied, or gated, by z_i on its way to the match interface where M_i is computed (Figure 11).

Because z_i changes only when the muscle is in a postural, or a passive state, terms γI_i and P_i typically represent the same position, or state of contraction, of the muscle group. Then inequality (18) says that the scale γI_i for measuring position I_i using inflow signals is larger than the scale $z_i P_i$ for measuring the same position using outflow signals.

When this happens, z_i increases until $M_i = 0$; viz., until outflow and inflow coordinates are equal.

On an occasion when the arm is passively moved by an external force, the inflow signal γI_i may momentarily be greater than the outflow signal $z_i P_i$. Due to past learning, however, the inflow signal satisfies

$$\gamma I_i = z_i P_i^*, \tag{19}$$

where P_i^* is the outflow command that is typically associated with I_i. Thus by (17),

$$M \simeq \frac{z_i}{\beta}(P_i^* - P_i). \tag{20}$$

By (13) and (20), P_i quickly increases until it equals P_i^*. Thus, after learning occurs, P_i approaches P_i^*, and M_i approaches 0 quickly, so quickly that any spurious new learning which might have occurred due to the momentary mismatch created by the onset of the passive movement has little opportunity to occur, since z_i changes slowly through time. What small deviations may occur tend to average out due to the combined action of the slow forgetting term $-\epsilon z_i$ in (16) and opponent interactions between agonist and antagonist motor commands.

Such adaptive coordinate changes seem to be of general importance in the nervous system, as well as of growing importance in neural computing. The learning of an invariant TPM is itself an adaptive coordinate change. Another adaptive coordinate change was developed in [4] to recode a visually activated target position in head coordinates into the same target position recoded into agonist-antagonist muscle coordinates. DAUGMAN [26] has developed a similar model for adaptive orthogonalization of a Gabor function decomposition of visual images, leading to efficient new schemes for image compression.

19. Concluding Remarks: Parallel Processing and Adaptive Coordinate Change

In summary, the present chapter has outlined recent results concerning how to learn an invariant target position map in head coordinates from combinations of retinotopic and eye position information; of how to learn an intermodality transformation from an eye-head target position map to a hand-arm target position map; of how to convert a hand-arm target position map into a synchronous arm trajectory under variable speed control; and of how to adaptively update the arm's present position command when the arm is passively moved by external forces.

Although these examples illustrate but a few of the neural network modules that are being developed and joined together to form models of complete sensory-motor systems, they already reveal several of the general design themes that repeat themselves, in various guises, during this type of research.

First there is the general theme of parallel processing: Each type of information is often used in multiple ways. For example, outflow information is herein shown to be used to move the arm, to update the active movement vector DV, to update the passive movement vector DV_p, and to provide a teaching signal to learn an intermodality map from eye-head TPM to hand-arm TPM, among other roles.

There is also the general theme of learning adaptive coordinate changes. This theme illustrates that there are several distinct types of learning in the nervous system—there is no single "engram". The chapter on ART [5] showed, for example, how learning can give rise to new sensory and cognitive codes. The present results have shown how learning can adaptively recalibrate one, or many, types of information organized into spatial maps into new spatial maps, thereby allowing dimensionally consistent information processing from multiple sampling sources to be cooperatively and autonomously computed within the unique milieu that evolves during the lifetime of each individual.

This research was supported in part by the Air Force Office of Scientific Research (AFOSR F49620-86-C-0037 and AFOSR F49620-87-C-0018) and the National Science Foundation (NSF IRI-87-16960). Thanks to Cynthia Suchta and Carol Yanakakis for their valuable assistance in the preparation of the manuscript and illustrations.

REFERENCES

1. D.H. Bullock, S.Grossberg: *Psychological Review*, **95**, 1988, p.49.

2. D.H. Bullock, S.Grossberg: In J.A.S. Kelso, A.J. Mandell, M.F. Shlesinger (Eds.), **Dynamic patterns in complex systems**, Singapore: World Scientific Publishers, 1988.

3. S. Grossberg: In D.M. Guthrie (Ed.), **Aims and methods in neuroethology**. Manchester: Manchester University Press, 1987, p.260.

4. S. Grossberg, M. Kuperstein: **Neural dynamics of adaptive sensory-motor control: Ballistic eye movements**. Amsterdam: Elsevier/North-Holland, 1986.

5. G.A. Carpenter, S. Grossberg: In H. Haken (Ed.), **Proceedings of the International Workshop on Neural and Synergetic Computers**, Schloss Elmau, Bavaria, 1988.

6. L.X. Cubeddu, I.S. Hoffmann, M.K. James: *Journal of Pharmacology and Experimental Therapeutics*, **226**, 1983, p.88.

7. M.L. Dubocovich, N. Weiner: In M. Kohsaka *et al.* (Eds.), **Advances in the biosciences, Vol. 37: Advances in dopamine research**. New York: Pergamon Press, 1982.

8. P.M. Groves, G.A. Fenster, J.M. Tepper, S. Nakamura, S.J. Young: *Brain Research*, **221**, 1981, p.425.

9. P.M. Groves, J.M. Tepper: In I. Creese (Ed.), **Stimulants: Neurochemical, behavioral and clinical perspectives**. New York: Raven Press, 1983, p.81.

10. D.M. Niedzwiecki, R.B. Mailman, L.X. Cubeddu: *Journal of Pharmacology and Experimenal Therapeutics*, **228**, 1984, p.636.

11. L. Siever, F. Sulser: *Psychopharmacology Bulletin*, **20**, 1984, p.500.

12. J.M. Tepper, S.J. Young, P.M. Groves: *Brain Research*, **309**, 1984, p.309.

13. S. Grossberg: *Cognitive Science*, **11**, 1987, p.23.

14. S. Grossberg: *Neural Networks*, **1**, 1988, p.17.

15. R.A. Anderson, G.K. Essick, R.M. Siegel: *Science*, **230**, 1985, p.456.

16. D. Zipser, R.A. Anderson: *Nature*, **331**, 1988, p.679.

17. J. Piaget: **The origins of intelligence in children**. New York: Norton, 1963.

18. S. Grossberg: In R. Rosen and F. Snell (Eds.), **Progress in theoretical biology**, Vol. 5. New York: Academic Press, 1978, p.233.

19. E. von Holst, H. Mittelsteadt: *Naturwissenschaften*, **37**, 1950, p.464.

20. A.P. Georgopoulos, A.B. Schwartz, R.E. Kettner: *Science*, **233**, 1986, p.1416.

21. F.B. Horak, M.E. Anderson: *Journal of Neurophysiology*, **52**, 1984, p.290.

22. W.D.A. Beggs and C.I. Howarth: *Quarterly Journal of Experimental Psychology*, **24**, 1972, p.448.

23. H.N. Zelaznik, R.A. Schmidt, S.C.A.M. Gielen: *Journal of Motor Behavior*, **18**, 1986, p.353.

24. A.P. Georgopoulos, J.F. Kalaska, R. Caminiti, J.T. Massey: *Journal of Neuroscience*, **2**, 1982, p.1527.

25. A.P. Georgopoulos, J.F. Kalaska, M.D. Crutcher, R. Caminiti, J.T. Massey: In G.M. Edelman, W.E. Gall, W.M. Cowan (Eds.), **Dynamic aspects of neocortical function.** Neurosciences Research Foundation, 1984, p.501.

26. J. Daugman: *IEEE Transactions on Acoustics, Speech, and Signal Processing*, **36**, 1988, in press.

Neural Nets for the Management of Sensory and Motor Trajectories

R. Eckmiller

Division of Biocybernetics, University of Düsseldorf,
Universitätsstr. 1, D-4000 Düsseldorf 1, Fed. Rep. of Germany

1. Introduction

We are presently witnessing the rapid development of a new interdisciplinary research field: Neural networks for computing alias Neuroinformatics (see, e.g.: ECKMIL-
LER, MALSBURG /1/). The concepts of brain function, cellular automata, artificial
intelligence, synergetics, self-organization, spin-glasses, and complex network topology are being merged in order to develop new information processing hardware as
well as software. This attempt to transfer concepts of brain function to artificial
neural nets for the design of neural computers demands several radical changes in
the approach:
a) The development of software for serial computers is being replaced by an iterative interaction between neural net topology and learning rules for self-organization;
and
b) Algorithmic representations of desired functions are gradually being replaced by
geometrical representations in neural nets with multi-layered or otherwise highly
ordered flexible topologies.

It is the object of this paper to emphasize one major aspect of information processing in biological nervous systems, namely the management of trajectories or space-time functions. Subsequently, a new concept for the management of trajectories in
artificial neural nets will be presented.

2. Sensory Trajectories versus Motor Trajectories

Sensory stimuli, be they auditory, visual, tactile, or vestibular, arrive at specific
sense organs or arrays of receptor cells and generate trajectories or space-time
functions of neural activity in real time. The available information can be decoded
as a pattern of stimulation intensity I at sensory location s as a function of time
t. Such an intensity-modulated trajectory $I(s,t)$ as depicted in Fig. 1, may occur
just once, but has to be stored for purposes of recognition, association, or later
generation of corresponding motor trajectories. Think for example of a young child's
ability to perform a vocalization on the basis of a previously heard new word without the need to make several false attempts and without even recognizing the meaning of this word. Location s in this scheme may correspond for example with retinal
location or with sound frequency (location on the basilar membrane). Stimulation intensity I is represented in Fig. 1 by the trajectory width. In order to store the
time parameter along the path s, we require the representation of time as spatial
location in a specialized neural map. According to my hypothesis sensory trajectories are being stored such that the spatial parameters are stored as a path on one

*) Supported by a grant from the Federal Ministry for Research and Technology
(BMFT), Fed. Rep. Germany.

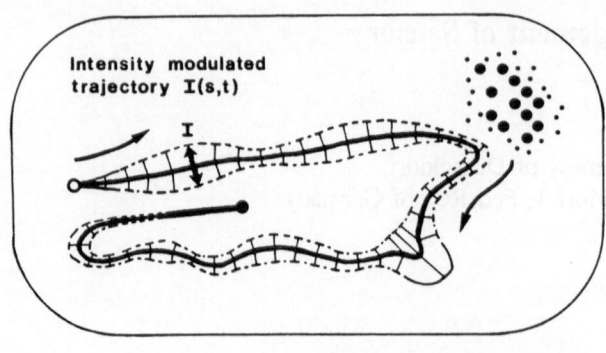

Fig. 1
Intensity-modulated sensory
trajectory I (s,t) on an
receptor array

Array of Receptor Cells

neural map, whereas the temporal parameters are stored as a separate path on another
neural map. Both neural maps can be thought of as 2-dimensional surfaces consisting
of neurons which have connections only to their neighbors.

Biological motor control systems for rhythmic time functions such as those controlling heartbeat, locomotion, and respiration (GRILLNER and WALLEN /2/, SELVERSTON
/3/) are assumed to be based on a small number of neurons. In these cases the time
parameter is represented by oscillations and time constants of neural activity within a small neural net that is capable of internally generating the required time
functions even without sensory feedback from muscles or joints. For quasi-ballistic
goal directed movements (e.g.: reaching, pointing, saccadic eye movements) various
mathematical and neural models have been proposed to explain the typical correlation
between amplitude, velocity profile, and duration of the movement (ARBIB /4/, BERK-
INBLIT et al. /5/, ECKMILLER /6/, HAKEN et al. /7/, HOGAN /8/, MILES and EVARTS
/9/).

A quite different kind of neural motor program generator is required for non-
rhythmic and non-ballistic smooth movements such as speech movements, voluntary limb
movements (dancing, drawing, or gesticulating), or pursuit eye movements. Biological
data on neural networks that act as function generators for such smooth movement
trajectories are scarce (BARON /10/, ECKMILLER /11/, GROSSBERG /12/, MORASSO and
MUSSA IVALDI /13/, SOECHTING et al. /14/).

3. Intelligent Robots as Federations of Various Neural Nets

The vertebrate central nervous system with its many distinct brain regions can be
considered to be a federation of 'special purpose' neural nets (e.g. visual cortex,
cerebellum). The hierarchy and communication lines within this federation as well as
the special purpose functions of its members evolve during early stages of ontogeny
on the basis of pre-set initial neural net architectures, learning rules for self-
organization, and learning opportunities.

One could argue that it is impossible to decode (and copy) an information processing mechanism which required millions of years to evolve. However, the rapidly growing research in the field of neural computers (ECKMILLER and MALSBURG /1/, FELDMAN
and BALLARD /15/) is based on the optimistic assumptions that:
a) It is possible to technically implement various initial neural net architectures
with the capability of self-organization;
b) It is possible to find and subsequently 'imprint' learning rules that lead to robust and rapidly converging self-organization of a neural net with initial architecture or even to changes in a previously established self-organization;

c) It is possible to generate and possibly even predict a desired information processing function by exposing an initial neural net architecture with its imprinted
learning rule to a set of learning opportunities, including experience of imperfect
performance in the learning phase;
d) Following a certain amount of learning experience a given neural net is able to
generalize from a limited set of correctly learned functions to an entire class of
special purpose functions (e.g. robot pointing to any visually marked point within
the entire pointing space independent of the initial robot position).

Intelligent robots are currently being conceived and designed that consist of various neural net modules with special purpose functions, such as: a) pattern recognition, b) associative memory, c) internal representation of patterns and trajectories, d) generation of motor programs, and e) sensory as well as motor coordinate
transformation (ECKMILLER /16/). Let's consider two typical tasks of intelligent robots:
1) 'Speak what you hear' as the ability to generate sound vocalizations that closely
match a newly presented unknown word, and
2) 'Draw what you see' as the ability to draw (or write) visually induced trajectories without seeing the drawing arm.
The modules eye and ear in Fig. 2 represent modules for visual and auditory trajectory detection (not necessarily including pattern recognition). Both sensory input
modules receive signal time courses, which occur as spatio-temporal trajectories at
the level of the corresponding sensory receptor array (retina or basilar membrane in
vertebrates). For example the vision module (eye) monitors the event of a letter 'b'
being written by a teacher on the x-y plane, whereas the hearing module (ear) is
thought to monitor the acoustic event of a spoken 'b'.

It is assumed here that each sensory module is connected with the internal representation module via a specific sensory coordinate transformation module (depicted
as three-layered structure). The function of these coordinate transformation modules
is to generate a normalized spatio-temporal trajectory to be stored in the internal
representation module. For sake of simplicity we might think of these internally
stored trajectories as paths in the 3-dimensional space of a neural net, along which
a particle, neural activity peak, or soliton can travel with constant or varying
speed. The internal representation module can store very large amounts of different

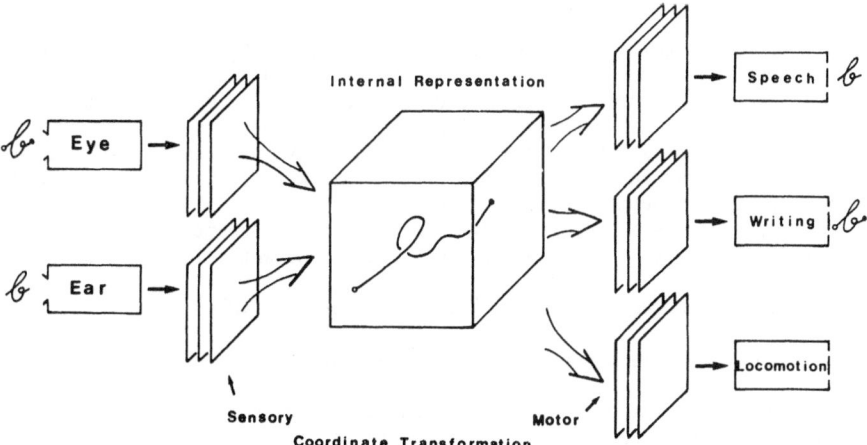

Fig. 2
Schema of an intelligent robot as a federation of various special purpose modules

trajectories, which could share neural elements in this 3-dimensional network. Certain elements of this hypothesis are similar to the engram hypothesis (LASHLEY /17/).

The subsequent generation of a corresponding movement trajectory for drawing a letter 'b' or speaking a 'b' is assumed here to require the activation of the appropriate stored trajectory. Please note that pattern recognition is not required for these two sensorimotor tasks. In fact, the pattern recognition process might operate on the basis of the internally stored trajectories, which are available for iterative recognition procedures after the end of the sensory event. For the purpose of motor program generation an activity peak travels along the stored path on the neural net of the internal representation module with a pre-defined velocity, thus generating various time courses for the spatial components of the desired trajectory.

The desired trajectory can be thought of as the 3-dimensional trajectory of the fingertips while drawing the 'b' or as the virtual center of contraction of all speech muscles while vocalizing a 'b'.

In such typically redundant motor systems specific motor coordinate transformation modules serve to transform the internally generated desired trajectory into a set of simultaneous control signals for the individual actuators (motors, muscles).

The schema of an intelligent robot in Fig. 2 emphasizes the notion that large portions of information processing in both primates and robots with neural net control involve handling of spatio-temporal trajectories for the purpose of detection, recognition, storage, generation, and transformation. The time parameter in these 'trajectory handling neural nets' is not defined by a central clock but is implicitly represented by the propagation velocity of neural activity within the neural net.

Obviously the presently available general purpose digital computers are not particularly well suited for trajectory handling, whereas special purpose neural nets are particularly ill equipped for most mathematical, logical operations. This principle difference between self-organizing neural computers and software-controlled digital computers should, however, not be interpreted as a problem but rather as indication for two application areas with minimal overlap.

4. Management of Motor Trajectories with a Neural Triangular Lattice

The generation of non-rhythmic and non-ballistic smooth velocity time courses requires a flexible neural net acting as a function generator. A neural net model (ECKMILLER /22/) for internal representation and generation of 2-dimensional movement trajectories including PEM is described below. This model accounts for sensory updating, prediction, and storage, and it is biologically (at least in principle) plausible though not based on experimental data. The key features of this neural function generator can be summarized as follows:
1) The velocity time course in one movement direction is represented by the trajectory of a neural activity peak (AP) that travels with constant velocity from neuron to neuron.
2) The neural net is arranged as a neural triangular lattice (NTL) on a circular surface.
3) The NTL output signal is proportional to the eccentricity of AP relative to the NTL center.
NTL Topology: Consider a large number of identical processing elements (neurons) with analog features. These neurons are arranged in a neural triangular lattice (NTL), in which they are connected only to their 6 immediate neighbors. Such a NTL with a radius of 50 neurons from NTL center to its periphery would consist of about 8,000 neurons. The connectivity strength c of tangential synaptic connections bet-

ween neurons located along concentric circles about the NTL center is slightly larger than c for radial synapses. Each neuron has a subthreshold potential P similar
to the membrane potential of biological neurons. P is always assumed to represent
the average of the potential values of all immediately neighboring neurons due to
continuous equilibration of possible potential gradients via neural connections. External potential changes can only be applied to the NTL center thus yielding a center symmetrical potential field. Such an input-dependent potential field can be
thought of as an elastic circular membrane whose center is being pushed up or down.
NTL Dynamics: A suprathreshold activity peak (AP) can be initiated (or extinguished)
only at the NTL center, and becomes superimposed on the potential field. For sake of
simplicity AP can be thought of as a neural action potential (RALL /19/) or a soliton (LAMB /20/), which travels away from the NTL center (following the potential
field gradient during a positive potential input) in the same angular direction (3
o'clock).

AP travels with constant propagation velocity from one neuron to one of its
neighbors, and its activity does not otherwise spread. Every time AP arrives at a
neuron, a decision has to be made concerning its next destination. Due to certain
transient constraints imposed on the adjacent synapses by the travelling AP, only
those 3 of the 6 neighboring neurons can be selected that are located in forward
(straight, right, or left) continuation of the trajectory. The decision as to which
of the 3 possible target neurons is selected, is based on a combined evaluation of
the connectivity strength c and the potential gradient dP/dR between the neighboring
neurons.
If the potential field P is flat throughout the NTL, AP will select the tangentially
adjacent neuron, since c_T is slighty larger than c_R for radial connections. In the
case of a positive potential input at the NTL center, however, the potential field
exhibits a radial gradient towards the periphery, which in effect pulls AP towards
the NTL periphery (equivalent to higher velocities at the NTL output). Similarly, a
negative input attracts AP towards the center.

A portion of the NTL together with AP (filled circle) and the last portion of the
memory trace of its trajectory is shown in Fig. 3. Horizontal connections in Fig. 3
are assumed to follow circles about the NTL center. AP had travelled from the left

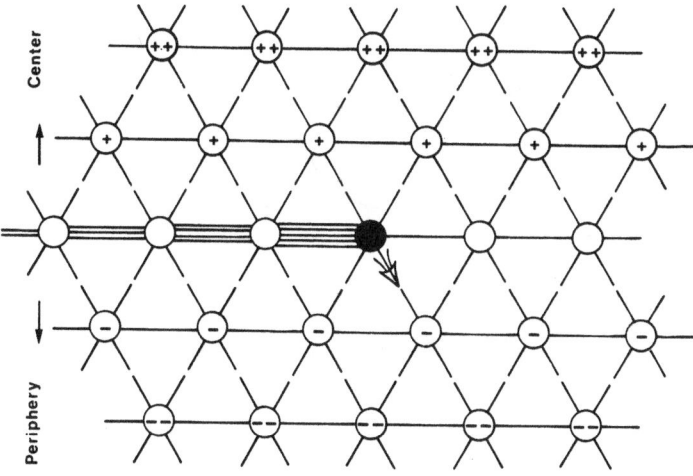

Fig. 3
Portion of neural triangular lattice (NTL) with a travelling activity peak (AP)

233

and is about to move towards the NTL periphery due to an assumed sudden positive step input at the NTL center.

A memory trace of the most recently traversed portion of the trajectory is created by means of a temporary increase of the connectivity strength as indicated by the number of connecting lines in Fig. 3. It is assumed that the memory trace, which gradually fades with a time constant of about 1 s, serves as a trajectory guide for subsequent movements, for example during post-pursuit eye movements after sudden target disappearance and for neural prediction during pursuit of a periodically moving target. Once a memory trace exists, it can be used to reduce the amount of time necessary for updating during periodical movements in one dimension (e.g. horizontal), which are generated by repeated creation of the same alternating velocity trajectories on the two velocity NTLs (Fig. 4).

The potential fields on the two velocity NTLs including the location of AP and the gradually increasing memory trace are shown in Fig. 4 at 4 successive times. AP trajectories always start and end in the NTL center, which corresponds to zero velocity. When the NTL input signal changes from zero to a positive or negative value, the topology of the two NTLs changes by equal amounts in opposite directions. It is

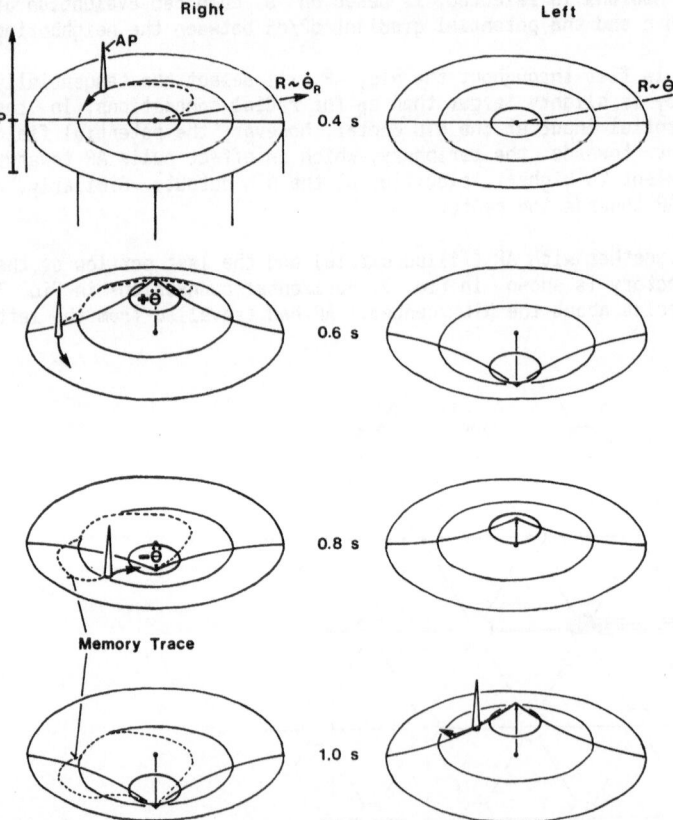

Fig. 4
Development of a trajectory of a travelling AP on a pair of NTLs encoding velocity to the right or to the left at four subsequent times. The membrane-like NTL surfaces indicate the potential fields P.

assumed here that AP had been initiated in the center of the NTL for velocities to
the right (left half of Fig. 4) at time zero and was travelling towards the periphery (starting in the 3 o'clock direction) due to a positive input. The memory trace
is indicated as a dotted line. At 0.4 s AP is travelling on a circle due to the flat
potential field (zero input). At 0.6 s AP travels towards the periphery in response
to a positive input and at 0.8 s towards the center due to a negative input. Please
note that the potential field of the other NTL is always identical except for the
sign and the absence of an AP. At 1.0 s AP had already reached the center of the NTL
and became extinguished there. Simultaneously AP became initiated at the opposite
NTL for velocities to the left (right half of Fig. 4) and is now travelling towards
the periphery. The radial location of AP is continuously being monitored by neurons,
which serve as eccentricity detectors.

Mechanism for learning and retrieval of trajectories:

The detailed synaptic connectivity pattern of 3 adjacent NTL neurons is shown in
Fig. 5 to describe the architecture of an element of this triangular lattice. NTL
neurons (large open circles with axon-like projections and synapses) are reciprocal-
ly connected via synapses. Two additional neurons with pre-synaptic synapses (synapse on another synapse), a pattern retrieval neuron (PR) and a velocity modulation
neuron (VM) have contacts with the NTL. The PR neuron belongs to a large population
of identical PR neurons for storage and retrieval of entire trajectories that always
begin and end in the NTL center. It is assumed that each PR neuron has excitatory
synapses (small circles) on each of the NTL synapses. In contrast, the VM neuron has

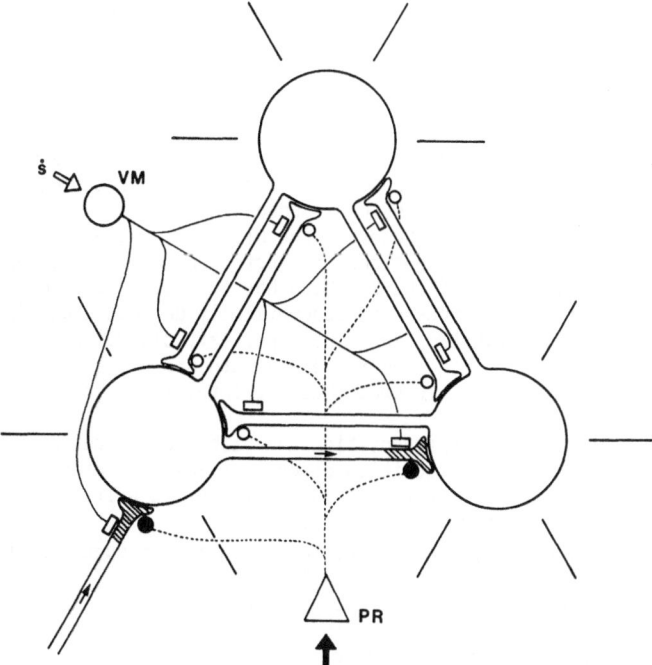

Fig. 5
Neural net architecture of three adjacent NTL neurons together with a pattern retrieval (PR) neuron and a velocity modulation (VM) neuron

inhibitory synapses (depicted as open blocks) on each of the NTL synapses in order
to control the propagation velocity v of AP on the NTL.

When a given trajectory has to be learned (stored), while being generated by means of input modulation of the potential field, one of the PR neurons (that had not
been used before) is first selected for this task, as indicated by the big arrow.

In the present example it is assumed that AP travels along a given trajectory
(from bottom left to right) including the two hatched NTL synapses in Fig. 5. The
connectivity strength of those few synapses which belong to the selected PR neuron
and have a pre-synaptic connection with the NTL synapses along the AP trajectory
(here, the two hatched synapses) becomes permanently increased. This process is assumed here to represent learning (memory) and is indicated by an enlargement of the
two corresponding PR synapses (filled circles). The synapse connectivity of the
other synapses as well as of all synapses of the non-selected PR neurons remains low
(except for those synapses that were involved in storing other trajectories). In
this way a single PR neuron stores one trajectory during the process of its first
generation on the NTL. The later retrieval of this permanently stored trajectory is
implemented by a brief activation pulse for the corresponding PR neuron. This activation pulse is assumed to yield a strong temporary enhancement of those NTL synapses that compose the desired trajectory, thus creating a memory trace. Immediately
afterwards AP becomes initiated in the NTL center. AP travels (although the potential input to the NTL center is zero) along this temporarily existing memory trace like in the groove of a record. Again the NTL output monitors the radial AP distance
during constant propagation velocity of AP and thereby generates the previously stored velocity time function R(t) proportional to s(t).

The concept of a motor program generator with NTL topology and a travelling AP
incorporates elements of soliton theory (LAMB /20/) and of cellular automata theory
(TOFFOLI and MARGOLUS /21/).

5. Application of NTLs for 2-Dimensional Motor Trajectories

The NTL function generator (Figs. 3-5) is equipped with mechanisms for storage and
retrieval of numerous smooth velocity time courses. A pair of NTLs is required to
generate 1-dimensional movements in both directions plus subsequent neural integra-
tors to generate various movement trajectories. However, the NTL concept can be further optimized to allow for generation of 2-dimensional trajectories. Let's consider
an intelligent robot with neural net modules (Fig. 2) which has to learn how to draw
visually monitored patterns. The proposed process is schematically demonstrated in
Fig. 6 by means of four diagrams (ECKMILLER /22/). The upper left diagram gives a
typical writing trajectory for letter 'b' in the x-y plane with start at the open
circle and stop at the filled circle. It is assumed that the intelligent robot has
the following modules available:
1) Vision module with retinotopic internal representation and with neural detectors
of target eccentricity relative to the fovea (position error), as well as of target
slip velocity and target slip acceleration. The target is the tip of a writing element, which the teacher uses to write (draw) on the x-y plane.
2) Velocity NTL with potential input at the NTL center as shown in the upper right
of Fig. 6. AP travels with constant velocity v=c.
3) Position NTL (lower right in Fig. 6) with a velocity modulation (VM) neuron at
the input to modulate the travel velocity of AP. In this case AP position on the NTL
is not monitored as radial distance from the NTL center but as horizontal (x^*) and
vertical (y^*) eccentricity. The VM neuron has a tonic activity,
which normally inhibits AP movement on the position NTL. However, the output of the
velocity NTL inhibits the VM neuron activity, thus yielding a propagation velocity
of AP on the position NTL proportional to the output signal of the velocity NTL.

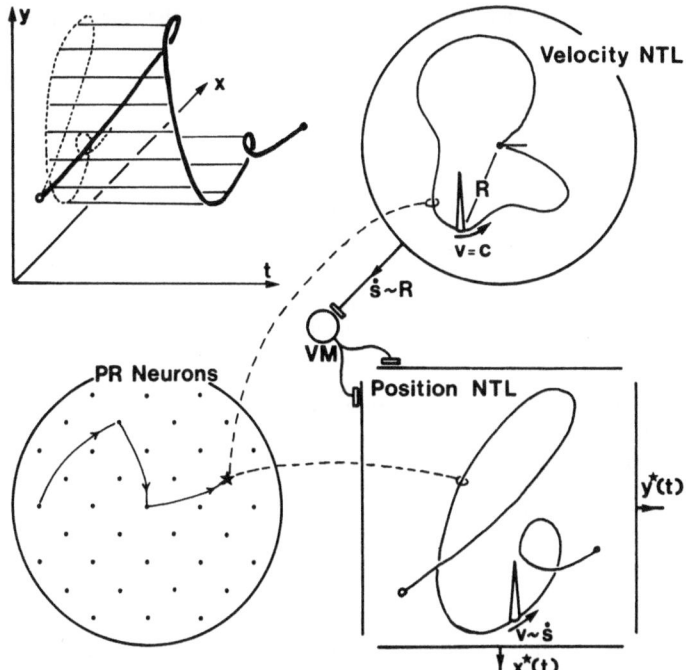

Fig. 6
Schema for generation of a drawing movement trajectory for a letter 'b' by means
of a velocity NTL and a position NTL

Learning and subsequent drawing or writing of a 2-dimensional trajectory such as
letter 'b' consist of the following functions:
a) Teacher draws letter 'b' with writing element in x-y plane.
b) Vision module detects acceleration $\ddot{s}$ of writing element in the direction of drawing trajectory s(t) and uses $\ddot{s}(t)$ to modulate the potential input of the velocity
NTL. This operation yields a velocity $\dot{s}(t)$, which trajectory is immediately stored
by a selected PR neuron (depicted as star in group of PR neurons; lower left in Fig.
6).
c) Upon completion of the teacher's drawing task, the vision module projects the entire pattern of letter 'b' onto the possition NTL, where the same selected PR neuron
(having synapses on both NTLs) stores the pattern.
d) If the robot has to draw the newly stored letter 'b' sometime later, the corresponding PR neuron (star) becomes briefly activated, thus generating the memory traces on both NTLs. The start location of AP on the position NTL is not the center but
the start position (open circle) of the trajectory for letter 'b'. The simultaneously travelling APs on both NTLs generate the appropriate horizontal and vertical components of the desired trajectory to drive a 2-dimensional drawing device (e.g. x-y
plotter).

Please note that this arrangement does not require a neural integrator and that
the spatial parameters of the trajectory are stored on the position NTL, whereas the
temporal aspects are stored separately on the velocity NTL. This important concept
can be compared with the situation of a car on a race track, in which the driver is
only told how to change his speed with time, but not how the track is formed. Only
when he actually drives along the race track (equivalent to a 2-dimensional path),

the spatio-temporal trajectory unfolds and can be monitored by two orthogonally placed eccentricity detectors.

6. Future Research on Neural Computers for Motor Control

The following considerations are based on the assumption that the schema in Fig. 2
is realistic for biological sensorimotor systems and is also feasible for robots
with neural net modules. The relevant modules for motor control are:
A) Internal representation of external space and of desired movement trajectories in
real time (abbreviated as: neural space net) and
B) Kinematics representation and transformation of desired movement trajectories into a set of simultaneous control signals for the corresponding actuators in real time (abbreviated as: neural kinematics net).

The research goal for the next decade is the development of neural space nets and
neural kinematics nets for the control of redundant, nonlinear, and non-orthogonal
motor systems. Such future systems should have the following main features:
1) Acquisition of the required knowledge regarding external space and kinematics by
means of self-organization during a learning phase.
2) Ability to consider obstacles within the grasping space for trajectory planning.
3) Adaptation to partial functional errors and defects or environment changes (e.g.
load, friction, etc.).

Progress in these developments will depend on our detailed knowledge of the corresponding modules in the vertebrate brain as well as on the availability of powerful simulation tools either as software for self-organizing virtual neural nets with
programmable architecture (e.g. neural net workstation), or as hardware for large
numbers of neurons and adaptive synapses with computer- controlled network topology.

Various research groups have recently begun to consider self-organization for
knowledge acquisition in functionally flexible neural nets by using a combination of
initial net topology, learning rules and learning opportunities (ALBUS /23/, BARHEN
et al. /24/, DAUNICHT /25/, ECKMILLER /16/, /26/, GROSSBERG /12/, KAWATO et al.
/27/, KHATIB /28/, KUPERSTEIN /29/). It can be expected that this approach will soon
expand and yield important advances not only in the area of industrial robots but
also in the field of prosthetics.

References

1. R. Eckmiller, C. v.d.Malsburg (eds.): Neural Computers (Springer,Heidelberg
   1988)
2. S. Grillner, P. Wallen In Ann. Rev. Neurosci. 8: 223-261 (1985)
3. A.I. Selverston (ed.): Model neural networks and behaviour (Plenum,
   New York 1985)
4. M.A. Arbib: In Handbook of Physiology, Section 1: The nervous sys,vol.II,
   Motor Control, part 2, ed. by J.M. Brookhart, V.B. Mountcastle, V Brooks (Bal-
   timore, MD: Williams & Wilkins) pp. 1449-1480
5. M.B. Berkinblit, A.G. Feldman, O.I Fukson: In Behav. and Brain Sci. 9:
   585-638 (1986)
6. R. Eckmiller: In IEEE Trans. on Systems, Man and Cybern. SMC-13: 980-989
   (1983)
7. H. Haken, J.A.S. Kelso, H. Bunz: In Biol. Cybern. 51: 347-356 (1985)
8. N. Hogan: In J. Neurosci. 4: 2745-2754 (1984)
9. F.A. Miles, E.V. Evarts: In Ann. Rev. Psychol. 30: 327-362 (1979)

10. R.J. Baron (ed.): In The cerebral computer - An introduction to the computational structure of the human brain (Lawrence Erlbaum Publ., Hillsdale, New Jersey 1987) pp. 402-452
11. R. Eckmiller: In Physiol. Rev. 67: 797-857 (1987)
12. S. Grossberg: In J. of Mathematics and Mechanics vol. 19, no.1: 53-91 (1969)
13. P. Morasso, F.A. Mussa Ivaldi: In Biol. Cybern. 45: 131-142 (1982)
14. J.F. Soechting, F. Lacquaniti, C.A. Terzuolo: In Neurosci. 17: 295-311 (1986)
15. J.A. Feldman, D.H. Ballard: In Cognitive Sci. 6: 205-254 (1982)
16. R. Eckmiller: In Neural Computers, ed. by R. Eckmiller, C. v.d. Malsburg (Springer, Heidelberg 1988) pp. 359-370
17. K.S. Lashley: In Symp. Soc. Exp. Biol. 4: 454-482 (1950)
18. R. Eckmiller: In J. Neurosci. Meth. 21: 127-138 (1987)
19. W. Rall: In Handbook of Physiology, The Nervous System I vol. 1, part 1, ed. by E. Kandel (1977) pp. 39-97
20. G.L. Lamb jr.: In Elements of soliton theory (John Wiley & Sons, New York 1980)
21. T. Toffoli, N. Margolus: In Cellular automata machines - A new environment for modeling (MIT Press, Cambridge, Mass. 1987)
22. R. Eckmiller: In Proc. IEEE First Int. Conf. on Neural Networks, vol. IV (SOS Publ., San Diego 1987) pp. 545-550
23. J.S. Albus: In Math. Biosci. 45: 247-293 (1979)
24. J. Barhen, W.B. Dress, C.C. Jorgensen: In Neural Computers, ed. by R. Eckmiller, C. v.d. Malsburg (Springer, Heidelberg 1988) pp.321-333.
25. W.J. Daunicht: In Neural Computers, ed. by R. Eckmiller, C. v.d. Malsburg (Springer, Heidelberg 1988) pp.335-344
26. R. Eckmiller: In Neural Networks Suppl., in press (1988)
27. M. Kawato, K. Furukawa, R. Suzuki: In Biol. Cybern. 56: 1-17 (1987)
28. O. Khatib: In Int. J. of Robotics Research vol. 5, no. 1: 90-98 (1986)
29. M. Kuperstein: In IEEE Int. Conf. on Robotics and Automation, vol. 3: 1595-1602 (1987)

Part IV

Optical Systems

Spatial Symmetry Breaking in Optical Systems

L.A. Lugiato[1], *C. Oldano*[1], *L. Sartirana*[1], *W. Kaige*[1], *L.M. Narducci*[2], *G.-L. Oppo*[2], *F. Prati*[3], and *G. Broggi*[4]

[1]Dipartimento di Fisica del Politecnico, Corso Duca degli Abruzzi 24, I-10129 Torino, Italy
[2]Physics Department, Drexel University, Philadelphia, PA 19104, USA
[3]Dipartimento di Fisica dell'Università, Via Celoria 16, I-20133 Milano, Italy
[4]Physik-Institut der Universität Zürich, Schönberggasse 9, CH-8001 Zürich, Switzerland

1. Introduction

One of the main issues raised by synergetics [1,2] is the emergence of spatial self-organization in nonlinear dynamical systems far from thermal equilibrium. Phenomena of this kind are often accompanied by the breaking of a spatial symmetry; for example, if an instability creates a spatial pattern in a previously homogeneous system, this corresponds to the spontaneous breaking of translational symmetry.

The formation of spatial structures has been the object of extensive investigations in such fields as nonlinear chemical reactions and developmental biology [1-4]. Here the instabilities that are responsible for the onset of spatial patterns arise from a diffusive mechanism and are usually referred to as Turing instabilities [5].

Optical systems are well known to produce temporal structures in the form of spontaneous oscillations of the regular and chaotic type [6,7]. Only very recently has a Turing instability been discovered [8-11] in an optical model. Here the resulting stationary pattern is produced by the interplay between diffraction and the nonlinear coupling among several transverse modes and not by a diffusion process. The spatial structures arise in the transverse profile of the electric field intensity and break the translational symmetry of the plane-wave configuration.

Recently the Turing instability of Ref. 8 was linked [12] to the phenomenon of spatial soliton formation discovered in earlier numerical investigations [13], and to the modulational instabilities analyzed in Ref. 14. The papers quoted in Refs. 8-11 describe the interaction of a nonlinear medium and an electromagnetic field inside two different kinds of resonators: one capable of supporting transverse modes of the cosine type, the other modes with a Gauss-Laguerre structure.

In this article we consider mainly passive systems with no population inversion. During the last decade passive optical systems have been the object of extensive theoretical and experimental investigations because of their propensity to display optical bistability [15-17] and their possible relation with optical devices that may operate as memories, transistors or logic gates. Hence optical bistability has come under close scrutiny in the areas of optical computing and optical data processing in the hope that it will open the door to the realization of an all-optical computer and the exploitation of massive parallelism. In this sense, perhaps, the optical computer may share some of the architectural features of neural networks. One cannot exclude that the transformations undergone by the transverse beam profile, as described in this paper, may prove useful to optical data processing and information technology.

We analyze also the phenomenon of spontaneous breaking of the cylindrical symmetry in the transverse configuration of an optical beam. In this case we consider an active, i.e. amplifying, system. In a laser operating with a symmetric single-mode Gauss-Laguerre configuration an increase of the pump parameter produces a spatial instability that induces the emergence of a radially asymmetric steady state pattern.

Springer Series in Synergetics Vol. 42: **Neural and Synergetic Computers**
Editor: H. Haken   ©Springer-Verlag Berlin Heidelberg 1988

2. The Kerr model with diffraction

We begin our discussion of spatial patterns in optical systems with the help of an especially simple model. We consider a Fabry-Perot cavity of length L (see Fig. 1) filled with a Kerr medium and driven by an external coherent field.

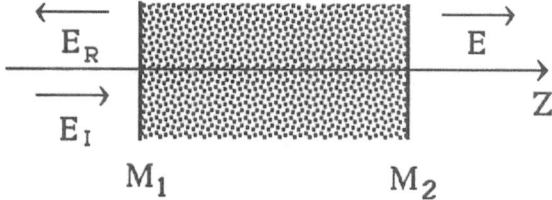

Fig. 1   Fabry-Perot cavity filled with nonlinear material. The mirrors M_1 and M_2 have transmittivity coefficient T. E_I, E, and E_R are the input, output and reflected fields, respectively.

The field internal to the cavity has the structure

$$\mathbb{E} = E \cos(k_z z) e^{-i\omega_0 t} + c.c. , \tag{1}$$

where ω_0 is the frequency of the injected field; $k_z = \pi n_z/L$ is the longitudinal wave number, and n_z is an integer. The envelope function E obeys the time-evolution equation

$$\frac{dE}{dt} = -\kappa\left\{E - E_I - iE\,(|E|^2 - \Theta)\right\} , \tag{2}$$

where $\kappa = cT/2L$ is the cavity damping constant, or cavity linewidth, and E_I is the amplitude of the input field which we assume to be real and positive for definiteness. The detuning parameter Θ accounts for the mismatch between the frequency of the input field and the cavity and for the linear, intensity independent part of the refractive index of the medium. The nonlinear contribution to the refractive index of the material is represented by the cubic term of Eq. (2). This model is valid in the uniform field limit which assumes that the transmittivity coefficient T of the mirrors is much smaller than unity [15].

From Eq. (2) we can obtain immediately the steady state equation linking the input intensity E_I^2 to the transmitted intensity $|E|^2$, i.e.

$$E_I^2 = |E|^2\left\{1 + (|E|^2 - \Theta)^2\right\} . \tag{3}$$

This equation is well known from the earliest days of optical bistability and it was proposed for the first time in Ref. 18. According to Eq. (3) the steady state curve linking $|E|^2$ and E_I^2 is single-valued for $\Theta < \sqrt{3}$ and S-shaped (i.e. bistable) for $\Theta > \sqrt{3}$.

Equation (2) is based on the plane-wave approximation and excludes the possibility of diffractive effects from the outset. In order to include diffraction we complete Eq. (2) as follows

$$\frac{\partial E(x,y,t)}{\partial t} = -\kappa\left\{E(x,y,t) - E_I - iE(x,y,t)\left[|E(x,y,t)|^2 - \Theta\right]\right\}$$

$$+ i\frac{c}{2k_z}\left(\frac{\partial^2}{\partial x^2} + \frac{\partial^2}{\partial y^2}\right) E(x,y,t) , \tag{4}$$

where x and y are the transverse coordinates. Because we assume that the input field E_I has a flat transverse profile, Eq. (4) allows stationary solutions that are homogeneous along the x and y directions and coincide, in fact, with the ones calculated using Eq. (2) as the starting point [see Eq. (3)].

In general, when E_I is a plane wave, it is also common to assume that the cavity field is independent of x and y, so that Eq. (4) automatically reduces to the form of Eq. (2). This assumption, however, is not always correct, as one can verify from the study of the linear stability properties of the homogeneous stationary solutions. The main problem with Eq. (2) is that it allows only those fluctuations that are uniform along the transverse directions. Random perturbations, on the other hand, do not have to be uniform at all, so that a more general linear stability analysis is needed using Eq. (4) as the starting point. The importance of diffraction in connection with the stability properties of plane-wave stationary solutions was already emphasized in Ref. 14.

The presence of second order space derivatives in Eq. (4) requires the specification of additional boundary conditions for our problem. For this purpose, we consider a cavity defined by four mirrors with transmittivity coefficient T<<1; two mirrors are orthogonal to the longitudinal z-axis and are separated by a distance L from one another; the other two are orthogonal to the x-axis and are spaced a distance b (see Fig. 2). The cavity is open along the y direction.

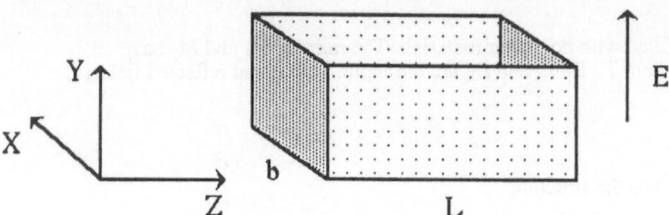

Fig. 2   Schematic representation of the cavity configuration discussed in this article. The four mirrors have reflection coefficients T<<1. The cavity is open along the y direction which is also the direction of polarization of the electric field.

We assume that both the input and the cavity field are linearly polarized along the y direction [other choices of polarization, leading again to a transverse modal configuration of the type $\cos k_x x$ as in Eq. (6a) are described in Ref. 8]. From this requirement and from the transversality condition, it follows that E is independent of y. In this case, Eq. (4) reduces to

$$\frac{\partial E(x,t)}{\partial t} = -\kappa\left\{E(x,t) - E_I - i\,E(x,t)\left[|E(x,t)|^2 - \Theta\right]\right\} + i\,\frac{c}{2k_z}\,\frac{\partial^2 E(x,t)}{\partial x^2}\ . \qquad (5)$$

Furthermore, we assume that the cavity can support modes of the type

$$\cos(k_x x)\cos(k_z z) \qquad\qquad (6a)$$

where

$$k_x = \frac{\pi n}{b}\ ,\quad k_z = \frac{\pi n_z}{L} \qquad\qquad (6b)$$

and n, n_z are nonnegative integer numbers.

Note that Eq. (5) fixes the value of n_z, while the integer n remains free to vary over the range 0,1,2,... As a result, Eq. (5) is a single longitudinal mode model, but it can account for an infinite number of transverse modes. The stationary solution, Eq. (3), corresponds to selecting n=0; however, as we shall show in the next section, solutions of this type can become unstable under appropriate conditions and allow the growth of transverse modes with n≠0.

3.   Stationary spatial patterns

In order to describe this instability, it is useful to focus on the cavity frequencies which are given by the usual expression

$$\omega = c\sqrt{k_x^2 + k_z^2}\ . \qquad\qquad (7)$$

The model (5) assumes the paraxial approximation $k_x \ll k_z$ so that Eq. (7) can be approximated by

$$\omega \approx ck_z + \frac{c}{2k_z}k_x^2 \qquad (8)$$

or, with the help of Eq. (6b), by

$$\omega \approx \pi n_z \frac{c}{L} + \pi^2 n^2 \frac{c}{2k_z b^2} \quad . \qquad (9)$$

Finally, if we take into account that $k_z = 2\pi/\lambda$, and if we introduce the Fresnel number

$$F = \frac{b^2}{\lambda L} \quad , \qquad (10)$$

we obtain the expression

$$\omega = \pi n_z \frac{c}{L} + a(n)\kappa \quad . \qquad (11)$$

In Eq. (11) we have defined

$$a(n) = \frac{\pi n}{2T\,F} \qquad n = 0,1,2,\ldots\ldots \qquad (12)$$

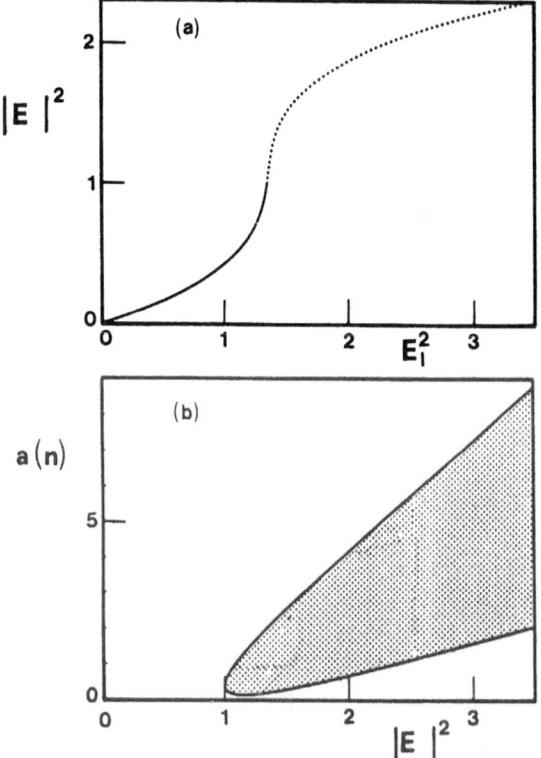

Fig. 3   (a) Steady state curve for $\Theta=1.6$. The broken part indicates the stationary solution that may become unstable. (b) The shaded region corresponds to the unstable domain in the plane of the variables a(n) [see Eq. (12)] and $|E|^2$.

The parameter a(n) represents the frequency difference between the n-th transverse mode and the resonant longitudinal mode, measured in units of the modal linewidth κ. Note that, on the frequency axis, the transverse modes lie to the right of the longitudinal mode. Because we assume that T<<1 and we want a(1) to be of the order of unity, we must require that F>>1.

We are now in a position to understand the origin of the transverse mode instability. If we consider first only the longitudinal cavity resonances, we note that they are very narrow (for T<<1) and well removed from one another because the modal width is much smaller than the mode spacing $2\pi c/2L$. As a result, the frequency ω_0 of the input field interacts only with the resonant cavity mode and the other longitudinal resonances do not affect the stationary state. If we now take into account also the transverse modes, we see from Eq. (11) that, when a(1) is of the order of unity, the nearest transverse resonances overlap the longitudinal resonant mode. This situation triggers a mode-mode competition which is at the origin of the destabilization of the homogeneous stationary solution.

Typical results of the formal linear stability analysis are illustrated in Fig. 3 for Θ=1.6. Consider, first, the steady state curve for the transmitted versus incident intensity (Fig. 3a): each point on the curve such that $|E|^2 \geq 1$ is unstable provided that for the corresponding value of $|E|^2$ at least one of the numbers a(n), n=1,2,... lies in the shaded region of Fig. 3b. This region, in other words, identifies the domain of the unstable transverse modes.

The special and interesting feature of this robust instability is that it does not lead to oscillatory behavior or to other dynamical effects, but it produces a <u>new stationary state</u>. To be more precise, the electric field in the new stationary configuration is still characterized by a single oscillation frequency that coincides with the incident frequency ω_0. Hence the input beam imposes its frequency to all the cavity modes that appear in the new stationary state.

In the neighborhood of the critical point we calculated the new stationary solution analytically using bifurcation theory [19], and found that the bifurcation is subcritical (i.e. the spatial structure

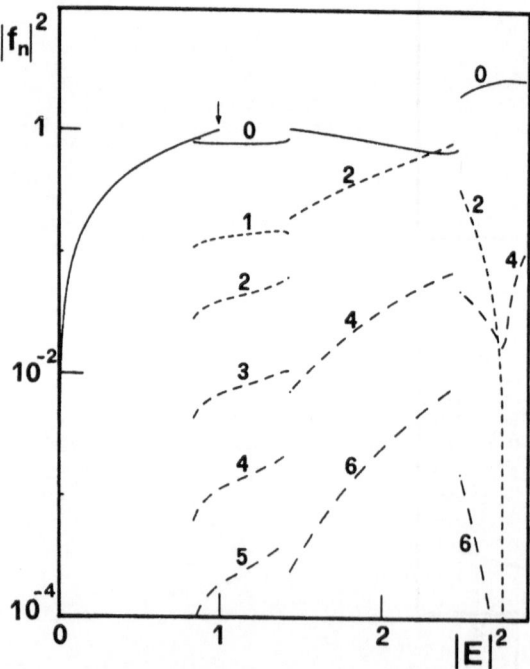

Fig. 4   Passive Kerr medium, Cartesian cavity configuration, Θ = 1.6. The intensities of the modes with n = 0, 1, ..., 6 are plotted as functions of the normalized output intensity of the homogeneous stationary solution. The arrow indicates the critical point.

emerges discontinuously at the critical point) for $\Theta > 41/30 \approx 1.4$. Figure 4 shows the result of a numerical calculation of the modal intensities $|f_n|^2$ as a function of the output intensity $|E|^2$ of the homogeneous stationary solution which is linked to the input intensity by Eq. (3). The modal amplitudes are defined by the equation

$$E(x',t) = \sum_{n=0}^{\infty} f_n(t) \cos(\pi n x') ; \qquad x' = x/b ; \tag{13}$$

in steady state the amplitudes f_n are time independent.

At the critical point there is a discontinuous jump in the intensity of the homogeneous mode, $n=0$, and the abrupt appearance of the transverse modes 1, 2,...,5. On decreasing the input intensity below the critical point, the system remains in the inhomogeneous state producing hysteresis, as is typical of subcritical bifurcations. If instead one increases the input power above the critical point over the interval of $|E|^2$ shown in Fig. 4, one finds two other values of the input intensity such that the transverse profile undergoes a discontinuous transition.

The analysis of the spatial instability formulated in Ref. 8 for a Kerr medium has been extended in Ref. 11 to the case of a two-level passive system. In this situation the model displays not only spatial but also temporal instabilities.

4. Passive systems, cavity with spherical mirrors

If we now insert the expansion (13) into Eq. (5) and take into account the orthogonality of the functions $\cos(\pi n x')$ over the interval $0<x'<1$, we obtain a set of ordinary differential equations for the modal amplitudes f_n:

$$\kappa^{-1} \frac{df_n}{dt} = - f_n + E_I \delta_{n,0} - ia(1)n^2 f_n + i \frac{1}{4(1+\delta_{n,0})} \left(\sum_{n'n''n'''}^{*} f_{n'} f_{n''} f_{n'''}^{*} - \Theta f_n \right) , \tag{14}$$

where the asterisk implies that the sum is restricted to terms with positive n''' and with $n''' = \pm n \pm n' \pm n''$; note that all the combinations of upper and lower signs must be included. The parameter $a(1)$ is given by

$$a(1) = \frac{\omega_1 - \omega_0}{\kappa} \tag{15}$$

where ω_0 and ω_1 are the frequencies of the modes $n=0$ and $n=1$, respectively; for the cavity shown schematically in Fig. 2, $a(1)$ coincides with the value given by Eq. (12) for $n=1$.

The Cartesian waveguide geometry of Fig. 2 is not very common with the possible exception of some investigations in the microwave frequency range. For this reason, we consider also a more standard cavity configuration with spherical mirrors in which the transverse modes are still discrete in spite of the absence of the lateral mirrors. For definiteness, we consider the ring cavity configuration of Fig. 5.

The cavity modes have the Gauss-Laguerre transverse structure [20,21]

$$A_{pl}(r,\varphi) = (\frac{r}{w_0})^l L_p^l (\frac{2r^2}{w_0}) \exp(- \frac{r^2}{w_0^2}) \begin{Bmatrix} \sin l\varphi \\ \cos l\varphi \end{Bmatrix} , \tag{16}$$

where $r = (x^2+y^2)^{1/2}$ and φ are the radial and angular transverse variables, respectively, w_0 is the beam waist parameter and L_p^l denotes the Laguerre polynomials, with $p,l = 0, 1,.....$ The cavity frequencies are given by

$$\omega = 2\pi n_z \frac{c}{L} + \frac{1}{2}a(1) (2p+l+1) , \tag{17}$$

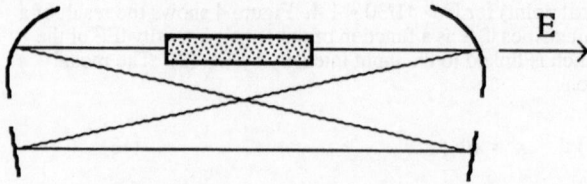

Fig. 5   Schematic representation of a ring cavity. The spherical mirrors have transmittivity coefficient T, the plane mirrors are ideal reflectors. E denotes the output field.

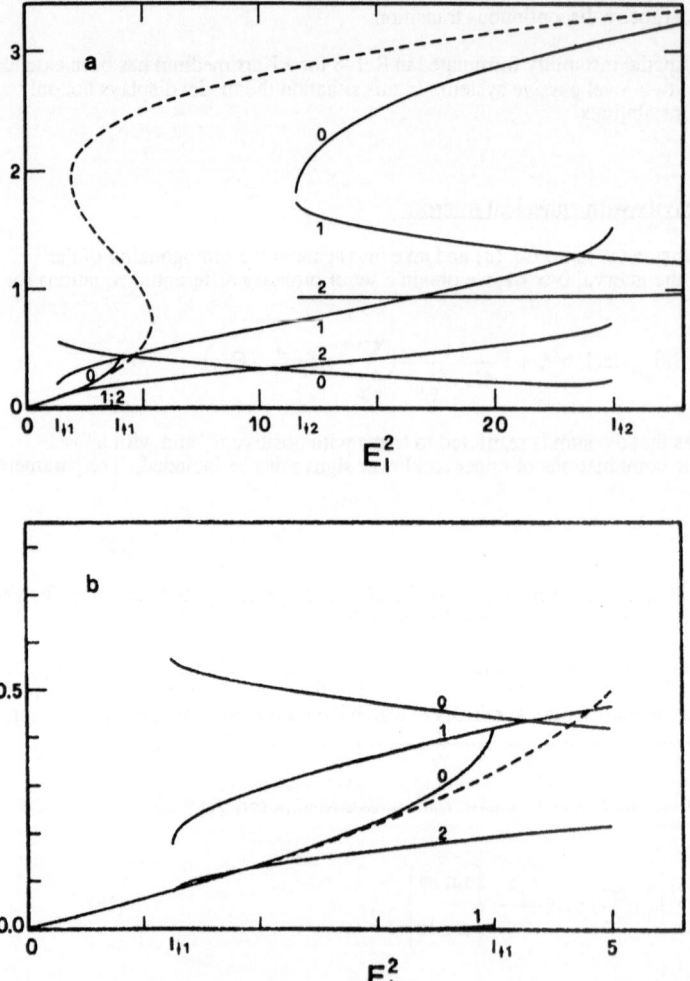

Fig. 6   Passive Kerr medium, cavity with spherical mirrors, $\Theta = 4$. (a) The broken line corresponds to the single-mode steady state curve (see text); the solid lines display the intensities of the modes $p = 0, 1, 2$ as functions of the input intensity. (b) Expanded version of part of figure (a).

where n_z has the same meaning as in Eq. (9), L is the cavity length and the constant a(1) depends on the exact geometrical details of the cavity [21]. For the moment, we assume cylindrical symmetry and therefore we expand the electric field in the following way :

$$E(r',t) = \sum_p f_p(t) \exp(-r'^2) L_p(2r'^2)$$ (18)

with $r' = r/w_0$. Assuming, as in the Cartesian case, that the Fresnel number $F = w_0^2/\lambda L$ is much greater than unity in such a way that $FT = O(1)$, one obtains [22] a set of dynamical equations for the amplitudes f_p of the form

$$\kappa^{-1} \frac{df_p}{dt} = -f_p + E_I \delta_{p,0} - ia(1)pf_p + i\left(\sum_{p'p''p'''} f_{p'}f_{p''}f_{p'''}^* \, c_{pp'p''p'''} - \Theta f_p\right),$$ (19)

where a(1) is still given by Eq. (15) and ω_0 and ω_1 are the frequencies of the modes p=0 and p=1, respectively. If we compare Eqs. (14) and (19) we note two main differences: (1) the term containing a(1), which arises from diffraction is proportional to n^2 in Eq. (14) and to p in Eq. (19); (2) the coefficients $c_{pp'p''p'''}$ of the nonlinear term in Eq. (19) couple all the amplitudes f_p. As a consequence, we have been unable to calculate analytically any stationary solutions of Eq. (19).

If, however, in Eq. (19), one neglects every amplitude except for f_0 and takes into account that $c_{0000} = 1$, one finds that $|f_0|^2$ obeys a steady state equation identical to Eq. (3). We shall call the solutions of this equation the "single-mode stationary solutions"; we must keep in mind, however, that these are not exact stationary solutions of Eq. (19), although they play a useful role in analyzing the actual stationary solutions and in studying the influence and behavior of the various transverse modes. Figure 6 illustrates the steady state behavior of the system for $\Theta = 4$; these results were obtained by solving numerically the time-dependent equations (19) with the three modes n = 0, 1, and 2. The broken curve traces the single-mode steady state; the solid lines show the intensities $|f_p|^2$ (for p=0,1,2) in the long-time regime.

For small values of the input intensity E_I^2 the steady state solution is practically of the single-mode type, the amplitudes f_1 and f_2 being negligible. For $E_I^2 = I_{\uparrow 1}$ one observes a discontinuous transition in which the modes p=1,2 acquire a sizable intensity. If at this point one decreases E_I^2 one finds hysteretic behavior, as expected. If, on the other hand, one increases E_I^2 beyond the value $I_{\uparrow 1}$, the intensity of the transverse modes p=1,2 becomes larger than that of the fundamental mode p=0 until one reaches the point $I_{\uparrow 2}$ where a second discontinuous transition brings the fundamental mode to dominate again.

Figure 6 displays two hysteresis cycles for $I_{\downarrow i} < E_I^2 < I_{\uparrow i}$ (i=1,2); this picture is very different from the standard hysteresis cycle of the single-mode solution because the bistability obtained here arises from discontinuous changes of the transverse profile.

5.   Spontaneous breaking of the cylindrical symmetry

In the case of the ring cavity with spherical mirrors we found transitions in the transverse spatial structure; strictly speaking, however, these transitions are not accompanied by a symmetry breaking. In this section we demonstrate that this type of cavity configuration can also generate a real symmetry breaking phenomenon: this is associated with the breaking of the cylindrical symmetry.

As shown in Eq. (17), different transverse modes with the same value of 2p+l are degenerate in frequency. We consider a ring laser with spherical mirrors in which the atomic line is exactly resonant with the three degenerate modes p = 1, l = 0 and p = 0, l = 2 [sine and cosine; see Eq. (16)]. We assume that all the other modes are quite removed in frequency from the atomic resonance, so that their dynamical contribution can be ignored and the problem can be reduced to the interaction of three modes. We assume that the pump profile has the Gaussian form

$$\exp\left(-\frac{r^2}{2R_p^2}\right),$$ (20)

where R_p is the pump waist. The system admits three distinct single-mode stationary solutions governed by the steady state equations

$$2C \int_0^{2\pi} d\varphi \int_0^\infty dr'\ r'\ \frac{A_{pl}^2(r',\varphi)}{1+A_{pl}^2(r',\varphi)\ f_{pl}^2} \exp\left(-2\frac{r'^2}{\psi^2}\right) = 1\ , \qquad (21)$$

where $r' = r/w_0$, A_{pl} is given by Eq. (16), f_{pl} is the amplitude of the mode (p,l) and $\psi = 2R_p/w_0$. The cylindrically symmetrical stationary state $p = 1, l = 0$ (Fig. 7a) has the lowest threshold. Hence, if we increase the pump parameter gradually, the laser starts oscillating with the symmetrical mode. We performed analytically [23] the linear stability analysis of this stationary solution obtaining two eigenvalue equations, one for the amplitude and the other for the phase fluctuations of the electric field. The latter equation predicts the emergence of an instability when the condition

$$2CI > 1 \qquad (22)$$

$$I = \int_0^{2\pi} d\varphi \int_0^\infty dr'\ r'\ \frac{A_{02}^2(r',\varphi)}{1+A_{10}^2(r')\ f_{10}^2} \exp\left(-2\frac{r'^2}{\psi^2}\right) \qquad (23)$$

is satisfied. The ratio of the instability threshold to the ordinary laser threshold is close to unity for $\psi \gg 1$. When the symmetric stationary solution becomes unstable, the asymmetric solutions are also unstable. The numerical solution of the dynamical equations show [23] that, if the system is initially in the symmetric state (Fig. 7a) and the pump parameter satisfies the instability condition (22), the system evolves into a new stationary state in which the three modes coexist and the cylindrical symmetry is broken (Fig. 7b). There is in fact an infinite number of asymmetric stationary solutions that can be obtained from one another by a simple rotation around the axis of the system; an initial fluctuation determines which of these solutions is approached by the system.

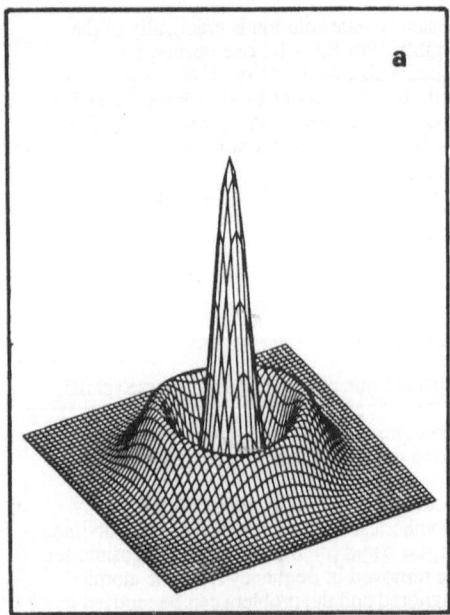 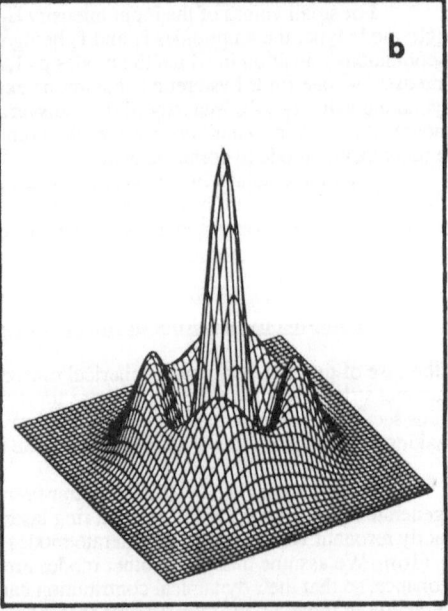

Fig. 7   Three-mode ring laser with spherical mirrors. Evolution from a cylindrically symmetric (a) to an asymmetric (b) steady state transverse configuration of the output intensity. The laser is 50% above threshold, $\psi = 1.5$ and $\kappa = \gamma_\perp = \gamma_\parallel$.

Similar phenomena involving the spontaneous breaking of the cylindrical symmetry have been discovered by numerical studies in Ref. 24 and have also been observed experimentally in four-wave mixing experiments [25] without a resonator and in a configuration involving a single mirror [26].

Acknowledgements

This work has been performed with the partial support of a NATO Collaborative Research Grant and in the framework of the EEC twinning project on "Dynamics of Nonlinear Optical Systems".

References

1. H. Haken, Synergetics - An Introduction, Springer-Verlag, Berlin, 1977.
2. H. Haken, Advanced Synergetics, Springer-Verlag, Berlin, 1983.
3. G. Nicolis and I. Prigogine, Self-Organization in Nonequilibrium Systems, Wiley, New York, 1977.
4. J.D. Murray, J. Theor. Bio. 88, 161 (1981).
5. A.M. Turing, Phil. Trans. Roy. Soc. London B237, 37 (1952).
6. N.B. Abraham, L.A. Lugiato, and L.M. Narducci, Eds., Feature Issue on Instabilities in Active Optical Media, J. Opt. Soc. Am. B2, January 1985
7. F.T. Arecchi, and R. Harrison, Eds., Instabilities and Chaos in Quantum Optics, Springer-Verlag, Berlin, 1987.
8. L.A. Lugiato and R. Lefever, Phys. Rev. Lett. 58, 2209 (1987).
9. L.A. Lugiato, L.M. Narducci, and R. Lefever, in "Lasers and Synergetics - a volume in honor of the 60th birthday of Hermann Haken", Springer-Verlag, Berlin 1987.
10. L.A. Lugiato, and R. Lefever, volume in honor of the 70th birthday of Adriano Gozzini, in press.
11. L.A. Lugiato and C. Oldano, Phys. Rev. A, in press.
12. A. Ouazzardini, H. Adachihara and J.V. Moloney, private communications.
13. J.V. Moloney and H.M. Gibbs, Phys. Rev. Lett. 48, 1607 (1982).
14. D.W. McLaughlin, J.V. Moloney and A.C. Newell, Phys. Rev. Lett. 54, 681 (1985).
15. L.A. Lugiato, "Theory of Optical Bistability", in Progress in Optics, Vol. XXI, edited by E. Wolf, North Holland, Amsterdam, 1984.
16. E. Abraham and S.D. Smith, Rept. Progr. Phys. 45, 815 (1982).
17. H.M. Gibbs, Optical Bistability: Controlling Light by Light, Academic Press, New York, 1985.
18. H.M. Gibbs, S.L. McCall and T.N.C. Venkatesan, Phys. Rev. Lett. 36, 113 (1976).
19. D. Sattinger, Topics in Stability and Bifurcation Theory, Springer-Verlag, Berlin, 1973.
20. A. Yariv, Optical Electronics , 3rd Edition, Holt, Rinehart and Winston, New York, 1985.
21. P. Ru, L.M. Narducci, J.R. Tredicce, D.K. Bandy and L.A. Lugiato, Opt. Comm. 63, 310 (1987).
22. L.A. Lugiato, L.M. Narducci and Wang Kaige, in preparation.
23. L.A. Lugiato, F. Prati, L.M. Narducci and G.-L. Oppo, in preparation.
24. J.V. Moloney, H. Adachihara, D.W. McLaughlin and A.C. Newell, in Chaos, Noise and Fractals, E.R. Pike and L.A. Lugiato, Eds., Adam Hilger, Bristol, 1987.
25. G. Grinberg, Opt. Comm. to be published.
26. G. Giusfredi, J.F. Valley, R. Pon, G. Khitrova and H.M. Gibbs, in J. Opt. Soc. Am. B, Feature Issue on Laser Dynamics, D.K. Bandy, J.R. Tredicce and A. Oraevsky, Eds., May 1988.

Cooperative Frequency Locking and Spatial Structures in Lasers

L.A. Lugiato[1], *W. Kaige*[1*] *L.M. Narducci*[2], *G.-L. Oppo*[2], *M.A. Pernigo*[2], *J.R. Tredicce*[2], *D.K. Bandy*[3], *and F. Prati*[4]

[1]Dipartimento di Fisica del Politecnico,
 Corso Duca degli Abruzzi 24, I-10129 Torino, Italy
[2]Physics Department, Drexel University, Philadelphia, PA 19104, USA
[3]Physics Department, Oklahoma State University, Stillwater, OK 74078, USA
[4]Dipartimento di Fisica dell'Università, Via Celoria 16, I-20133, Milano, Italy

1.   Introduction

Lasers have enjoyed from their very beginning a preeminent spot in the physical sciences not only for their obvious and by now ubiquitous technological role, but also as a testing ground for new and far reaching ideas. Haken's felicitous characterization of the laser as the trailblazer of Synergetics was advanced in recognition of the laser's conceptual contributions to our present understanding of pattern formation and cooperative phenomena [1-3]. The early focus of theoretical laser studies was on symmetry breaking phenomena in the time domain, and the field of laser instabilities that grew in response to the pioneering works of the early 1960's came to be identified closely with the study of output pulsations and dynamical chaos [4].

Investigations of spatial symmetry breaking effects flourished independently and in fact, well before the advent of the laser, had already become an important component of chemistry and developmental biology [5-7]. From the early 1970's Synergetics provided the essential impetus for the unification of these apparently unrelated problems with the development of the appropriate mathematical language and, even more so, with the identification of the central issues: By what mechanisms do entirely different systems undergo spontaneous symmetry breaking, and develop spatial and temporal patterns with a stability of their own? What causes these patterns to grow in complexity until chaos, the ultimate complexity, eventually dominates? Why, as the original symmetry is degraded through a hierarchy of transformations, does a new deeper form of symmetry, the scale invariance, emerge where only disorder seems to be present?

While the search for a unified explanation of these and other questions continues from within many branches of science, the study of optical interactions is providing additional evidence for the existence of an underlying unity of symmetry breaking phenomena in space and time. The recent discovery of a Turing instability in an optical system [8] has stimulated investigations of the spontaneous formation of spatial and temporal patterns in passive and active optical devices [9-11]. A companion article in this book contains an overview of our current understanding of this problem in passive systems [12]; here the focus is on the behavior of the theoretical and experimental side of their active counterparts.

The essential motivation for these studies came from the realization that diffraction effects in nonlinear optical systems can be responsible for mode-mode coupling and for the growth of previously unexcited modes of the radiation field [8], in much the same way as diffusion effects in chemical and biological systems produce permanent concentration gradients and spatial inhomogeneities [13]. In a laser the competition among transverse cavity modes is responsible for several new dynamical features [14] that are hidden or, perhaps better, frozen in the traditional plane-wave models. Transverse modes in a cavity have different modal volumes and are coupled to the active population with varying efficiency; they suffer different diffraction losses because of their different transverse cross sections, and have a frequency distribution that depends in a

---

*   Permanent address: Department of Physics, Beijing Normal University, Beijing, People's Republic of China.

Springer Series in Synergetics Vol. 42: **Neural and Synergetic Computers**
Editor: H. Haken   ©Springer-Verlag Berlin Heidelberg 1988

sensitive way on the geometry of the resonator. In particular, their frequency spacing, hence their degree of overlap, can be controlled by suitable variations of the radii of curvature of the mirrors which leave the cavity free spectral range constant. Thus, the interaction among transverse modes can be varied without significant competition from neighboring longitudinal modes which would complicate the picture.

As we learn from elementary laser physics, the resonant frequency of an empty cavity mode changes somewhat in the presence of an active medium because of the frequency pulling effect. If ω_A represents the center frequency of the atomic transition and ω_C the resonant frequency of an empty cavity mode, a single-mode laser operates in steady state with a carrier frequency

$$\omega_L = \frac{\kappa\omega_A + \omega_C}{\kappa + 1} \quad , \tag{1.1}$$

where κ is the linewidth of the cavity mode measured in units of the atomic linewidth $\gamma_\perp$. Thus, it is normally assumed that if two or more laser modes are excited, the resulting output intensity will display a beat pattern caused by the interference among the pulled frequencies of the oscillating field components. This picture, in fact, is supported by numerical solutions of the plane-wave Maxwell-Bloch equations [15].

A notable twist develops if one takes transverse effects into consideration because under appropriate conditions a remarkable locking phenomenon takes place by which the modal frequencies drift into a common value [14]. The long term solution of the laser equations becomes time independent as in the case of the single-mode configuration, i.e. the beat pattern disappears. At the same time, each mode is characterized by a different transverse profile so that the output field configuration carries the imprint of the coherent superposition of the various modal patterns. In this way, stationary spatial inhomogeneities are born. This effect is not a symmetry breaking phenomenon in the strict sense of the word if the laser operates with cylindrically symmetric modes, because even with a large frequency spacing one always observes a small mixture of higher order transverse components, but it becomes a real symmetry breaking effect if one accounts also for possible variations of the field profile around the polar angle [12].

In this article we summarize the main findings of a study of laser dynamics with a fairly realistic resonator model and use these results to interpret experimental tests carried out with a CO_2 laser of the Fabry-Perot type after suitable modifications of the cavity design to allow a continuous variation of the effective radius of curvature of the mirrors. Our theoretical formulation evolves from the standard framework of the Maxwell-Bloch equations under the assumptions of cylindrical symmetry of the dynamical variables and with the restriction that the transverse mode spacing is of the order of the cavity linewidth, a configuration that enhances the dynamical interaction of the transverse modes [14].

2.   The mathematical description of the model

The starting point of our considerations is the set of Maxwell-Bloch equations for the amplitudes $f_p(\eta,\tau)$ of the cavity modes [p=0,1,2,...], the atomic polarization $P(\rho,\eta,\tau)$ and population difference $D(\rho,\eta,\tau)$

$$\frac{\partial f_p}{\partial \eta} + \frac{1}{v}\frac{\partial f_p}{\partial \tau} = i\frac{\delta\Omega}{v}f_p - \alpha\Lambda \int_0^\infty d\rho\, \rho\, A_p^*(\rho,\eta)\, P(\rho,\eta,\tau) \tag{2.1a}$$

$$\frac{\partial P}{\partial \tau} = -\left[FD + (1 + i\Delta)P\right] \tag{2.1b}$$

$$\frac{\partial D}{\partial \tau} = -\gamma\left[-\frac{1}{2}\left(F^*P + FP^*\right) + D - \chi(\rho,\eta)\right] \quad . \tag{2.1c}$$

These are identical to those derived in Refs. 16,17 apart from minor rescaling of the variables and parameters. Here we have set

$$\eta = \frac{z}{\Lambda}, \quad \rho = \sqrt{\frac{\pi}{\Lambda\lambda}} \ r, \quad \tau = \gamma_\perp t, \quad v = \frac{c}{\Lambda\gamma_\perp} \ , \tag{2.2}$$

where λ is the wavelength of light, Λ is the length of the ring cavity, z is the longitudinal and r the radial coordinate, $\gamma_\perp$ is the polarization relaxation rate and γ is the population relaxation rate, in units of $\gamma_\perp$. The constant α denotes the unsaturated gain of the active medium per unit length. The field and polarization envelopes $F(\rho,\eta,\tau)$ and $P(\rho,\eta,\tau)$ are related to the usual slowly varying variables $F(\rho,\eta,\tau)$ and $P(\rho,\eta,\tau)$ by

$$F(\rho,\eta,\tau) = e^{-i\delta\Omega\tau} \ F(\rho,\eta,\tau) = e^{-i\delta\Omega\tau} \sum_p A_p(\rho,\eta) \ f_p(\eta,\tau) \tag{2.3a}$$

and

$$P(\rho,\eta,\tau) = e^{-i\delta\Omega\tau} \ P(\rho,\eta,\tau) \ , \tag{2.3b}$$

where $\{A_p(\rho,\eta)\}$ denote the empty cavity modes [18]. The parameter $\delta\Omega$ measures the unknown offset between the operating frequency of the laser in steady state and the frequency of the TEM_{00} mode that lies nearest to the center of the atomic line and acts as the reference frequency; Δ is defined as the difference $\delta_{AC} - \delta\Omega$ where δ_{AC} is the detuning between the center of the gain line and the cavity reference frequency. The detuning parameters are all measured in units of $\gamma_\perp$. Finally, $\chi(\rho,\eta)$ denotes the equilibrium population difference. Equations (2.1) are supplemented by the boundary conditions

$$f_p(-\frac{1}{2}1,\tau) = R \ e^{-i\delta_p} \exp[i\delta\Omega\gamma_\perp \frac{\Lambda-L_A}{c}] f_p(\frac{1}{2}1, \tau - \gamma_\perp \frac{\Lambda-L_A}{c}) \ , \tag{2.4}$$

where $1 = L_A/\Lambda$, $\delta_p = (\omega_p - \omega_0)\Lambda/c$, L_A is the length of the active medium and $R=1-T$ is the power reflectivity coefficient of the mirrors [see Fig. 1].

In order to focus on the essential features of this problem, we assume the limits

$$\alpha L_A \ll 1, \quad T \ll 1, \quad \text{with } \frac{\alpha L_A}{T} \equiv 2C = \text{arbitrary} \ . \tag{2.5}$$

In addition, we require that

$$\frac{\gamma_\perp\Lambda}{c} \ll 1, \quad \text{with } \kappa = \frac{cT}{\Lambda\gamma_\perp} = \text{arbitrary} \ . \tag{2.6}$$

Conditions (2.5) and (2.6) insure that the dynamics of the system is governed by the single longitudinal mode that is nearest to the atomic line (the so-called resonant mode) as all other longitudinal modes lie well outside the atomic linewidth. We select the radius of curvature of the

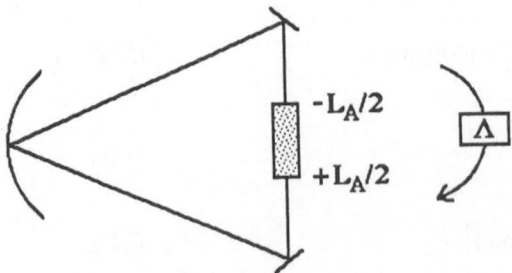

$-L_A/2$

$+L_A/2$

Λ

Fig. 1   Schematic representation of the ring cavity. The length of the resonator is Λ and the active medium is contained in the region $-L_A/2 < z < L_A/2$.

spherical reflector and the length of the resonator in such a way that several higher order transverse modes fall in the vicinity of the resonant mode, in the sense implied by the condition

$$\delta_p \equiv \frac{\omega_p - \omega_0}{c/\Lambda} \ll 1, \quad \text{with } a_p \equiv \frac{\delta_p}{T} = \text{arbitrary} ,$$
(2.7)

where ω_p denotes the frequency of the transverse mode of order p (p=0 labels the resonant mode) and a_p measures the frequency separation between the p-th and the resonant mode in units of cavity linewidth. Condition (2.7), together with the requirement T<<1, implies that the beam waist is nearly uniform along the atomic sample and that the transverse modes, in the region occupied by the active medium, have the structure

$$A_p(\rho,\eta) \cong A_p(\rho,0) = \frac{2}{\sqrt{\eta_0}} \exp[-\frac{\rho^2}{\eta_0}] L_p(\frac{2\rho^2}{\eta_0}) ,$$
(2.8)

where $(\eta_0)^{1/2}$ is the beam waist and $L_p(x)$ is the Laguerre polynomial of order p and the indicated argument. In order to take advantage of the simplifications afforded by the conditions (2.5) and (2.6), it is convenient to introduce the new independent variables

$$\eta' = \eta$$

$$\tau' = \tau + \gamma_\perp \frac{\Lambda-L_A}{cl}(\eta' + \frac{1}{2}l)$$
(2.9)

and the field modal amplitudes $\varphi_p(\eta',\tau')$ defined by

$$f_p(\eta',\tau') = \varphi_p(\eta',\tau') \exp\left\{ - \left[\ln R - i\delta_p + i\delta\Omega \gamma_\perp \frac{\Lambda-L_A}{c}\right] \frac{\eta'+\frac{1}{2}l}{l} \right\} .$$
(2.10)

The reason, as usual [17], is that the new field variables obey standard periodicity boundary conditions

$$\varphi_p(-\frac{1}{2}l,\tau') = \varphi_p(\frac{1}{2}l, \tau')$$
(2.11)

and become independent of η' in the limit (2.5).

In this way the equations of motion become ordinary time-dependent differential equations and under the conditions (2.5) and (2.6) the field equations take the form

$$\frac{\partial \varphi_p}{\partial \tau'} = - \kappa(1+ia_p-i\Delta)\varphi_p - 2C\kappa \int_0^\infty d\rho \, \rho \, A_p(\rho) \, P(\rho,\tau')$$
(2.12)

while the atomic equations retain the form given in Eqs.(2.1b,c). In arriving at Eq.(2.12), for convenience, we have chosen $\delta\Omega$ as the frequency shift predicted by the standard mode pulling formula, i.e.

$$\delta\Omega = \frac{\kappa\delta_{AC}}{1+\kappa} = \kappa\Delta .$$
(2.13)

Note that this selection of $\delta\Omega$ implies that, in steady state, the time derivatives of the field and of the atomic variables do not vanish because, in general, the actual frequency shift of the stationary state is not given by Eq.(2.13). After a final change of variables

$$\xi = \frac{\rho}{\sqrt{\eta_0}}, \quad \psi_p = \frac{\varphi_p}{\sqrt{\eta_0}}$$
(2.14a)

$$A_p(\rho) \rightarrow B_p(\xi) = 2 \, e^{-\xi^2} L_p(2\xi^2)$$
(2.14b)

and with the modal expansion

$$
\begin{bmatrix} F(\xi,\tau') \\ P(\xi,\tau') \\ D(\xi,\tau') \end{bmatrix} = \sum_{p=0}^{\infty} B_p(\xi) \begin{bmatrix} \psi_p(\tau') \\ p_p(\tau') \\ d_p(\tau') \end{bmatrix} ,
$$

(2.15)

the dynamics of the system is governed by the modal equations

$$
\frac{d\psi_p}{d\tau'} = -\kappa_p \psi_p - \kappa\, i\, (a_p - \Delta)\psi_p - 2C\, \kappa\, p_p
$$

(2.16a)

$$
\frac{dp_p}{d\tau'} = -\sum_{qq'} \Gamma_{pqq'}\, \psi_q d_{q'} - (1+i\Delta)p_p
$$

(2.16b)

$$
\frac{dd_p}{d\tau'} = -\gamma \left[-\frac{1}{2} \sum_{qq'} \Gamma_{pqq'} \left(\psi_q^* p_{q'} + c.c. \right) + d_p - d_p^{(0)} \right] ,
$$

(2.16c)

where the mode-mode coupling coefficients are defined by

$$
\Gamma_{pqq'} = \int_0^{\infty} d\xi\, \xi\, B_p(\xi)\, B_q(\xi)\, B_{q'}(\xi)
$$

(2.17a)

and

$$
d_p^{(0)} = \int_0^{\infty} d\xi\, \xi\, B_p(\xi) .
$$

(2.17b)

In deriving Eq.(2.16a) we have introduced a phenomenological damping coefficient to simulate the losses introduced by the finite diameter of the laser tube for the different transverse modes. For definiteness (and somewhat arbitrarily) we have chosen

$$
\kappa_p = \kappa\, (1+\beta p^4), \quad (p=0,1,2,\dots)
$$

(2.18)

where β is a constant. In addition, we have assumed a uniform pump profile $[\chi(\xi,\eta)=1]$ in the region occupied by the active medium so that the equilibrium modal populations are given by the simple formula $d_p^{(0)} = (-1)^p$. Equations (2.16) represent the starting point of our dynamical studies. In steady state we set

$$
\begin{bmatrix} \psi_p \\ p_p \end{bmatrix} = \exp(-i\varepsilon\kappa\tau') \begin{bmatrix} \psi_p^{st} \\ p_p^{st} \end{bmatrix}
$$

(2.19a)

$$
d_p = d_p^{st}
$$

(2.19b)

where $\varepsilon\kappa$ is the frequency shift of the true steady state from the frequency given by the approximate mode pulling formula Eq.(2.13), and obtain

$$
(1+\beta p^4)\psi_p^{st} + i\, (a_p - \varepsilon - \Delta)\psi_p^{st} = -2Cp_p^{st}
$$

(2.20a)

$$
\left[1+i\, (\Delta - \varepsilon\kappa) \right] p_p^{st} = -\sum_{qq'} \Gamma_{pqq'}\, \psi_q^{st} d_{q'}^{st}
$$

(2.20b)

$$
d_p^{st} = d_p^{(0)} + \frac{1}{2} \sum_{qq'} \Gamma_{pqq'} \left(\psi_q^{st*} p_{q'}^{st} + c.c. \right) .
$$

(2.20c)

The structure of the mode-mode coupling coefficients (2.17a) introduces some quantitative differences relative to the steady state solutions of the Cartesian model discussed in Ref. 11. However, important qualitative similarities persist. First of all, for sufficiently large values of the radial mode spacing (i.e. $\kappa a_1 \sim 1$) steady state solutions exist which are essentially of the single-mode type, in the sense that one modal amplitude, e.g. p=n, dominates over all the others. The approximate analytic expression of these "single-mode" steady state solutions is given by

$$ x_n^2 = \frac{1}{I_n} \left\{ \frac{2C}{1+\beta n^4} - 1 - (\Delta - \epsilon \kappa)^2 \right\} , \tag{2.21} $$

where $x_n = \text{Re}(\psi_n)$, $y_n = \text{Im}(\psi_n)=0$,

$$ \epsilon = \frac{a_n}{1+\kappa} \tag{2.22a} $$

and

$$ I_n = \sum_{q'} \Gamma_{nnq'} \Gamma_{q'nn} = 4 \int_0^\infty dx\, e^{-2x} (L_n(x))^4 . \tag{2.22b} $$

For general values of the parameters we have solved Eqs.(2.20) by successive iterations, using the single-mode solutions as the starting approximations for large values of a_1 and then reducing the frequency spacing between transverse modes "adiabatically". In resonance and for large values of a_1 we found stationary solutions which are very nearly of the TEM$_{00}$ type, as expected. For small values of κa_1, instead, significant mode mixing develops and a considerable distortion of the transverse intensity profile becomes apparent [Fig.2]. This is the result of the fact that several modal amplitudes have become

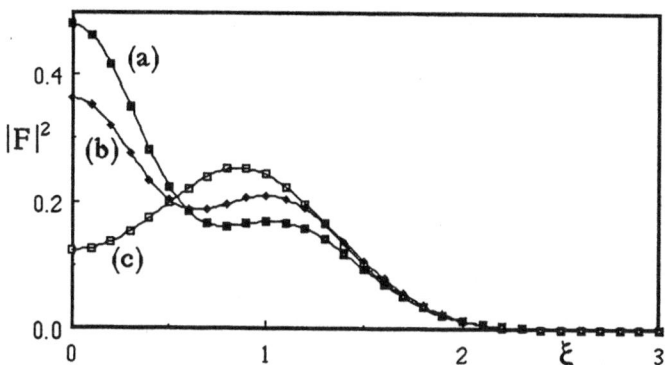

Fig. 2   Transverse profile of the stationary output intensity for 2C=1.2, β=0.05, Δ=0.0, κ=0.1, and (a) $a_1 = 0.1$, (b) $a_1 = 0.05$, (c) $a_1 = 0$.

comparable to one another [the actual number of significantly excited modes is controlled by the effective diffraction losses; here, for the selected values of β, the dominant modes are p=0,1,2].

The stationary modal equations, of course, cannot yield information on the stability of the solutions or on the course taken by the system in its evolution. We have investigated this aspect of the problem with the help of the time-dependent modal equations (2.16) whose numerical solutions have shown the existence of varied and complicated behaviors. Here, we limit ourselves to a survey of a few typical results. Figure 3a provides a global overview of a series of numerical runs in which we varied the frequency spacing of the transverse modes, holding every other parameter fixed. In this figure the single solid lines denote the steady state values of $|\psi_0|$ for different values of a_1 when the solutions approach a stationary state, while the shaded regions represent the domains where self-pulsing solutions develop; these domains are bounded by the maxima and minima of the oscillating moduli of the field modes to give an idea of the modulation depth of the

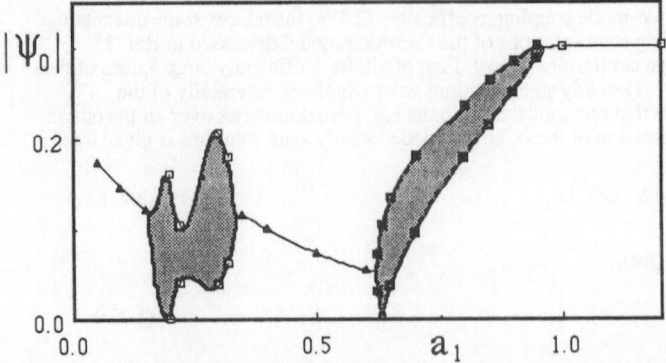

Fig. 3a   A global overview of the solutions of Eqs. (2.16) for different values of a_1 and $2C = 1.2$, $\kappa = 1$, $\beta = 0.005$, $\gamma = 0.05$, and $\Delta = 0.15$. The stationary values of $|\psi_0|$ are indicated by the single solid lines; the domains where time-dependent pulsations exist are shown by the shaded areas which are bounded by the maxima and minima of the oscillating $|\psi_0|$.

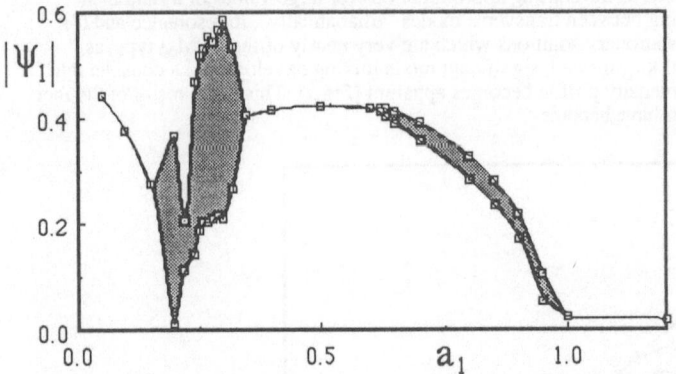

Fig. 3b   A global overview of the solutions of Eqs. (2.16) for different values of a_1 and $2C = 1.2$, $\kappa = 1$, $\beta = 0.005$, $\gamma = 0.05$, and $\Delta = 0.15$. The stationary values of $|\psi_1|$ are indicated by the single solid lines; the domains where time-dependent pulsations exist are shown by the shaded areas which are bounded by the maxima and minima of the oscillating $|\psi_1|$.

solutions; in the regions where dynamical chaos develops, we have used the averages of several maxima and minima. Figure 3b shows the complementary data for mode p=1.

For large values of a_1 the long time solutions are stationary and essentially of the TEM$_{00}$ single-mode type. Over this range of parameters the higher order modes are locked to the oscillation frequency of the lowest order solution in a way that is reminiscent of a laser with an injected signal, when the injected signal strength is large with respect to the single-mode laser field. For decreasing values of a_1 the amplitude of the p=1 mode increases, as shown clearly in Fig. 3b, and the frequency locking is destroyed. Thus, in the range $0.65 < a_1 < 1$, regular time-dependent oscillations develop with a frequency given by the mode pulled intermode spacing (See Fig.4). The strength of the mode p=1 continues to grow for decreasing a_1 until it becomes the dominant contribution to the total cavity field, for $a_1 < 0.65$. In this region a new locking regime ensues where the laser carrier frequency is given by the mode pulled value of the frequency of mode p=1.

The next region of unstable oscillations develops when a_1 becomes smaller than approximately 0.35, in response to the growth of the p=2 modal component which is now getting

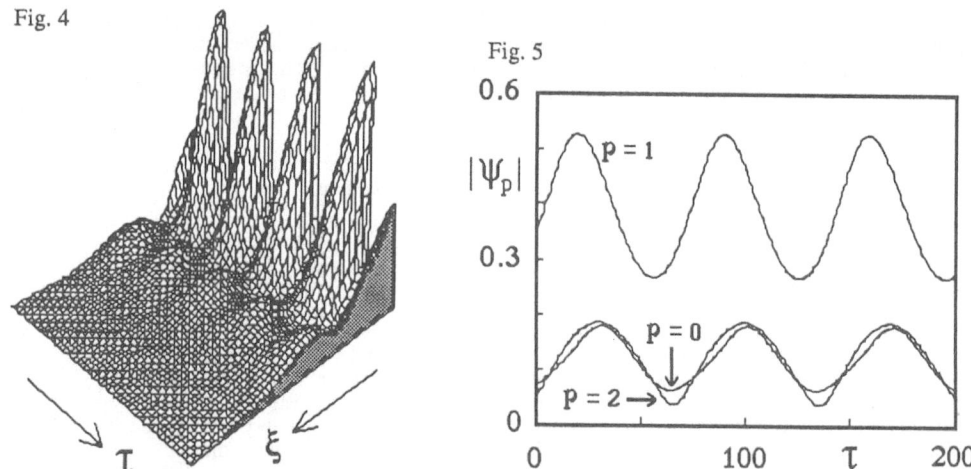

Fig. 4

Fig. 5

Fig. 4   Periodic oscillations of the output intensity for $2C = 1.2$, $\kappa = 1$, $\beta = 0.005$, $\gamma = 0.05$, $\Delta = 0.15$, and $a_1 = 0.7$.

Fig. 5   Periodic oscillations of the moduli of the modes p=0,1,2 for $a_1 = 0.32$. These three modal amplitudes are now of comparable strength. The parameters for this simulation are $2C = 1.2$, $\kappa = 1$, $\beta = 0.005$, $\gamma = 0.05$, and $\Delta = 0.15$.

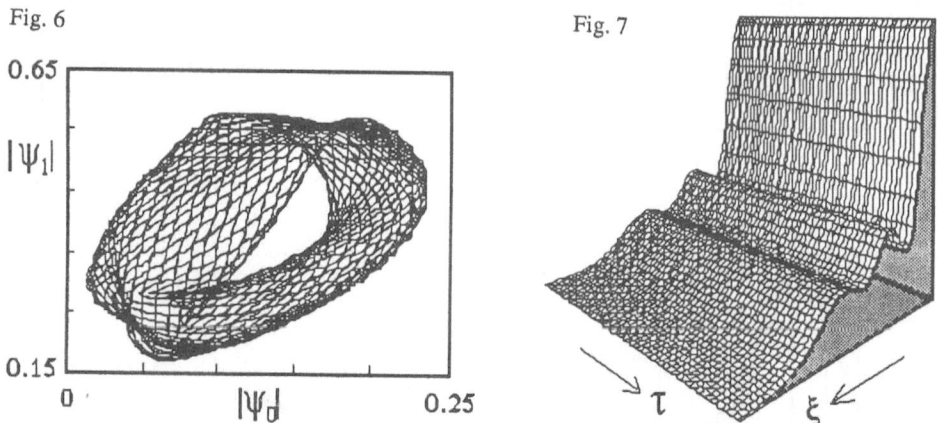

Fig. 6

Fig. 7

Fig. 6   Projection of the trajectory traced by the moduli $|\psi_p|$ for p=0,1,2 in the plane of $|\psi_0|$ and $|\psi_1|$. The trajectory in the full tridimensional space moves on the surface of a torus. The parameters are the same as in Fig. 5 except for a_1 which equals 0.29.

Fig. 7   An example of a frequency locked solution for $a_1 = 0.1$. The remaining parameters are the same as in Fig. 5.

closer in frequency to the resonant mode. Here the situation is more complicated because, in addition to the regular oscillations (see Fig. 5), we observe also the appearance of chaotic solutions perhaps induced by the growth of the p=2 modal component.

This region is also characterized by small domains of quasiperiodicity in which the moduli of the dominant modes (p=0,1,2) move on the surface of a torus in three-dimensional space (See Fig.6), as one often expects when the varying frequencies of oscillation become commensurate.

Below a certain well-defined threshold value (in this simulation, for a_1 less than approximately 0.2) the system enters a new locked state in which several modes oscillate synchronously with comparable strength. In this case the output field intensity becomes time independent and the radial profile develops considerable modulation. An example of a frequency locked solution is shown in Fig. 7.

The complex interplay between temporal and spatial variations is currently the subject of extensive investigations which will be discussed more fully elsewhere.

3. Experimental results

The theoretical analysis of the previous section emphasizes the important role played by the frequency spacing κa_1 between adjacent Gauss-Laguerre modes of the resonator. This parameter depends on the free-spectral range of the cavity and on the radius of curvature of the output couplers. In order to hold the free spectral range constant and to vary the radius of curvature of the mirrors in a controllable way we built a resonator of the Fabry-Perot type with plane output couplers and inserted two lenses inside the cavity. The effective radius of curvature of the mirrors depends on the spacing between the lenses which we can vary quite accurately with the help of a micrometer driven translation stage. With this arrangement we

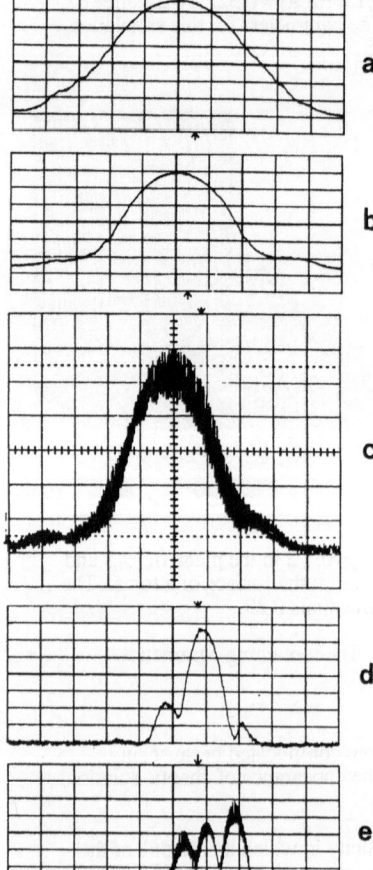

a

b

c

d

e

Fig. 8   Experimental radial profile of the output intensity. The sequence from (a) to (e) corresponds to a progressively longer radius of curvature of the output mirror.

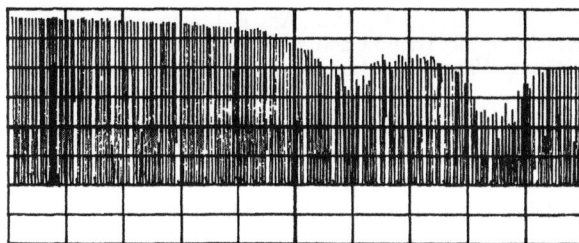

Fig. 9   For each setting of the radius of curvature of the output coupler, the vertical bars
          indicate individual measurements of the average laser power, which is highest when
          the laser operates stably. The left hand side of the figure corresponds to the largest
          modal spacing; the right hand side to the smallest (near planar configuration).

have been able to reproduce qualitatively many features predicted by the theoretical scan of Fig.
3. In Fig. 8 we show a number of observed spatial profiles of the laser output, obtained by
sweeping the beam across a HgCdTe detector. We begin by considering a nearly confocal
cavity configuration with a maximum separation of about 50 MHz between transverse modes
and a cavity linewidth of about the same magnitude (approximately 60 MHz). In this case the
radial intensity distribution is practically Gaussian and constant in time (see Fig. 8a). Upon
varying the cavity configuration from confocal to nearly planar with a continuous sweep of the
effective radius of curvature of the mirrors, the output intensity begins to show a measurable
distortion from the original Gaussian profile (Fig. 8b) which is then followed by oscillations
affecting the entire profile (Fig. 8c). As the transverse modes become progressively closer in
frequency, the laser output becomes stable again and the radial distribution approaches a shape
that is typical of the p=1 mode, as predicted by the theory.

A further increase in the radius of curvature of the mirrors leads to another region of
instability (Fig. 8d) with a more complicated structure which evolves eventually into the stable
configuration shown in Fig. 8e. An interesting aspect of this scan is that the unstable regions
are characterized by a lower average output power (Fig. 9) as compared to the stable domains.
Presumably this effect is related to the mode-mode competition and to the asynchronous nature
of their oscillations away from the cooperative frequency locking states.

This work was partially supported by a NATO Collaborative Research Grant, by the
EEC Twinning Project on "Dynamics of Nonlinear Optical Systems" and by a Joseph H.
DeFriees grant of the Research Corporation and by the Samuel Roberts Noble Foundation of
Oklahoma.

References

1.   H. Haken, Synergetics - An Introduction, Springer-Verlag, Berlin, 1977.
2.   H. Haken, Advanced Synergetics, Springer-Verlag, Berlin, 1983.
3.   H. Haken, Rev. Mod. Phys. 47, 67 (1975)
4.   See for example N.B. Abraham, P. Mandel and L.M. Narducci, in Progress in Optics,
      Vol. XXV, edited by E. Wolf, North Holland, Amsterdam, 1988.
5.   A.M. Turing, Phil. Trans. Roy. Soc. London B237, 37 (1952).
6.   G. Nicolis and I. Prigogine, Self-Organization in Nonequilibrium Systems, Wiley, New
      York, 1977.
7.   J.D. Murray, J. Theor. Bio. 88, 161 (1981).
8.   L.A. Lugiato and R. Lefever, Phys. Rev. Lett. 58, 2209 (1987).
9.   L.A. Lugiato, L.M. Narducci, and R. Lefever, in Lasers and Synergetics - a volume in
      honor of the 60th birthday of Hermann Haken, Springer-Verlag, Berlin 1987.
10   L.A. Lugiato and C. Oldano, submitted for publication.

11. L.A. Lugiato, C. Oldano and L.M. Narducci, in J. Opt. Soc. Am. B, Feature issue on Laser Dynamics, edited by D.K. Bandy, J.R. Tredicce and A. Oraevsky, to appear.

12. L.A. Lugiato, C. Oldano, L. Sartirana, Wang Kaige, L.M. Narducci, G.-L. Oppo, F. Prati and G. Broggi, this volume.

13. See also the important contributions by J.V. Moloney and H.M. Gibbs, Phys. Rev. Lett. 48, 1607 (1982); D.W. McLaughlin, J.V. Moloney and A.C. Newell, Phys. Rev. Lett. 54, 681 (1985); A. Ouazzardini, H. Adachihara and J.V. Moloney, private communication.

14. L.A. Lugiato, G.-L. Oppo, M.A. Pernigo, J.R. Tredicce, L.M. Narducci and D.K. Bandy, Opt. Comm., to be published.

15. L.A. Lugiato, L.M. Narducci, E.V. Eschenazi, D.K. Bandy and N.B. Abraham, Phys. Rev. A32, 1563 (1985).

16. L.A. Lugiato, F. Prati, D.K. Bandy, L.M. Narducci, P. Ru and J.R. Tredicce, Opt. Comm.64, 167 (1987).

17. L.A. Lugiato, F. Prati, L.M. Narducci, P. Ru, J.R. Tredicce and D.K. Bandy, Phys. Rev. A (in press).

19. P. Ru, L.M. Narducci, J.R. Tredicce, D.K. Bandy and L.A. Lugiato, Opt. Comm. 63, 310 (1987).

Index of Contributors